Conservation ar
Development

Conservation and development share an intertwined history dating back to at least the 1700s. But what are the prospects for reconciling the two, and how far have we come with this project? This book explores these questions through a detailed consideration of the past, present and future of the relationship between conservation and development. Bringing to bear conceptual resources from political ecology, social-ecological systems thinking and science and technology studies, *Conservation and Development* sets this relationship against the background of the political and economic processes implicated in environmental degradation and poverty alike. Whilst recognising that the need for reconciling conservation and development processes remains as compelling as ever, it demonstrates why trade-offs are more frequently encountered in practice than synergies. It also flags alternative visions for conservation and development obscured or ignored by current framings and priorities.

Bringing together policy and theory, *Conservation and Development* is an essential resource for undergraduate and postgraduate students and a useful reference for researchers in related fields. Each chapter contains a reading guide with discussion questions. The text is enlivened by a number of new case studies from around the world. A must-read for anyone interested in understanding the history, current state and projections for future shifts in the relationship between conservation and development.

Andrew Newsham is Lecturer in International Development at the Centre for Development, Environment and Policy, SOAS (University of London). He brings practical and theoretical insights from political ecology and the sociology of scientific knowledge to a research agenda which encompasses environment and development in Southern Africa and South America. His research can be broadly characterised by its focus on a) the relationship between conservation and development and b) climate change adaptation and development.

Shonil Bhagwat is an academic at the Open University, UK. He has training in ecology and forestry from the University of Oxford. His current research engages key environmental concerns: biodiversity conservation, agriculture and food security, and environmental change. He addresses these concerns through a blend of theoretical and methodological influences drawn from the social and natural sciences.

Routledge Perspectives On Development

Series Editor: Professor Tony Binns, *University of Otago*

Since it was established in 2000, the same year as the Millennium Development Goals were set by the United Nations, the *Routledge Perspectives on Development series* has become the pre-eminent international textbook series on key development issues. Written by leading authors in their fields, the books have been popular with academics and students working in disciplines such as anthropology, economics, geography, international relations, politics and sociology. The series has also proved to be of particular interest to those working in interdisciplinary fields, such as area studies (African, Asian and Latin American studies), development studies, environmental studies, peace and conflict studies, rural and urban studies, travel and tourism.

If you would like to submit a book proposal for the series, please contact the Series Editor, Tony Binns, on: jab@geography.otago.ac.nz

Published:

Third World Cities, 2nd edition
David W. Drakakis-Smith

Rural–Urban Interactions in the Developing World
Kenneth Lynch

Environmental Management & Development
Chris Barrow

Tourism and Development in the Developing World
Richard Sharpley & David J. Telfer

Southeast Asian Development
Andrew McGregor

Postcolonialism and Development
Cheryl McEwan

Conflict and Development
Andrew Williams & Roger MacGinty

Disaster and Development
Andrew Collins

Non-Governmental Organisations and Development
David Lewis and Nazneen Kanji

Cities and Development
Jo Beall

Gender and Development, 2nd Edition
Janet Momsen

Economics and Development Studies
Michael Tribe, Frederick Nixson and Andrew Sumner

Water Resources and Development
Clive Agnew and Philip Woodhouse

Theories and Practices of Development, 2nd Edition
Katie Willis

Conservation and Development

Andrew Newsham and
Shonil Bhagwat

Routledge
Taylor & Francis Group

LONDON AND NEW YORK

First published 2016
by Routledge
2 Park Square, Milton Park, Abingdon, Oxon OX14 4RN

and by Routledge
711 Third Avenue, New York, NY 10017

Routledge is an imprint of the Taylor & Francis Group, an informa business

© 2016 Andrew Newsham and Shonil Bhagwat

British Library Cataloguing in Publication Data
A catalogue record for this book is available from the British Library

Library of Congress Cataloging in Publication Data
A catalog record for this book has been requested

ISBN: 978–0–415–68780–5 (hbk)
ISBN: 978–0–415–68781–2 (pbk)
ISBN: 978–1–315–69477–1 (ebk)

Typeset in Times New Roman and Franklin Gothic
by Cenveo Publisher Services

Printed and bound in Great Britain by
TJ International Ltd, Padstow, Cornwall

Andy dedicates this book to his dad, Peter Newsham, for encouragement, inspiration and insatiable curiosity.

Shonil dedicates this book to his son Dominic who arrived when this book was being written. Hope this book makes a small contribution to equipping his generation with ideas and thoughts that are needed to address grand challenges in environment and development.

Contents

Figures, tables and boxes

Figures

Tables

Boxes

Acknowledgements and dedications

It is not possible to trace all of the ways in which our interactions with others contribute to the writing of a book. As such, acknowledgements sections are inevitably incomplete, and perhaps a level of generalised gratitude to the universe, as it were, is always appropriate. Clearly, however, the presence of particular people has been key to the completion of this book. Together, therefore, we would like to thank, first and foremost, the editorial team at Routledge – Sarah Gilkes, Faye Leerink and Andrew Mould for their support, flexibility and understanding, especially when we requested leeway with deadlines (not a one-off event!). We are also grateful to the series editor Tony Binns for his very helpful comments, as well as to the anonymous reviewers who gave insightful, constructive feedback on early versions of chapters and the first draft. A longer list of people who have provided inspirational, intellectual and practical contributions include: Jennifer Constantine, Rosaleen Duffy, Sam Geall, Dabo Guan, Maymay Knight, Peter Newell, Peter Newsham, Alison Ormsby, Claudia Rutte, Alex Shankland, Tom Tanner and Frauke Urban.

Andy would like to thank: first of all, family (Dad, Nick, Anna, assorted aunties, uncles and cousins, you know who you are!) and friends who have pulled me through the hard, sad times that have been the backdrop to much of the writing with countless acts of love and kindness. I've loved the regular episodes of fun, exercise, chocolate and utter exhaustion with my nephew and niece, Barney and Lucie! I'm deeply grateful to and perpetually in awe of Jenny for her huge heart and vast reserves of empathy. I've been blessed by the presence of Agnes, Akshay, Ana, Ben, Carrie, Ewan, Hannah, Marisa and Maymay. I honestly don't know what I would have done without you all.

Shonil would like to thank: family and friends who have always been there and without whose help and support it would have been impossible to dedicate time to writing this.

Every effort has been made to trace and contact copyright holders for their permission to reprint material in this book. The publishers would be grateful to hear from any copyright holder who is not acknowledged and will undertake to rectify any errors or omissions in future editions of this book.

Acronyms and abbreviations

3iG	International Interfaith Investment Group
ARC	Alliance of Religions and Conservation
AKDN	Aga Khan Development Network
ANT	Actor-Network Theory
BBC	British Broadcasting Corporation
BR	Biosphere Reserve
CBD	Convention on Biological Diversity
CBD PoWPA	Convention on Biological Diversity Programme of Work on Protected Areas
CBNRM	Community-Based Natural Resource Management
CBT	Community-Based Tourism
CBTE	Community-Based Tourism Enterprise
CFCs	Chlorofluorocarbons
CI	Conservation International
CII	Conservation International Indonesia
CITES	Convention on International Trade in Endangered Species of Wild Fauna and Flora
CO_2	Carbon Dioxide
CO_{2e}	Carbon Dioxide or Equivalent Greenhouse Gases
COP	Congress/Conference of Parties (the name for the international meetings at which signatories to United Nations conventions gather)
CPR	Common-Pool Resource
CPRM	Common-Pool Resource Management

DDVE	Development Dialogue on Values and Ethics
DFID	(UK) Department for International Development
DMG	Daureb Mountain Guides
DMGA	Daureb Mountain Guides Association
DNA	Deoxyribonucleic Acid
DRC	Democratic Republic of Congo
ESAV	Ecosystem Service Assessment and Valuation
EU	European Union
FAO	(United Nations) Food and Agriculture Organization
FSC	Forest Stewardship Council
FUNGIR	Management and Regional Research Foundation (Argentina)
GATS	General Agreement on Trade and Services
GDP	Gross Domestic Product
GIS	Geographic Information Systems
GM	Genetically Modified
GRN	Government of the Republic of Namibia
IBP	International Biological Programme
IAASTD	International Assessment of Agricultural Knowledge, Science and Technology for Development
ICCAs	Indigenous and Community Conserved Areas
ICDP	Integrated Conservation and Development Project
ICSU	International Council of Scientific Unions
IFOAM	International Federation of Organic Agriculture Movements
IIRSA	Integration of the Regional Infrastructure in South America
ILO	International Labour Organisation
IMF	International Monetary Fund
IOPN	International Office for the Protection of Nature
ISKON	International Society for Krishna Consciousness
ITK	Indigenous Technical Knowledge
IUBS	International Union of the Biological Sciences
IUCN	International Union for the Conservation of Nature
IUPN	International Union for the Protection of Nature
MAB	Man and the Biosphere
MDGs	Millennium Development Goals
MEA	Millennium Ecosystem Assessment
MSY	Maximum Sustainable Yield
NACSO	Namibian Association of CBNRM Support Organisations
NCCL	Nhambita Community Carbon Livelihoods
NGO	Non-Governmental Organisation
OECD	Organisation for Economic Co-operation and Development
PA	Protected Area

PSAH	Pagos por Servicios Ambientales Hidrológicos (Payments for Hydrological Environmental Services Programme, Mexico; Spanish acronym used here)
PES	Payments for Ecosystem Services
PWS	Payments for Watershed Services
RA	Resilience Alliance
REDD(+)	Reducing Emissions through Avoided Deforestation and Degradation
SCPE	Scientific Committee for Problems of the Environment
SES	Social-Ecological System
SLCP	Sloping Land Conversion Programme in China
SPWFE	Society for the Preservation of the Wild Fauna of the Empire
TBNRM	Transboundary Natural Resource Management
TEEB	The Economics of Ecosystems and Biodiversity
TEK	Traditional Ecological Knowledge
TEV	Total Economic Value
TNC	The Nature Conservancy
UAV	Unmanned Aerial Vehicle
UN	United Nations
UNCHE	United Nations Conference on the Human Environment
UNDESA	United Nations Department of Economic and Social Affairs
UNEP	United Nations Environment Programme
UNESCO	United Nations Educational, Scientific and Cultural Organization
UNFCCC	United Nations Framework Convention on Climate Change
UNFPA	United Nations Population Fund
UNSCCUR	United Nation Scientific Conference on the Conservation and Utilisation of Resources
UNU	United Nations University
UNWTO	United Nations World Tourism Organization
USD	United States Dollar
USAID	United States Assistance for International Development
VCM	Vicious Circulation Model
WCC	World Council of Churches
WCED	World Commission on Environment and Development
WCRP	World Conference on Religions for Peace
WCS	World Conservation Strategy
WCMC	World Conservation Monitoring Centre
WFDA	Women, Faith, and Development Alliance
WFDD	World Forum for Development Dialogue
WHC	World Heritage Convention

WILD	Wildlife Integration for Livelihoods Diversification
WDPA	World Database on Protected Areas
WTO	World Trade Organization
WWF	World Wide Fund for Nature

Introduction

Why look at conservation and development together?

Simply put, the relationship between conservation and development has been fraught with conflict because the environmental implications of predominant development trajectories are profound. Yet, conservation and development have to somehow work together. The so-called 'forced marriage' between conservation and development has been a topic of academic debates and policy discussion for well over four decades; although in some sense the two have been intertwined since at least as far back as the 1700s (Grove 1995). There is a spectrum of positions taken: some commentators contend that conservation and development have entirely different objectives and they should be kept separate, while others argue that conservation and development cannot work without each other. Trying to do conservation and development simultaneously may have fallen out of fashion, but it has not stopped being necessary. Indeed, the need for it may be greater now than ever.

Current development trajectories do not appear to be tending toward a state of global environmental well-being. Instead, as this book will explore, we seem to be in an era that some have christened 'the Anthropocene' (Crutzen, 2000), that is, an age in which human activity exerts an even greater influence on the global environment than do natural processes occurring largely independently of humans. This framing may underestimate the potential for 'natural variability' to harm human beings and societies (for instance, in the shape of an asteroid colliding with the planet, or a change in El Niño Southern Oscillation which alters the frequency and distribution of storms and cyclones). But it requires a lot of optimism not to worry that our impacts upon the global environment will not at some point rebound upon our global efforts at development. Therefore, some kind of conservation of the environment is indispensable for our development prospects, whether we define that in terms of economic growth, or in terms of a notion of well-being which to a greater or lesser extent is generated through processes independent of economic activity.

This book looks at the relationship between conservation and development. It is not a book which introduces in great detail and depth the 'worlds' of conservation and development separately; there are plenty of other books which do that either for conservation or for development. Instead, this book focuses on their intersection. Some prior acquaintance with either, therefore, is helpful, but our account does not require specialist knowledge of its readers.

There are numerous reasons why conservation and development are better explored together than separately. One key reason is that both are rooted in the same historical processes of European imperial expansion and the intensification of increasingly global trade. Underpinning these processes was what might be called the 'commodification of nature' (Castree, 2008): making more and more of nature into products which can be traded in markets; and coming to understand nature more and more in terms of its potential to be turned into tradable products. The economic activity and growth which is generated by the commodification of nature remains at the heart of dominant development trajectories. The conservation imperative emerged from a concern with the environmental impacts of the projection of colonial power and increasingly globalised trade. Thus, the commodification of nature renders the relationship between conservation and development as one of necessary interdependence, given that the prospects for economic growth are intricately bound up with the health of the planet's biodiversity.

Another key reason why it is not possible to understand conservation or development in isolation from each other is because it can be very difficult to separate out nature from society. We might choose to retain this artificial distinction for analytical purposes, perhaps, but it remains artifice. The extent of human influence on the planet now appears to be so great that it may not be possible to find 'natural' environments, if these are defined as being free from the presence or influence of humans. Bill McKibben some decades ago christened this tendency 'the end of nature' (1989). Nor does nature appear to behave as we thought it did, being both less stable and less predictable than we had previously envisaged. For these reasons, it may be more useful to conceive of the relationship between nature and society as a complex social-ecological system whose resilience to shocks and disturbances is determined by social and ecological processes.

Thinking along these lines has profound implications for the kinds of conservation and development that we try to do.

Calls to reconcile or integrate conservation and development became louder in the last decades of the twentieth century. The economic post-Second World War boom stimulated a popular wave of environmentalism which documented and raised concerns about environmental destruction caused by increasing (Northern) prosperity. This in turn opened the question of how sustainable the development process fuelling such prosperity could be if it was causing such damage to the environment on which it depended. In consequence, more standard conservation instruments, such as protected areas and international treaties with legally binding national conservation targets, were supplemented by efforts to do conservation outside of protected areas in ways which could support development objectives, such as community conservation in its various guises. The push for sustainable development and taking conservation beyond the protected area were accompanied, overall, by an ever-greater emphasis on the role of economic incentives, either through leveraging capital in markets or through state compensation, as a means of ensuring the conservation of valued and indeed economically useful biodiversity. As a result, mechanisms like payments for ecosystem services, and discourses around the green economy, have become the tools and concepts of choice for trying to reconcile conservation and development in the 21st century.

How do we explore the relationship between conservation and development?

First and foremost, we consider the relationship between conservation and development not just to be one of interdependence, but one of trade-off. For some time, perhaps especially between the 1970s and the 1990s, there was quite some hope that efforts to conserve nature could be reconciled with efforts to bring development to impoverished and marginalised parts of the world. The discourse was about – and sometimes continues to be about – synergies, win-win situations, even triple wins. Concepts such as integrated conservation and development were championed and highlighted at global summits on sustainable development. However, perhaps a better way to understand the relationship is in terms of trade-offs that we do not manage to reconcile. It is not that win-win situations are impossible; it is, rather, that against the background of current political and economic priorities, they are highly improbable. Broadly speaking, this trade-off manifests in the extent to which environmental considerations, and ways of relating with nature which are not so bound up in commodifying it, tend to be seen as less important than, and therefore traded off against efforts to ensure development primarily through a trajectory of continually attempting to increase economic growth. Arguably, it is inherent in neoliberal capitalism – still our global political and economic system – to discount (or at the very least not to account fully for) the value of nature, even though capitalism is built upon the commodification of nature. Furthermore, for as long as this prioritisation of economic growth pertains, the relationship between conservation and development will always be a trade-off, because there is not the political space to consider other alternatives.

This is not to say that development is never traded off against conservation objectives. In this book, we document instances in which local livelihoods have been prohibited and sometimes criminalised, in which people have been removed from the places they live in, and in which huge restrictions are placed upon land use. These are by no means automatic processes, the inevitable result of the establishment of a protected area or a biological corridor. The changes to these forms of intervention over the course of the twentieth century, and in particular the greatly heightened awareness of the need to reduce, or even, preferably, eliminate the human costs of conservation, have made genuine differences to the well-being of many people living in areas where conservation activities are implemented. Nevertheless, a plan for the conservation of a particular landscape would look very different from a plan for its economic development. The process for determining which of these plans will prevail is deeply political, and inevitably there are winners and losers from the outcome of such processes, because a) people sometimes pursue opposing objectives, and b) some people are better placed than others to achieve their objectives.

Second, this book takes the position that conservation and development are about something bigger than the narrow confines of 'conservation' and 'development' as we often think and write about them. The relationship between conservation and development is shaped by a variety of wider issues – from globalisation, neoliberalism, population-consumption dynamics through to global environmental governance. This book therefore puts efforts to realise conservation and development simultaneously in that broader context. We do so as a corrective to the predominant tendency to focus on rural areas high in biodiversity and poverty, rather than the underlying processes that drive either (Adams, 2013). The book has broadly three parts:

1 Histories of conservation and development
2 Conservation and development in the broader global context
3 Conservation and development in practice

Together, these sections and chapters make a case for conservation and development that strives to go beyond the confines of 'conservation' and 'development' as we tend to think and speak of them. When we set the relationship between conservation and development against this broader background, they bring up a variety of historical, contemporary, global, local, socio-economic and cultural issues that this book explores. A more thoroughgoing and less superficial effort to reconcile conservation and development must engage, of necessity, with these issues and processes.

We have tried to write this book in the way we think it most likely to be read: as a resource to be 'dipped into' rather than read from cover to cover. Each chapter therefore works as a stand-alone piece, and does not require readers to have read the rest of the book in order to make sense of it. We flag at pretty much every opportunity where more information can be found in other chapters on specific areas of coverage, for those wishing to find out more, but tend to give enough of an introduction on that area for this to be optional, rather than necessary. Having said that, the chapters are informed by and all speak to the central agenda of themes and approaches we have highlighted here in the introduction:

1 the importance of social-ecological systems thinking and political ecology in particular (but also other resources like cultural and spiritual values, environmental history, science and technology studies, common pool resource theory, etc.) for understanding the relationship between conservation and development;
2 recognising that conservation and development unfold against a background of a globalised, neoliberal, capitalist political and economic system;
3 taking conservation and development beyond a focus on rural biodiversity and poverty intersections.

Part 1: Histories of conservation and development

Part 1 is about the histories of conservation and development and comprises two chapters. Between them, they stitch together a longer view of the origins of conservation and development, with a consideration of how the need to reconcile them came to be seen as a global imperative.

Chapter 1, 'Conservation and development in historical perspective', presents a historical overview of the co-emergence of conservation and development, stemming from the massive expansion in European empire and global trade from the eighteenth century onwards which sought to turn nature into commodities. The scientific revolution that followed provided the tools for documenting the environmental impacts of this expansion, and in many ways the impetus for minimising these impacts. The international conservation movement from the mid-nineteenth century onwards was dominated in its origins – and continues to be dominated by – Europe and the USA. Within conservation, two broad strands can be identified as different ways of responding to what we would these days call 'development': preservationism and wise use, which both drew upon an increasingly broad spectrum of emerging environmental sciences to further their objectives. Preservationism sought to draw a line between humans and wilderness, fearing that the latter would succumb to the reckless proclivities of the former. Wise use similarly started from an awareness of anthropogenic environmental disturbance, but saw the possibility of redemption through rational, scientific management regimes. These two different philosophies and modes of practice continue to run deep in conservation today and prompt a host of views on what the relationship between conservation and development should be.

Chapter 2, 'Sustainable development in the twentieth and twenty-first centuries', charts the path for the sustainability agenda, crowned in 1972 with the Stockholm conference and the establishment of the United Nations Environment Programme. This is relevant to the increasing desire to join up conservation and development. It sets out the concerted, popular efforts to force a reckoning between global development and its resulting environmental discontents between 1972 and 2002, when the World Summit on Sustainable Development was held in Johannesburg. This period played host to the Rio conference in 1992, perhaps the twentieth century's biggest display of political commitment to global environmental objectives, some of them legally binding. Yet the precedent set was one of repeated disappointment, stemming from the inability and/or unwillingness of governments to act upon commitments made. This sense of disappointment, at least for environmentalists, only intensified in the decade between 2002 and 2012, the year in which the Rio+20 conference was held. This conference sought to take stock of what had occurred over the twenty years since the first Rio meeting. In so doing, it was an attempt to regain momentum behind a sustainable development agenda that, even with the looming threat of global environmental catastrophe posed by climate change, had stalled.

Part 2: Conservation and development in the broader global context

Part 2 introduces important elements of contemporary social and political theory which guide the thinking throughout the book about the relationship between conservation and development. It also explores the various global processes and the dynamics which, together, provide the broader background against which the relationship between conservation and development is more properly understood, and with which efforts to reconcile conservation and development need to engage more fully. Its six chapters each take on different global dynamics and processes with profound implications for conservation and development.

Chapter 3, 'Conservation, development and the relationship between nature and society', introduces the two broad sets of theoretical resources that underpin our approach to understanding the prospects for reconciling conservation and development. In order to do this, the chapter suggests, we need first to get to grips with the underlying relationship between nature and society. This is, in a sentence, because the nature that is available to conserve and the development paths we choose to follow are contingent upon the processes through which nature and society are co-produced. We use three key sets of conceptual lenses to bring the relationship between nature and society into sharper focus: social-ecological systems theory; social constructivism applied to accounts of nature and society; and political ecology.

Chapter 4, 'Conservation and development in an era of global change', starts by outlining the unprecedented ecological, economic and social changes that have prompted the discussion on global change. It picks up two contemporary themes that have come forward: (1) 'the perfect storm', a metaphor used to define the synergistic effects of the unprecedented ecological, economic and social changes on the environment; and (2) 'planetary boundaries', a concept that puts forward the limits within which humanity should live in order to avert catastrophic environmental consequences of human actions. Both these themes bring up a wide range of issues relevant to conservation and development. The chapter argues that the discussion on climate change does bring together many of these issues, but it perhaps does not have the same moral underpinning as sustainability, the need to live within our own means. It explores the implications this has for the desire to join up conservation and development.

Chapter 5, 'Conservation and development in the context of globalisation and neoliberalism', examines conservation and development in relation to prevalent economic paradigm and political systems. Since the end of the Second World War, the world is increasingly connected in its economic and political systems. Conservation and development have to operate in a world that is increasingly reliant on free market capitalism. As such, different ways of connecting markets with conservation and development have been put forward. Viewing nature as a tradable commodity has given rise to market-based mechanisms for conservation and development. However, the global economic crisis that started in 2008 has shown that markets are volatile, raising concerns for purely economic justifications for conservation of nature. Management of nature as a common pool resource, beyond the state and the markets, can safeguard natural resources from the vulnerabilities in the global economic system. The level of such management often has to be 'local' as opposed to 'global' and this raises the question of whether 'localisation' of the economy might be a way forward to make economic and ecological systems more resilient.

Chapter 6, 'Global environmental governance', puts conservation and development in the context of an environmental governance landscape shaped chiefly by an attempt to get beyond state-centric understandings of how nature is governed. Contemporary environmental governance is characterised by the emergence of new actors, institutions and networks. In particular the private sector and civil society have become much more visible actors in processes of governing the environment. Hybrid actors, which cannot easily be categorised as state, private or civil society but which incorporate elements of all three, have also become increasingly prevalent and influential. In addition to the more conventional governance structures of hierarchy and markets, network and adaptive governance arrangements have also emerged, as well as Foucauldian perspectives expressed principally in the idea of 'environmentality'. These are characterised by more fluid and flexible approaches; although their efficacy and influence have yet to be comprehended fully. What is increasingly clear from the literature on environmental governance is that the mechanisms of the market, in tandem with the ideology of neoliberal capitalism, have gained much ground. Some important conservation actors appeared to have embraced (or at least accommodated) aspects of neoliberal capitalism as a means through which to achieve their objectives. However, a number of commentators maintain that these forms of political and economic activity continue to be the root cause of environmental degradation and uneven, deeply inequitable forms of development.

Chapter 7, 'Population, consumption, conflict and environment', takes stock of the relationships between the growing human population and the environment. Demographers are now less ambitious in their claims about the role of population growth, and much better at linking it to other variables which intersect and also play important roles in environmental change. Furthermore, development trajectories based on unequal distribution of wealth are implicated both in the major challenges for environmental conservation of the twenty-first century and in the higher likelihood of poor people, in parts of the world such as China, being exposed to adverse effects of environmental degradation and contamination. The relationship between population, environment and conflict is highly contested and the research on it somewhat inconclusive, but so far, none of the scholarship on these issues has identified population as a potential driver of conflict around the illegal wildlife trade.

Chapter 8, 'Management of natural resources in a globalised world', takes stock of the impacts of globalisation on natural resource management. Going beyond population growth as an explanation for the depletion of natural resources, this chapter argues that it is the increasingly globalised nature of natural resource consumption that leads to its depletion. It focuses on agriculture, fisheries, forestry and wildlife as they are on an increasing trajectory of globalisation. These resources originate in one geographic location but are exploited to fulfil the demands of people in another geographic location.

The examples of 'localisation' of resource use demonstrate that a greater participation of local communities can be beneficial to the conservation of natural resources as well as development. The chapter also examines challenges for local governance in a world that continues to be increasingly globalised.

Part 3: Conservation and development in practice

Part 3 explores the most common varieties of conservation and development intervention. It comprises five chapters, each introducing and exploring different approaches to doing conservation and development together, as well as assessing experiences with them and examining the current state of the art.

Chapter 9, 'Protected areas, conservation and development', starts off by looking at the growth of protected areas ever since the establishment of the Yellowstone National Park in 1872. These areas have diverse management objectives, which are reflected in six protected categories designated by the International Union for Conservation of Nature, IUCN. Protected areas, however, are not well-equipped to meet the objectives of conservation and development simultaneously. As such, newer categories of protected areas have come up, where natural resources are managed jointly by local communities and the state. These new categories show promise in conserving biodiversity whilst also providing alternative livelihoods to people who depend on natural resources.

Chapter 10, 'Integrated conservation and development through community conservation', examines community-based conservation as the key concept through which the desire to integrate conservation and development has been expressed. The rehabilitation of 'community' was prompted by a re-evaluation, both amongst conservation and development thinkers and practitioners, around the abilities of local people and institutions to conserve, as well as the development benefits that could be derived from community conservation. The enthusiasm for integrated conservation and development translated into a substantial number of interventions, across the globe. However, the majority of integrated conservation and development projects were not deemed to deliver enough of either. The idea of the 'win-win' outcome had by the turn of the century been displaced by a more sober series of reflections on the trade-offs entailed by trying to do conservation and development simultaneously. Whilst integrated conservation and development projects (ICDPs) as an approach have become less influential, there continues to be broad agreement that we do need to reconcile conservation and development, and that the 'communities' bearing the costs of conservation must be better treated. Whether the approaches that aim to continue this conceptual project have learned the lessons of ICDP experiences remains to be seen.

Chapter 11, 'Ecotourism, conservation and development', examines the birth of ecotourism as a way of integrating conservation and development. Both conservation and development constituencies share a strong interest in ecotourism and through such initiatives efforts are made to improve the conservation status of an area and deliver benefits to local people often through the use of market mechanisms. Whilst ecotourism in many ways is one of the most lucrative examples of market-based conservation and development, it is also constantly at risk of providing cover for 'greenwash'; although the advent of ecotourism certification schemes have sought to tackle this problem. Because of its commitment to free markets and trade, some commentators see ecotourism as a way of 'neoliberalising nature', that is, of bringing nature more and more under the dominant control of market mechanisms. As long as ecotourism remains, in Martha Honey's oft-cited phrase, 'the road less travelled', it risks co-option by a broader tourism industry which is very far from internalising environmental objectives into its core model, or to contributing to reducing poverty and inequality (Honey, 2008).

Chapter 12, 'Conservation and development in the context of religious, spiritual and cultural values', examines the challenges and opportunities in linking religion with conservation and development. The world's religions have made a contribution to environmental conservation and sustainable development through faith-based non-governmental organisations. Partnerships between secular conservation and development organisations and faith-based groups, however, are not always easy. Such partnerships face various challenges, including differences in worldviews, conflict between identities, and the attitudes and behaviour of religious groups that may not be favourable to conservation and development. Despite a possible overlap of values, these incompatibilities can often cause tensions between secular organisations and religious groups. A number of examples, however, suggest that faith-based groups are starting to address these incompatibilities. This chapter suggest that partnerships with faith groups might be valuable because these groups can enhance public support for conservation and development. While secular organisations need to work with faith groups on the basis of shared ethical or moral values, identifying effective ways to strengthen the linkages between secular organisations and faith groups is also necessary.

Chapter 13, 'Payments for ecosystem services in the context of material and cultural values' applies cultural values to ecosystem services. Ecosystems services are provisions that humans derive from nature. Ecologists trying to value ecosystems have proposed five categories of these services: preserving, supporting, provisioning, regulating and cultural. While this ecosystem services framework attributes 'material' value to nature, sacred natural sites are areas of 'non-material' spiritual significance to people. The chapter asks whether we can reconcile the material and non-material values. Ancient classical traditions recognise five elements of nature: earth, water, air, fire and ether. The chapter demonstrates that the perceived properties of these elements correspond with the ecosystem services framework. Whilst the two can be reconciled, the 'elements of nature' framework is more suitable to make a case for conservation of sacred natural sites because it can be attractive to traditional societies whilst being acceptable to Western science. The chapter goes on to examine practical application of this framework for designing payments for ecosystem services, which aim to achieve both the conservation of nature and the alleviation of poverty.

Chapter 14, the conclusion, identifies three key themes and trends pertinent to future efforts to reconcile conservation with development. Two of these trends, around conservation and conflict, and around resilience, conservation and development, are already receiving increasing attention. However, the third theme speculatively considers what the relationship between conservation and development might look like if it paid more attention to the 'urban transition': a fundamental, ongoing transformation which is going to be critical to the prospects of both. We offer this in view of our critique that current efforts to achieve conservation and development objectives simultaneously do not address the underlying drivers of biodiversity loss and poverty, and suggest it as an area to focus on more in the future.

PART 1
Histories of conservation and development

① Conservation and development in historical perspective

- The early roots of conservation
- Managing global nature: science and conservation
- The consolidation of international conservation

Introduction

From at least the seventeenth century onwards there has been an increasing sense of worry that there is a limit to what we can take from, or do to, our natural environment (Grove, 1995). It is a narrower history of the concern with limits which we present in this chapter, as opposed to the much broader history of human thought on nature. It is the concept of limits that gave rise to, and continues to galvanise, global environmental movements. It remains the essential tension which has so frequently set the clamour for development against the conservation imperative.

This chapter sketches the historical roots of concerns for the environment, and surveys the conservation measures enacted as a result of these concerns. It provides examples of their genesis from Africa, Asia and the Americas. The importance of Western thinking – from the USA and Europe – generated in relation to Western contexts is, of course a very important part of this history. Yet environmental historians have shown that contemporary international conservation intervention emerges from "the kind of homogenising, capital-intensive transformation of people, trade, economy and environment" that drove European colonial expansion (Grove, 1995: 2).

The chapter begins by looking at the early roots of conservation, and the emergence of 'preservationist' and 'wise use' schools of thought.

It then charts the ways in which an environmental conscience emerged partly in the tumult and heat of European colonial expansion. Finally, it maps the consolidation of the international conservation movement.

The early roots of conservation

The commodification of nature and the creation of environmental concern

When Richard Grove published *Green Imperialism* in 1995, he was at pains to counter the impression that environmentalism derived historically from Northern responses to the local impacts of industrialisation, and that conservation interventions originated broadly in North America. Through his efforts and those of other environmental historians, we are now aware of a much richer, more dispersed history of the relationship between conservation and what these days we called development, as impelled by European colonial expansion. Yet a common element underlies both the home-grown reactions to industrialisation in countries like Britain and the wider arc of colonial encounters: the 'commodification of nature' (Beinart and Hughes, 2008; Castree, 2008). Turning natural 'raw material' into tradable commodities did not just transform landscapes: it fuelled colonial expansion, which in turn often (though not automatically) had an amplifying effect on global trade. It was the increasing scale at which this self-reinforcing process was occurring that triggered environmental concern, wherever it was observed. In this way, the engagement of conservation with capitalism – which has been the subject of attention in relation to the latter decades of the twentieth century and the start of the twenty-first (see chapters 5 and 6) – is not new. Rather, both conservation and development are the product of a capitalist economic model.

What is more specific to the North is its historical dominance of international environmentalism (Guha, 2000; Adams, 2009),[1] even processes of imperial expansion fundamentally shaped the environmental conscience that gave rise, in the twentieth century, to the formation of an international conservation architecture. In this picture, countries such as the USA, the UK and Germany figure centrally. Some aspects of the historical development of environmental thinking and conservation movements in these countries were driven by the national setting. Yet the kinds of conservation that emerged were heavily influenced by people and events across the world, even as they became truly international in their reach (Guha and Martínez-Alier, 1997; Guha, 2000). The UK and USA in particular were very important – though by no means the only – settings in which two sometimes opposed, sometimes overlapping impulses emerged. The first impulse was fuelled, in the eighteenth and nineteenth centuries, by a counter-reaction to the industrial revolution, and manifested itself in both the 'back to the land' movement of the UK and the 'wilderness' movement of the USA. The second impulse, part ambivalent and part pragmatic about these processes of intensification and transformation, sought to govern them through rational, scientific management to curb wasteful, destructive tendencies.

'Back to the land' and 'wilderness': the beginnings of US and UK conservation

In the nineteenth century, concern grew over the effects of urbanisation, the intensification of agriculture to serve urban markets and the proliferation of industries turning nature into commodities to trade in domestic and international markets. In the UK setting, Romantic poets – most famously Wordsworth but also others such as John Clare – decried the adverse environmental effects of a proliferation of coal mines, textile mills, shipyards and railway lines. They scorned what they saw as city dwellers that, however literate or sophisticated, had lost the elemental connection with nature abounding in even the humblest shepherd (Marsh, 1982). Central to the Romantic view was the sheer aesthetic beauty of nature, accompanied by an acute horror of its being despoiled by industrial pollution, and devalued by an inexorably expanding urban culture. But as the fascination with rural ways of living denotes, Romantics also favoured a particular relationship between humans and nature, one that they deemed industrialisation to have disrupted. In this there was more than an element of social realism: industrialisation did indeed change the structure of the agrarian economy in ways which undermined the basis of peasant farming (Williams, 1973). As wealthy landowners enclosed erstwhile communal grazing pastures in order to intensify production for urban markets, many peasants were forced to move to towns and cities in search of other ways to make a living (Williams, 1973).

Near-contemporary commentators such as John Ruskin, Edward Carpenter, and later William Morris, all took their cue from

Wordsworth, condemning what Guha terms the 'desacralisation' of nature (2000); that is, viewing the natural environment only as raw, exploitable material. They added to the Romantic view a keener, grittier awareness of the adverse impacts on human health of the industrial revolution. Where others saw progress – indeed something akin to what these days we would call development – they saw human misery: impoverished workers leading lives shortened by factory conditions, insanitary urban dwellings, smog-blackened air and contaminated drinking water. They rejected industrialisation and the creeping materialism accompanying it, advocating instead a rural life characterised by simplicity, self-sufficiency and a more harmonious relationship with nature (Marsh, 1982).

The Romantic movement was a critical source of inspiration for the societies that sprung up from the 1840s onwards, which intended to conserve a variety of things perceived to be at risk from industrialisation. These included:

- the Scottish Rights of Way Society, formed in 1843, and the Commons Preservation Society formed in 1865, which both sought to defend rights of way across/access to increasingly enclosed rural and urban landscapes;
- the Lake District Defence Society formed in 1883, dedicated to preserving the natural beauty of the area;
- The National Trust, dedicated originally to the preservation of both beautiful landscapes and architecture (though better known these days for the latter). (Guha, 2000)

A point in common with all of these societies was that humans and human creations were an integral part of what they wanted to conserve. The pre-industrial human–environment relationship was admired, and much value conferred on the intimate, intricate agro-environmental knowledge possessed by rural farming populations, regardless of their social standing. Yet this way of thinking found relatively little favour in key international conservation efforts driven by the North in much of the twentieth century, especially in relation to attempts to establish protected areas in the South. It was not until the 1970s that a dramatic resuscitation of the value of local knowledge would become a mainstream feature of international conservation (see Chapter 10). In many ways this can be explained by the power and prevalence of the wilderness 'ideal' that characterised US conservation thinking from the nineteenth century onwards, and particularly associated with the works of one of conservation's most enduringly influential thinkers, John Muir.

John Muir, though born in Dunbar, Scotland, is better known for his writings on places and events in the USA, particularly as they related to the Sierra Mountains of California. He founded the Sierra Club, one of the USA's most enduringly influential conservation organisations and provided the inspiration for the UK's John Muir Trust. He was a highly vocal commentator on the levels of deforestation driven by the expanding agricultural frontier and the trading that accompanied it. His fear that forests might disappear drove his passion for wilderness, that is, for landscapes which were not transformed by the forces of economic dynamism. He was a lover of nature for its own sake, a defender of the right of all species, not just humans, to persist (Cohen, 1984). He prized, moreover, the spiritual value of a wild landscape, which he felt was a necessary connection for people to maintain with nature, and which, he believed, underpinned the burgeoning camping and trekking industry catering to urbanites seeking distance from city life (Fox, 1985). Yet for urban sophisticates to be able to appreciate nature, it was necessary for it to be left 'untouched' by human hand, and it was this sentiment which drove Muir's insistence on a wild landscape as one free from human habitation. Where Wordsworth had considered the shepherd to be 'part of the scenery', a feature of the beauty of England's Lake District, Muir recoiled at the idea of grazing in his Sierra (though his ire was aimed as much at logging and mining as it was at livestock farming). He advocated strongly that the USA's nascent system of national parks should be guarded by the military, to ensure human 'disturbance' was kept at bay (Guha, 2000). These ideas would come to be echoed by commentators at the end of the twentieth century (for instance Terborgh, 2004 – see Chapter 10). In this, John Muir was a precursor of the 'siege' mentality which still runs strong in some conservation circles in the twenty-first century.

The establishment of Yellowstone in 1872, the first national park in the world, heralded the institutionalisation of the preservationist tendency in conservation, of cutting off 'wild' landscapes from the reach of permanent human settlement, and in particular the consumptive economic activities associated with it. This tendency would, as we shall see later on in the chapter, become one of the USA's most powerful conservation exports. A part of this model which has become equally enduring is the 'wilderness tourism' that was as – if not more – important than strictly ecological criteria in the creation of

Figure 1.1 *John Muir by Carleton Watkins, c.1875*

the USA's network of national parks (Runte, 2010). This notion of separating humans from wild landscapes, other than for the purposes of recreation or study, was very much contingent upon a particular moment in US cultural history. The notion of wilderness provided a means by which a settler society could invent its own history and cultural identity separate from those of the Native Americans whose presence it had fought to displace (Nash, 1973). In the notion of wilderness were also the foundations for a national culture which could rival the ancient regimes of Europe. They might not have the longstanding heritage, culturally or architecturally, of a France or a Greece, but they could hold their natural wonders, which *had* existed over a much longer timeframe, to surpass anything to be found in Europe (Nash, 1973). Critically, moreover, they could enforce a separation of people from landscapes much more easily than could be done in smaller, more heavily-populated European countries.

Managing global nature: science and conservation

Whilst the preservationist ethos was a central tenet of conservation in the USA, it ran in parallel – and indeed often in competition with

– another school of thought, one which did not start from the premise of prohibiting all forms of consumptive use of nature. Instead, it was proposed – and still is – that the key to the conservation of what human societies need from and value in the environment is the wise use of the natural resources available to society (Ekirch, 1963; Nash, 1973). What it shared with preservationism was a growing awareness that there were environmental limits to ever-expanding, consumptive economic activity: nature's bounty was no longer, even by the eighteenth century, considered boundless, nor its ruination beyond the reach of humankind. There was room, too, for aesthetic and spiritual engagements with nature. Yet the notion of wise use of nature made for a much more pragmatic stance towards consumptive economic activity. Squandering the natural resource base for the sake of ever-expanding trade was (and still is) seen from this perspective as a false economy.

Nevertheless, destructive as humans could be, eighteenth- and nineteenth-century commentators in Europe often lauded them as capable of great feats of creativity and redemption, the key to which was the harnessing of scientific knowledge. Just as looming environmental catastrophe was foretold by scientific observation, scientific management held out the prospect of living in greater harmony with nature. This thinking was at the heart of one of the seminal works in US environmentalism: *Man and Nature: or, Physical Geography as Modified by Human Action*, by George Perkins Marsh. First published in 1864 and still in print, it is difficult to think of a book which has been more influential within US environmentalism, even if figures such as Muir, Henry David Thoreau, Aldo Leopold and Lewis Mumford have made comparably weighty contributions.

As environmental historians have been documenting for some decades, it is important to recognise that the work of Perkins Marsh should be set in the wider context of the meteoric rise of scientific managerialism, which itself was intricately bound up with the project of European imperial expansion (Grove, 1995; Beinart and Hughes, 2008). It was this push for territory and trade which made scientific conservationism a truly global phenomenon. Colonialism provided a platform for ecological learning and assessment of the environmental impacts of trade. It also facilitated access to the altered environments which gave rise to an early wave of apocalyptic discourses of ecological collapse – which prefigure those of modern-day environmentalists. The implications of deforestation for soil erosion in Venezuela were famously captured by Alexander von Humboldt (Walls, 2009), whilst the forest conservation science of Pierre Poivre, Philibert Commerson and Bernardin de Saint-Pierre on the topical island of Mauritius gives them a strong claim to be "the pioneers of modern environmentalism" (Grove 1995: 9). Yet there was often a receptivity to the idea that short-term capital gain should sometimes be curbed for the sake of averting ecological crisis, which could be perceived as a threat to the longer-term economic and political security of the colonial state (Beinart and Hughes, 2008). Moreover, the emergence of national states with centralised administrations in the nineteenth century enabled scientific prescriptions to form the basis for government intervention into land access at the national level (Guha, 2000).

Professionalising and globalising forestry science

One of the best-known examples of the character and reach of scientific conservationism is the emergence of the state forestry sector as a global phenomenon. By the eighteenth century Germany had emerged as pre-eminent in forestry science, by virtue of the quantitative methods Prussian state foresters had devised for calculating growing stock and yield (Lowood, 1990). These calculations, which estimated with a high degree of accuracy the yields that could be expected of given species, formed the basis of the 'sustained yield' forestry model, which would become a very influential German export. Forestry schools in the German tradition were established across many different parts of Europe, including Austria, Poland, Russia and France. Yet German influence was not limited to Europe: as Franz Heske's work has shown, the thinking was disseminated much more widely through colonial expansion (Heske, 1938). Germany sent forestry experts to its own colonies, but they were also in demand from other colonial powers. They assisted the Dutch in establishing state control over forests in their colonial possessions, for instance for the extraction of teak from Java in the 1840s. At this point Britain, which lacked its own forestry schools, sent students to France and Germany for training. Britain also followed the Dutch example when it came to its colonies. From 1866 it appointed three successive former German Inspectors Generals of Forests to run the Imperial Forest Department in the British Raj: Dietrich Brandis, Wilhelm Schlich and Bertold von Ribbentrop (Heske, 1938). Its modern-day counterpart is the Indian Forest Service. What goes around comes around, and in later years, the same Dietrich Brandis served as mentor to Gifford Pinchot, the man selected to establish the first US forestry service in 1900, super-vising Pinchot's education in Europe and continuing to advise him following his return to the USA (Pinchot, 1998 [1908]).

As it transpires, the concept of sustained yield forestry did not always travel with great success even when measured on its own terms – at least when applied in the tropics. Guha (2000) suggests that India's forests were, by the end of the twentieth century, in a worse state than when the Indian Forest Service's forerunner, the Imperial Forest Service, was created in 1867. Ecologists have since questioned the notion of equilibrium in nature upon which sustained yield is based (Berkes, Colding and Folke, 2008; Gunderson and Holling, 2002). Nor should the geo-political purposes served by establishing conser-vation bureaucracies be forgotten. Establishing strictly protected forest reserves was also a means of consolidating state power, and one which often met with fierce local resistance which sought to wrest back local control over forests (Gadgil and Guha, 1995). Moreover, it is important to keep in mind that European forestry science was not always a mere blueprint exported from Germany or France and applied as a fully-formed doctrine in colonial territories overseas. As Grove (1995) reminds us, Dietrich Brandis himself acknowledged that the Indian Forest Service was also substantially influenced, when first set up, by pre-colonial forest usage systems deployed, for example, in Sind and Maratha (1995). This theme of local knowledge changing – and being changed by – conservation-oriented scientific knowledge and practice is not restricted to India: it is, in fact, a common feature of colonial encounters globally (Grove, 1995, but see also Beinart and Hughes, 2008). Chapter 10 demon-strates how the question of *whose* knowledge is privileged conserva-tion theory and practice continues to resonate in the twenty-first century. What we may not realise is that the history of this question is as old as that of conservation itself.

Caribbean plantations, environment and empire

A recurring theme in this chapter is the weight of influence carried by particular individuals in the history of conservation, as well as their loudly-voiced fears regarding the environmental impacts of expanding economic activity, be they preservationist or conservationist in tone. Yet the roots of contemporary conservation and development cannot be reduced solely to the places, persons and happenings which brought about opposing conservation and development 'camps', glaring at each other across the divide. They are embedded in the interwoven relationships between environment, empire and trade, which were mutually shaped by, and shaping of, each other. A compelling example of the push-and-pull character of these relationships is found in the plantation economies of the Caribbean and the Americas.

The plantation economies were central to an 'Atlantic order' that evolved roughly between the seventeenth and nineteenth centuries. This order was driven partly by shifting asymmetries of power between Europe, Africa and America, but also by environment and disease (Beinart and Hughes, 2008). The concentrated, intensive model of plantation production emerged to cater for European consumer demand primarily for cotton, coffee, tobacco and sugar. Yet the capacity to produce the quantities necessary to meet this growing demand depended on:

a) securing access to good sites for plantations; and
b) coercing sufficient labour to work in the plantations.

For the land access requirements, tropical islands, such as Barbados, Jamaica or Tobago, served very well, as they offered large areas of often-fertile land with easy access to the water-bound modes of transportation that literally kept imperial trade afloat (Beinart and Hughes, 2008). The crops grown in the plantations were not indigenous: sugar cane, for instance, came from New Guinea, spread to India through trade routes and spawned hybrid varieties along the way, prior to its introduction to the Caribbean (Diamond, 1998). In this way, environmental factors weighed heavily not just by providing the products for consumption but also by affecting the geographical distribution of production and, thereby, the geography of empire (Crosby, 2004). It was precisely because it was not possible to grow plantation crops in Europe that the Portuguese, Spanish, Dutch and British all sought suitable colonial possessions for their production on the other side of the Atlantic.

For the labour requirement, the plantations have become infamous for the use of African slaves, shipped across from West Africa. Paul Lovejoy (2000) suggests that as many as 11–12 million slaves were ferried across to the Americas between 1450–1750. Towards the end of this period as many as 80,000 slaves were transported each year, with brutal, heart-rending, cross-continental consequences. In this regard, the Atlantic slave trade followed the distribution of production across empires. The rationale for it was driven not just by political economy but also by the same environmental considerations that focused the European imperial gaze in the direction of the Caribbean. After all, it would have made more sense for Europeans to focus efforts on establishing plantations in West Africa itself, given its agro-ecological potential for staple commercial crops (Crosby, 2004). However, the presence and distribution of diseases like malaria and dengue in West Africa, driven by ecological factors, were one reason why it was so difficult for Europeans to establish a more predominant position in West Africa. They did not have immunity to these diseases and therefore maintaining a military presence or other forms of settlement was more difficult (Crosby, 1986). Other factors – such as the militarily strong and well-organised polities of West Africa (Eltis, 2000) – must also be considered, but the role played by ecological factors in impeding Europeans from establishing plantations in West Africa should not be underestimated.[2] They thereby contributed to making it necessary for slaves to be taken across the Atlantic instead.

Once in the Caribbean, environment and disease factors had devastating impacts on human demography. Not only did the pre-colonial Carib populations suffer the onslaughts of Spanish and British invaders, but African slaves and traded goods transmitted infectious diseases such as smallpox and tropical staples (at least in Africa and Asia) including malaria and yellow fever (Kiple, 1984). The pivotal role played by smallpox in the defeat of the Aztecs – who had no immunity – by the forces of the Spanish conquistador Hernán Cortés, is well known. Yet on the mainland, at least some non-immune populations in remote places, or at higher elevations, were spared. The fate of the Carib Arawak populations, extinguished by colonisation and principally tropical disease, was if anything even more extreme. Over time, slaves with immunity to African and European diseases – generally through exposure in childhood – were biologically favoured, in contrast both to Arawak and European settler populations (ibid.). This relentless, grim selection process determined the broad demographic characteristics of the Caribbean that remain broadly unchanged in the twenty-first century.

Whilst, then, this brutal epoch in Caribbean history demonstrates unequivocally the massive repercussions of environmental factors for imperial economic development, it also demonstrates the converse. The composition of flora was radically altered by the introduction of exotic species (Watts, 1987; Kiple, 1984). Both deforestation and soil erosion occurred on a dramatic scale on islands such as Barbados and Grenada (Richardson, 1997). One measure of the environmental toll taken is that sugar production in Barbados in the late eighteenth century was less than a third of what it had reached a century earlier (Carrington, 2002). Another measure is that by the late eighteenth century, Barbados was importing timber from Tobago, as a result of the scale at which forest clearance had occurred (Carrington, 2002).

However, one result of the environmental impacts of the plantations was the emergence of colonial forest protection regimes on St Vincent and Tobago (Grove, 1995). The very forces that transformed Caribbean environment and society were the same forces that generated the conservation imperative. If Kiple's account of the long-term environmental ramifications is to be believed, however, this early conservationist mentality, and the legislation that accompanied it, were at best insufficient (Beinart and Hughes, 2008).

Quagga-mire: European settlement, hunting and trade in Southern Africa

In his engaging book *Against Extinction* (2004), William Adams explains how the fate of the Quagga, an erstwhile cousin of the zebra hunted out of existence by the 1870s, symbolises an elemental driving force of the conservation movement: to prevent (human-caused) extinction. Hunting has played a contentious, double-edged role in the history of conservation, perhaps nowhere more centrally than in Africa. The thronged masses of migratory game herds encountered by early Dutch settlers of South Africa in the 1650s had dwindled by the nineteenth century, much reduced by subsistence and sport hunting carried out by Boer – and other European-descent – settler groups (Adams, 2004). The agricultural frontier that extended progressively northwards proved intolerant of competition from game – or people who happened not to be white settlers, for that matter (MacKenzie, 1988). Two crucial factors, then, driving the exponential increases in wildlife hunting were, first, the colonial expansion of the frontier, partly with a view to, second, the trade of wildlife in international markets (Beinart and Hughes, 2008). The effects of both trends were facilitated and amplified by the introduction of the firearm, which transformed hunting as much for pre-colonial as for European settler populations (MacKenzie, 1988).

Figure 1.2 *Quagga*

Yet concerns surrounding the fate of wildlife in Southern (and other regions of) Africa cannot be heaped solely on Boer (and other European-descent) farmers. By the 1850s, it was common for wealthy Europeans – typified by the Victorians – to set out on exploratory journeys, developing a seemingly insatiable appetite for the kill as they went (Mackenzie, 1988). Back in Britain, books with titles such as *The Wild Sports of Southern Africa* (Harris, 1850, cited in Adams 2004) found a ready market for exotic tales from distant continents. Yet as awareness of the scale of loss of game populations grew, so too did the desire to reign in the bloodthirsty tendencies, ushering in a new breed of hunter turned conservationist; the 'penitent butchers', as one memorable characterisation had it (Fitter and Scott, 1978).

Looming fears of a wave of extinction of popular game such as lion, elephant, rhino and giraffe united influential men such as Edward North Buxton in a new cause: preservation. Buxton was a wealthy Victorian businessman and an acolyte in the early struggles of the British conservation movement in the colonies. His travels in British East Africa and Somaliland – in which he himself hunted – persuaded him of the need for stricter action to preserve game. In his book *Two African Trips: with notes on big game preservation* (Buxton, 1902), he would argue not against hunting *per se*, but against indiscriminate killing which showed little or no regard for the health of the population being hunted. A 'true sportsman' respected his prey and knew when, therefore, to put the gun away. The issue, then, was not one of prohibition but rather of encouraging the 'right' kind of hunting, of curbing excessive tendencies. This framing of the hunting problem, propagated by Buxton and other leading commentators of the era such as Henry Seton-Karr, fingered the culprits even as it exonerated the 'true sportsman'. In the first instance, these were amateur sportsmen and naturalists whose desire, respectively, for trophies or specimens outweighed their consideration for the survival of the species. But settler farmers were also singled out for their share of the blame. Presaging debates that ran the length of the twentieth century, grounded in what we now call human–wildlife conflict, the farmers themselves resented damage to crops and the dangers posed by wild animals. Moreover, they were unwilling to give up an accessible, virtually free source of food which had, some argue, in effect subsidised the costs of European colonisation and settlement (McKenzie 1988). Yet, for John McKenzie (1988), the age-old Victorian prejudice against subsistence hunting became embedded in the conservationist 'mindset', and extended increasingly beyond the European settler farmer to the pre-colonial populations more broadly. Some argue that such prejudice was itself the cause of the decimation of wildlife numbers in Southern Africa. Commentators such as Thomas (1995) and Murombedzi (1992) maintain that in the case of modern-day Zimbabwe, the arrival of Europeans, and the systems of colonial administration put in place towards the end of the nineteenth century, displaced pre-colonial resource use regimes which had regulated hunting much more effectively. It was not until the latter half of the twentieth century that this attitude towards subsistence hunting in conservation circles would change; chapters 9 and 10 explore the contemporary legacy for local people of this position. In fact, as conservation grew as a movement, hunting would remain on the 'fault line' of conceptions of conservation as fencing off parts of the natural world, to be protected against human use, and conservation as the wise, scientifically rational use of natural resources for the prosperity of current and future generations.

Buxton, meanwhile, was more than an author: he was also a man of considerable influence. He was one of the founding members of the Society for the Preservation of the Wild Fauna of the Empire (SPWFE) in 1903 (Adams, 2004). The significance of the Society, as even its name suggests, is that it linked the preservation agenda explicitly to the project of empire. Not only did its gaze fall upon Britain's colonial possessions – many of them recently 'acquired' in Africa not long before the turn of the century – but it saw colonial administrations as the proper agents of preservation. In 1906, the Society called upon the British government to establish a network of protected areas in East Africa, based on the US model (Adams, 2004; McCormick, 1989). As this example demonstrates, whilst other conservation organisations had emerged in national settings, the SPWFE was the first to look beyond country borders and embrace a more global focus. It was in key respects a gentlemen's club that found it easy to concentrate on influencing the British government because it had so many links to it, both publically and privately (Fitter and Scott, 1978).

Fledgling colonial administrations were increasingly aware of growing fears of an extinction spasm (Adams, 2004). By 1900, in both Southern and East Africa, a number of colonies had passed legislation governing hunting, introducing stipulations on what could be hunted, in what season and by whom (MacKenzie, 1988). The other kind of initiative that colonial administrations were increasingly setting their sights on by the end of the nineteenth century was the game reserve. At first favoured more in German than British East Africa, the idea of a physical space in which to preserve endangered species from all hunting activities would become one of the cornerstone activities of the conservation world (Prendergast and Adams, 2003; Neumann, 1996). Sabie, the first game reserve in Africa, was established in 1892 (Graham, 1973), whilst the first national park came into being, courtesy of King Albert in the former Belgian Congo, in 1925 (Boardman, 1981).

However, at least in some cases, despite accommodations made towards the preservationist cause, colonial policy remained attentive to the demands of settler constituencies seeking to expand the agricultural frontier. For example, at the beginning of the twentieth century in erstwhile Rhodesia, the colonial government launched a tsetse fly clearance programme, which set up a hunting zone later extended all around the periphery of the Zambezi tsetse belt, an area of 10,000 km^2. By 1961, this programme had resulted in the slaughter of 750,000 game animals, with a view to ensuring that settlers and their livestock remained free of the sleeping sickness (trypanosomiasis) disease spread by the tsetse fly (Ford, 1971).

The consolidation of international conservation

Hunting in Southern Africa was one amongst a number of concerns which gave impetus to the internationalisation of conservation. Indeed, the consternation generated by the tsetse clearance programmes was cited as one of the factors justifying the signing of the 'Convention on the Preservation of Flora and Fauna in their natural state', ratified by most of the colonial powers in 1934 (McCormick 1989). As we have seen, the emergence of an environmental conscience, as well as a set of conservation measures grounded in botanical, forestry, soil and other sciences, emerged over centuries. These were born of trade and colonisation, of European encounters with geographies, peoples, knowledge and practices previously unknown to them. The rise of an international conservation architecture in organisational terms, from the late nineteenth century onwards, sprang from this many-splintered set of experiences from around the world. And yet it was dominated in many ways by Northern actors, both in terms of the countries in which international conservation organisations were founded and often the countries which were held up as providing good examples to practise in other parts of the world. This section cannot do justice to the intricacies and institutional complexity of the emergence of this architecture, but it provides a sketch of key moments, interventions and characteristics.

The pre-First World War and inter-war periods

The elitist beginnings of the establishment of conservation initiatives in the nineteenth century, in which powerful, well-connected individuals held substantial influence over key government actors, were just as evident at the start of the twentieth. In some instances, conservationists found sources of support right at the very top. In the USA, private conservation efforts found a ready champion in President Theodore Roosevelt; albeit in the sense of wise use, as opposed to the preservationism of John Muir. Gifford Pinchot, who had been so important in grounding the newly-established US Forestry Service in the doctrine of wise use, was a key advisor to and political ally of Roosevelt during his presidency (1901–09), and would go on to become a US senator in his own right, reflecting his desire to bring environmental matters into the broader social and political spectrum (Miller, 2013). One of his highest profile achievements in terms of

Figure 1.3 *Gifford Pinchot*

setting a national US conservation agenda was his organisation of the White House Conference of Governors on Conservation, held in May 1908. Whilst the conference was to some extent characterised by state resistance to the idea of federal control over natural resources within their boundaries, a key recommendation was the creation in each state of a conservation commission which would work with other state commissions and also the federal government. The publicity surrounding the conference was credited with making conservation a matter of national public debate in the USA for the first time (McCormick 1989).

Roosevelt and Pinchot were also keen to put conservation on the international stage, because they felt that it was not an issue that could be dealt with solely at the national level. The next logical step in this direction was an international conservation meeting which could speak to this agenda and burnish the environmental legacy of the Roosevelt presidency. Shortly before he left office, Roosevelt called a world conservation congress, which was to be held at The Hague, and invitations to 58 countries had been issued by the time he left office; half of the countries invited had accepted. With the

arrival of William Taft to the presidency, the meeting was can-
celled, and Pinchot removed from his post as chief of the US Forest
Service (Miller, 2013).[3] Roosevelt and Pinchot were denied this
ambition, but had demonstrated that there was international appetite
for a broad discussion about global conservation, and that the USA
would be a key interlocutor in any international processes
(McCormick, 1989).

Environmental events in the USA were also highly influential in
policies pertaining to colonial possessions. One well-known example
is the environmental catastrophe that came to be known as the Dust
Bowl, which affected the Great Plains (a vast region of flatlands in
the USA and Canada, occupying a corridor of land west of the
Missouri River and east of the Rocky Mountains). Between 1934 and
1937, this region was very badly affected by over 200 dust storms,
which, amongst other impacts, caused widespread soil erosion across
16 states, and left the USA reliant on wheat imports (Worster, 1987).
In its report into the causes of the Dust Bowl disaster, the Great
Plains Committee laid the blame on agricultural practices, such as
ploughing long, straight furrows, clearing fields of vegetation, overde-
pendence on cash crops and the destruction of native grasses (Worster, 1987). In the USA, the magnitude of the impacts dispelled the
idea, arguably too prevalent within 'wise use' conservation circles
that nature could be easily shaped to serve the pursuit of economic
gain (Worster, 1987). However, the episode also received significant
attention abroad, for instance, in British colonial agricultural policy in
Eastern and Southern Africa, where awareness of the Dust Bowl's
effects fuelled concerns about the effects of agriculture on the soil
(Anderson, 1984; Beinart, 1984).

Meanwhile, plans were afoot in Europe to create an international
conservation organisation – touted not least by national organisations
dedicated to birdlife conservation, whose work on bird migration was
tangibly international in character (Boardman, 1981). The idea first
came to prominence in the International Congress for the Preservation
of Nature, which took place in Paris in 1909; in 1913, seventeen
European countries met in Berne, becoming signatories to the 'Act of
Foundation of a Consultative Commission for the International
Protection of Nature' (Holdgate, 1999). Much of the credit for the
creation of this body was owing to Swiss conservationist Paul Sarasin. Like US efforts before them, these energetic beginnings at first
came to very little, stopped in their tracks by the outbreak of the First

World War the following year. The Commission's importance would become more in terms of the precedent it set than in terms of its work as a functioning international conservation body (Holdgate 1999). By 1928, however, momentum had been re-built, culminating in the creation of the Office International de Documentation pour la Protection de la Nature, established at the assembly of the International Union of the Biological Sciences (IUBS) (Boardman 1981). Dutch conservationist P.G. Tienhoven assumed the work of Sarasin, heading up the Office and securing its funding from the Dutch government. It was consolidated in 1934 as the International Office for the Protection of Nature (IOPN), and given a wider mandate, its fundamental object now being 'to work internationally for the progress of nature protection' (article 2 IOPN 1936, cited in Boardman 1981).

Post-Second World War

As had happened with the First World War the efforts towards greater international cooperation on conservation matters, which were beginning to be channelled through the IOPN, dissipated as a result of the outbreak of the Second World War (Adams 2004, 2009; Boardman 1981; Holdgate 1999; McCormick 1989). Yet one of the consequences of the War was a much more energetic, international order, bent on cooperation rather than conflict; unsurprising, given the terrible costs paid by millions around the world. The replacement of the League of Nations with the United Nations Organization, and the enthusiasm that accompanied it, provided a much more conducive context in which to pursue an international conservation agenda (McCormick 1989). At a world conference on nature protection, in Basle in 1946, organised by the Swiss League for the Protection of Nature and British biologist Julian Huxley, there emerged a majority in support of significant institutional change, as a means for pursuing this agenda (Boardman, 1981). It is not surprising that it was deemed fitting to make the United Nations the vehicle for such changes, though at first glance it might seem odd that UNESCO, the United Nations Educational, Scientific and Cultural Organization, was chosen as its institutional home. Not all UNESCO member states were at first convinced that international conservation should be part of its remit. Nor was it clear in what relation this new organisation should stand to the International Organization for the Protection of Nature. Nevertheless, after Julian Huxley was installed as UNESCO's first Director-General, these doubts were overcome (or at least sidelined), and a

'Provisional Union for the Protection of Nature' was tabled at the UNESCO general conference of 1947 (Holdgate, 1999). A year later, at a conference in Fontainebleau, the constitution for IUPN, the International Union for the Protection of Nature (Holdgate, 1999), was formally adopted. Reflecting the ambiguities over the proper role of governments that all along had pervaded the efforts to establish this international conservation organisation (Boardman, 1981), it consisted of a somewhat irregular melange of governmental and non-governmental actors of national and international ilk (ibid.). IUPN's *raison d'être* was to promote:

● wildlife preservation and the natural environment
● public awareness of the importance of conservation
● international conservation research
● conservation legislation, both national and international
    (Holdgate, 1999).

Early forays of international conservation organisations into development

All that was now required was a plan of action. This emerged largely from the 'International Technical Conference on the Protection of Nature', which the IUPN held concurrently with the UN Scientific Conference on the Conservation and Utilisation of Resources (UNSC-CUR). These meetings took place at Lake Success, in New York State (Boardman, 1981). The outcomes may not quite have lived up to the name of the conference venue, but these meetings produced a work programme that was to guide the construction of the international conservation agenda throughout the 1950s. Moreover, it was proposed that the IUPN and other UN agencies should work with development agencies with a view to conducting ecological impact surveys for development projects (McCormick, 1989). As Adams (2009) notes, this was one of the first expressions, within the context of the international conservation organisation, of the need to consider the environmental implications of development processes.[4]

The call to engage with broader development processes became stronger throughout the 1950s, prompting in 1956 a re-naming of the IUPN to the International Union for the Conservation of Nature and Natural Resources (Holdgate, 1999). Specifically, the 'wise use' brand of conservation, as advocated by Gifford Pinchot, and which had found strong purchase in the USA, had gained in influence relative to the preservationist leanings of many working within the field of international conservation (Holdgate, 1999). Yet this was not simply a case of shrewd institutional politics and lobbying from US conservationists – though the increased involvement of the USA in international conservation from the 1950s onwards was a marked development in the post-war period (Adams, 2009). Partly, it was a result of the rise of ecology, which emphasised, long-term, systems-wide approaches to understanding the workings of nature (Worster, 1987). This meant folding a 'flora and fauna' perspective into the broader purview of 'planet and people' and, thereby, a host of issues around resource use, population dynamics. Just looking to preserve parcels of natural landscapes was to miss an essentially global picture, to which conservationists must attend in its entirety. Looking at this bigger picture meant enquiring much more into social and economic considerations. In its 10th General Assembly, held in New Delhi, a new mandate for the IUCN was proposed: 'the perpetuation and enhancement of the living world – man's natural environment – and the natural resources on which all things depend' (Holdgate, 1999: 108). This more ambitious goal, in pointing out humankind's dependency on nature, was a precursor to the definitions of sustainable development that would become common currency towards the end of the 1970s. What it started to convey was the growing sense of concern over the magnitude of adverse environmental impacts resulting from human activity, the sense that the 'living world' itself could be in jeopardy. This preoccupation would form the backdrop against which ever more vocal calls for greater sustainability in human use of nature would be made.

It is as well to remember, though, that even if a vision for international conservation was well-established by the end of the 1960s, it was at best only partially translated into practical action. As Boardman notes, in the 1940s and 50s, there was not enough known about the status of flora and fauna, threatened or otherwise, across many countries, so as to be able prioritise interventions (Boardman 1981). Nor was there a cadre of professionals with the knowledge and skills to apply the recommendations generated by conservation thinking even where they were available (ibid.). If the workings of international conservation seem slow to us today, it is at least in part because they got off to a slow start.

Some of the crucial efforts to remedy the information deficit sought further to consolidate the links between conservation and development, not least through the establishment of international scientific organisations which sought to ascertain, in global terms, 'the biological basis of man's welfare' (Worthington, 2009: 5). One attempt to bring ecological and biological insights, both theoretical and practical, to the world of international development, took the form of the International Biological Programme (IBP). Advanced by the International Union of Biological Sciences, under the auspices of the International Council of Scientific Unions (ICSU), the IBP comprised seven broad research groups, one of which was terrestrial conservation (Worthington, 2009). This exercise in research planning and coordination was accompanied by various meetings, from the 1950s onwards, all sharing the same broad objective of uniting important actors from across the world to discuss global environmental topics (Dahlberg, 1985). Moreover, out of the International Biological Programme emerged the Scientific Committee for Problems of the Environment (SCPE), part of whose remit was to look at environmental problems and strengthen scientific understanding of them in developing country contexts (Worthington, 2009).

Summary

- Conservation and development have always been 'joined at the hip', as it were. The desire to conserve nature from destruction, and the documenting of environmental harm, crops up, of course, throughout recorded history. However, the narrower concern that human actions could exceed natural limits actually arises as a response to – and is a by-product of – the massive expansion in European empire, trade and economic activity which sought to turn nature into commodities, from the eighteenth century onwards.

- The scientific revolution that contributed so much to the spread and intensification of this process also provided the tools for documenting the environmental impacts of this expansion, and in many ways the impetus for those who wanted to avoid or minimise these impacts.

- The emergence of conservation and the ever more frequent tendency to mention it along with development cannot, therefore, be attributed to the rise of the international conservation architecture, from the mid-nineteenth century onwards. Yet whilst environmental historians have deeply enriched our understanding of the beginnings of conservation, no-one would deny that international conservation as a movement was dominated in its origins – and continues to be dominated by – Europe and the USA.

- The making of the international conservation architecture, largely in the twentieth century, was a slow-burn process, interrupted by two world wars and sidelined by a preoccupation with recovery and economic prosperity whose environmental impacts would only be challenged later on.

- Within conservation, two broad strands can be identified as different ways of responding to what we would these days call 'development': preservationism and wise use, which both drew upon an increasingly broad spectrum of emerging environmental sciences to further their objectives.

- Preservationism sought to draw the line between humans and wilderness, fearing that the latter would succumb to the reckless proclivities of the former.

- Wise use similarly started from an awareness of – often bordering on horror at – human-caused environmental disturbance, but envisaged the possibility of redemption through the instigation of rational, scientific management regimes.

- These two different philosophies and modes of practice continue to run deep in the conservation universe, and to prompt a host of views on what the relationship between conservation and development should be.

Discussion questions

1. With which kinds of conservation – preservationist or wise use – are you most familiar? To what extent are they related directly or indirectly to development considerations in the contexts you know best?

2. What are the implications for contemporary conservation of its roots in 'Western' ways of thinking and projects of Imperial and trade expansion?

3. How similar and different do you think the international conservation architecture is, in the early decades of the twenty-first century, to how it was at the beginning of the twentieth?

4. How similar or different are global economic development models today from the ones presented here, which were rooted in an ever increasing level of the commodification of nature?

Notes

1. This dominance is somewhat diminished at the start of the twenty-first century.

2. Perhaps the bigger question is why West Africans did not themselves establish plantations – see Beinart and Hughes (2008) for posited explanations.

3. Pinchot's dismissal from office was unsurprising, given that one of his other claims to fame in American history was the bequeathing of the set of publicity techniques he used to brief – with divisive efficacy – against President Taft. These were seen as so effective that US bureaucrats adopted them as a model in manipulating information (Miller, 2013).

4. As Adams also notes, this was also an early instance of a rhetoric which came far in advance of the changes envisaged: implementation of these early environmental impact assessments would not occur for another 20 years.

Further reading

Adams, W. M. (2004) *Against Extinction: the Story of Conservation*, London, Earthscan.

Beinart, William and Lotte Hughes (2007) *Environment and Empire*, Oxford University Press.

Boardman, R. (1981) *International Organisations and the Conservation of Nature*, Bloomington, IN, Indiana University Press.

Guha, Ramachandra (2000), *Environmentalism: A Global History*. New York: Longman

Marsh, G. P. 1965. *Man and Nature*, Harvard U.P; Oxford U.P. Also available to read for free on Google Books

McCormick, J. (1989) *The Global Environmental Movement: reclaiming paradise*, London, Belhaven Press.

Online resources

Environmental History Journal – http://www.environmentalhistory.net/ [accessed 29 January 2015]

Environmental History resources – http://www.eh-resources.org/press [accessed 29 January 2015]

The Sierra Club – http://www.sierraclub.org/about [accessed 29 January 2015]

George Perkins Marsh – http://history1800s.about.com/od/americanoriginals/a/ GeorgePerkinsMarshbio.htm [accessed 29 January 2015]

George Perkins Marsh Online Research Center – http://cdi.uvm.edu/collections/ search.xql?start=0&term1=george%20perkins%20marsh&field1=ft&rows=25& fq=parent_facet:%22George%20Perkins%20Marsh%20Online%20Research%20 Center%22 [accessed 29 January 2015]

The international Union for the Conservation of Nature (IUCN)

http://www.iucn.org/ [accessed 29 January 2015]

http://www.iucnredlist.org/ [accessed 29 January 2015]

http://worldparkscongress.org/ [accessed 29 January 2015]

Society for the Protection of the Fauna of Empire – www.fauna-flora.org/wp-content/ uploads/setting-up-of-FFI_1904.pdf [accessed 29 January 2015]

2 Sustainable development in the twentieth and twenty-first centuries

- The onset of popular environmentalism, 1962–1972: a spur to sustainability thinking
- 1972–1992: the birth of the sustainability paradigm
- The rise and fall of sustainable development, 1992–2012

Introduction

Whilst, by the 1950s, an international conservation architecture had been established, an international focus on reconstruction and reconsolidation, dominated the post-Second World War international agenda, leaving little space for environmental politics. In the 1960s, conservation was given much greater impetus after environmentalism exploded into (Northern) public consciousness, spearheaded by commentators who were raising the alarm about the destructive impacts of the accelerated industrialisation that characterised the twentieth century. The idea that there were ecological limits to growth was gaining traction, albeit not as rapidly as the development processes it targeted.

The establishment of the international conservation architecture, in tandem with this popular environmental push, sparked the 'official' birth of the sustainability agenda, crowned in 1972 with the Stockholm Conference on the Human Environment and the establishment of the United Nations Environment Programme. This 1972 conference paved the way for a series of 'Earth Summits', in Rio in 1992, Johannesburg 2002 and most recently again in Rio in 2012, which have provided a framework for international negotiations around sustainable development.

An exploration of these events leads the chapter to identify two key tendencies:

1   The multilateral ideals, and the international institutions that emerged to put them into practice, after the Second World War, have proved unequal to the task of garnering sufficient agreement and engendering the necessary collective action to tackle environmental problems robustly.

2   When push has come to shove, safeguarding the prospects for short-to-medium term economic growth has been deemed more important than environmental concerns by the countries – North and South – whose economies have the biggest environmental impacts.

The onset of popular environmentalism, 1962–1972: a spur to sustainability thinking

The early roots of popular environmentalism

As Chapter 1 explains, the institutional architecture of international conservation was slow in the making, and the conservation agenda correspondingly marginal. Even though the international conservation institutions that had been created were very much a product of the multilateralism that came out of the quest to build a safer and more peaceful world, environmental concerns paled into insignificance relative to the task of rebuilding. Moreover, the international conservation architecture remained an elitist affair, which infrequently forayed into the realm of popular concern (Adams, 2009). Yet the pace quickened in the 1960s, and not solely – or perhaps not even primarily – owing to these international conservation structures. Conservation was given new impetus by the outbreak of a popular wave of environmentalism, which in many ways took its cue from the lack of heed paid to the environmental impacts of the overwhelming focus on reconstruction that had characterised the post-war period (Guha, 2000).

One of the core characteristics of the popular environmental movement that emerged was that its focus was much broader than the 'preservationist' vein in conservation thinking, as it sought to engage with the environments people lived in, not just those characterised by high levels of biodiversity which were to be hived off from all human use. The environmental movement's message was much more explicit about society's relationship with the environment, and how some of the practices that impelled the development of industrial society were coming back to haunt it. In this sense, it shared an intellectual heritage with the 'wise use' school of conservation, looking at the wider ramifications of natural resource use, and how these could be regulated and reformed. This was not, however, always conscious: as Hay points out, 'the modern environmental movement set out on a path it thought to be wholly novel, the subsequent discovery of precursors being an unlooked for surprise' (2002: 26).

The backdrop to this popular wave of environmentalism, at least as it manifested in the US and Europe, was the prizing open of the existing political space by the New Left (Rome, 2003; Rootes, 2004). Its genesis was, in other words, entwined with that of the range of political and social movements that rose to prominence in the 1960s, which in various ways presented a critique of the status quo and a challenge to it through the means of mass protest and activism. Environmentalism profited as much as any movement from the development of the mass media, which made it easier to communicate environmental messages, whilst direct mail marketing and polling techniques made it much easier to build mass memberships (Mitchell, Dunlap and Mertig, 2014). However, environmentalism was not equally well-linked to all social movements. In the USA, for instance, there were strong links to the peace movement. 'Going back to nature' was also a core value for the hippy movement in its rejection of liberal, capitalist democracy. Yet environmentalism was conspicuously distanced from the civil rights movement, some of whose proponents saw it as a source of competition for scarce resources, and as an elitist white middle-class dominated project that was not concerned with poverty and inequality. Justice demanded that these be addressed; concerns with wilderness paled by comparison. Indeed, only in the 1980s would the environmental mainstream become more concerned with the extent to which issues pertaining to poor and/or specific ethnic groups mapped onto higher exposure to environmental contamination, via the concept of environmental justice (Brulle, 2000). In contrast, in the global South, considerations of justice, access to resources and exposure to pollution were always central to what might, to paraphrase Guha and Martinez-Alier (1997), might be called the 'environmentalism of the belly'.

Environmental disasters

In the twenty-first century, environmental disasters are an established topic for news stories, activist campaigns and government oversight, and very much a part of the mainstream of current affairs. When, in April 2010, an explosion on the Deepwater Horizon drilling rig killed 11 people and leaked 4.9 million of barrels of oil into the Gulf of Mexico, it became a test of US president Obama's authority and a turning point in the fortunes of BP, the company ultimately responsible for it.[1] At the Fukushima Daiichi nuclear power station, Japan, in the aftermath of the Tōhoku earthquake and tsunami in 2011, the radiation leak, the evacuation of the surrounding area and the frantic efforts at stabilising the three reactors which experienced meltdown all received wall-to-wall international press coverage.[2] As Chapter 1 has shown, the adverse environmental consequences of trade, empire, and industrialisation have long been a cause for concern. Yet the very public and widespread coverage of environmental disasters characterising these contemporary examples has its roots in the 1960s.

The events at Fukushima are simply, in fact, a more recent example of the consternation that the onset of the atomic age brought about, a cruel awakening ushered in by the devastating power of the nuclear bombs dropped on the Japanese cities of Hiroshima and Nagasaki, towards the end of the Second World War in 1945. In the nuclear arms race that ensued – in the first instance largely between the USA, Soviet Russia and Britain – testing comprised a fundamental stage in the development of atomic weapons. The increasing anxieties around the nuclear fallout produced by such tests has been held up as one of the earliest examples of an environmental problem publically acknowledged to be of global proportions (McCormick, 1989).

Oil spills, likewise, became a public issue in the post-war period, made much likelier by the substantially greater quantities of oil tankers (Gundlach, 1977) that were transporting the commodity around the world, and the corresponding increase in the size of the shipping vessels. In the 1960s, incidents such as the sinking of the Torrey Canyon tanker off the coast of south-western England, and the blowout in 1969 on a Union Oil Company platform near the coastal city of Santa Barbara, California, prompted public debates on how rises in personal affluence had not been matched by the public provision of environmental and health protection. It was the missing link in the post-war search for prosperity, a point made pithily by 'new liberal' J.K. Galbraith in his enduringly influential book, *The Affluent Society* (1998 [1958]).

The environmental 'prophets of doom'

One of the key catalysts of popular environmental concern was a mushrooming of books which sought to take scientific accounts of environmental problems to a wider audience, and which in the process captured the attention of the major political actors of the day. The most influential of such publications, now a canonical text in the history of conservation, often credited with kick-starting the global environmental movement (i.e. Fox, 1985), was Rachel Carson's *Silent Spring* (1963). This book offered an account of the threat posed by the 'contamination of man's total environment' (p. 2), through the use of pesticides – iconically DDT – in large–scale agricultural production

Figure 2.1 *Rachel Carson*

in the USA. She argued that pest-controlling chemicals had been allowed to leach into the environment in such quantities that they had been transferred to the food chain, both for animals and humans, but with little real understanding of their effects. These 'elixirs of death' (the title of Chapter 3), she maintained, were potentially so harmful that unfettered use of them could 'still the song of birds' (p. 10), to which the silence of the book's title alluded.

In addition to quickly assuming best-selling status and being published in 15 countries by the end of 1963, *Silent Spring* attracted some highly influential followers, notably John F. Kennedy, and was attributed with being a key factor leading to the introduction of the Pesticide Control Act in the USA in 1972 (Fox, 1985). Unsurprisingly, given the interests at stake in the agricultural sector and the very direct line of attack adopted by Carson, her thesis was quickly embroiled in controversy. Carson was frequently derided by critics as an amateur with a poor grasp of the science on which she based spurious, overblown claims. In particular, her central thesis that the harm caused by the insecticide DDT outweigh any benefits (including as an anti-malarial treatment) is highly contested. The polemic surrounding her work remains unresolved: it has simultaneously been voted one of the 25 greatest science books of all time[3] and given an 'honourable mention' in the list of contenders for the 'Ten most harmful books of the 19th and 20th centuries'.[4] (Both of these sources come with the health warning that they may say as much about those conferring the awards as they do about *Silent Spring* itself.)

However contentious, *Silent Spring* set the tone for a number of works published over the course of the 1960s and into the early 70s, which in different ways, and through different issues, collectively expressed the idea that the environmentally destructive tendencies of industrialised development were looking increasingly likely to undermine humanity's common future. More than anything, the worry was that there were limits to the Earth's capacity to absorb our rapidly-accelerating, post-war quest for ever greater economic expansion and prosperity. When a critical mass of thinkers started to suggest that these limits might be reached sooner rather than later, and with globally catastrophic ramifications, they came to be known as the 'prophets of doom'.

Population, technology and pollution were three key themes singled out by this pessimistic turn within the popular environmental agenda

(McCormick 1989). They remain at the heart of contemporary debates around squaring conservation and development. In the late 1960s and early 1970s perhaps the best-known commentators spearheading them were Paul Ehrlich, Barry Commoner and Garrett Hardin.

Concerns about population have a venerable history, long pre-dating but most frequently associated with Thomas Malthus and his *Essay on the Principle of Population* (Malthus, 1999 [1803]). The central thesis in Malthus's work was that (exponential) human population growth would, if left to its own devices, eventually outstrip (arithmetical) food production (see Chapter 7 for more detail). In the late 1960s, in the context of increasing human population growth at a global level, Paul Ehrlich, brought these arguments into public debate with his book *The Population Bomb* (Ehrlich, 1975 [1968]). He maintained that the scale of this growth would quickly provoke widespread famine by the end of the 1970s. Ehrlich argued that the best – indeed the only truly effective – environmental protection was birth control. He decried the degradation and other environmentally adverse impacts he saw as the result of efforts to ensure that agricultural production responded to population increases. By the mid 1970s, three million copies of *The Population Bomb* were in circulation.[5] Like Rachel Carson, Ehrlich excelled at reaching a popular audience, though over time, and especially when the levels of famine he projected did not come to pass, his work would become just as controversial as hers.

Garret Hardin, an ecologist at University College Santa Barbara, was another scholar firmly in the neo-Malthusian mould. His most famous work explored what he viewed as the inexorable logic that impelled humans to treat the environment and its resources so irresponsibly. He captured this with the notion of the 'Tragedy of the Commons' (Hardin, 1968), a term which remains in frequent contemporary use. The Tragedy of the Commons is Hardin's scenario of a pasture open to as many herders as can gain access to it. From the viewpoint of an (instrumentally rational) individual herder, seeking to maximise his or her benefits, it is best to have as many animals grazing the pasture as possible. However, if all herders pursue this strategy, they will degrade their commonly held resource base, which will become less and less productive over time. In the absence of regulation to control access, the resource base becomes wholly depleted. Individuals receive all of the benefits accrued from putting more animals out to graze, but bear only a small share of the costs of resource degradation, albeit one that increases over time. This dynamic provides the incentive for self-interested behaviour, also known as the 'free rider problem', tragically at odds with the maintenance of a finite resource base. Hardin's fear was that the same logic applied to population growth, and saw in this the makings of global tragedy, which led him, like Ehrling, to become a strong advocate of population control measures. He also explored the ethics of coming to terms with life in a limited world – most provocatively captured in the title of his paper, 'Lifeboat Ethics: the Case Against Helping the Poor' (1974). The essentially Malthusian approach taken by Erlich and Hardin has been staunchly critiqued in recent decades (see Chapter 7), but the fear of runaway population growth remains with us both in public debate and policy spheres.

Quantifying the limits to growth

Common to all of these thinkers was a rehabilitation of the longstanding notion, evident also in Chapter 1, that human activity could exhaust the natural asset base, smothering its capacity for regeneration. That there were biophysical limits that we looked to be exceeding was not new, but the scale at which human activity was propelling us towards them, as well as the magnitude of the consequences, was what the 'prophets of doom' added to the existing picture. Their legacy includes those questions, so familiar to us now, of whether or when fossil fuels would run out, whether we could produce so much pollution that our future (and current) prospects might be irrevocably damaged, and how many other forms of life would become extinct as a result of our settlement and consumption patterns, in the period which would come to be known as the Anthropocene (see Chapter 4). As McCormick (1989) points out, none of these ideas was taken as capable of explaining the whole of the environmental crisis that seemed to loom. Problems of population, hunger and pollution were instead better seen as symptoms of the underlying 'disease': the kind of economic growth which was, in Northern countries at least, heightening prosperity in the post-war period – the very engine of progress itself – would also be our undoing.

With this primary cause identified, the challenge became one of quantifying the size of the impacts of economic growth, and the timeframes within which biophysical limits would be reached.

The projections of scholars such as Ehrlich, were a partial attempt to do precisely this. But a broader, more systematic exercise was deemed necessary, one which could take advantage of innovations in computer modelling in order to project into the future the likely trajectories for the 'end of growth'. The most enduringly influential example of these efforts is the programme of work that emerged from what became known as the 'Club of Rome'.

Spearheaded by Italian management consultant Aurelio Peccei, the Club formed in 1968, comprising a mix of scientists, economists, pedagogues and industrialists. The Club's findings were reflected in and popularised by *The Limits to Growth* (Meadows, Randers and Meadows, 1972). This study employed the computer modelling of social systems pioneered by Jay Forrester to generate twelve potential scenarios for the 'end' of growth, with modest consumption and population levels achieved at one end of the scale to catastrophic crash at the other. All twelve scenarios put the end of economic growth at some point in the twenty-first century. Having set up the contrast between needless catastrophe and the rosy future that humankind could enjoy if shaken from complacency, the aim of the book was to make the case for a global effort to bring human impact on the environment to within critical ecological boundaries. In this sense, whilst it built on the neo-Malthusian logic that infused the work of the 'doomsters', it was also characterised by an optimistic view of the ability of humans, guided by the wisdom of science, to rise to the existential challenge that Club of Rome members felt themselves to be chronicling.

Not all were readily convinced of the need to impose limits on growth, chiefly owing to doubts about the quantity and quality of the data fed into Forrester's models. It came under particular attack from a group of thinkers who came to be known as the 'cornucopians' or 'doomsters' (see also Chapter 7). They held to the belief that there would be no resource crisis, on account of a different reading of the continued abundance of the earth's resources, and a seemingly unshakable faith in the ability of humans to adapt and innovate in times when scarcity loomed (see Simon and Kahn, 1984 as an example of supreme technological optimism). Moreover, 'doomsters' have in recent decades been challenged frequently on the grounds that neither the level of environmental catastrophe they initially projected, nor the absolute depletion of heavily traded natural resources, have come to pass (i.e. Peet, Robbins & Watts, 2010).

Indeed most governments remain opposed to the idea that there is a limit to growth, which has not shaken off its characterisation as radical, almost subversive. *The Limits to Growth* was, nevertheless, influential in important places, and as one of the key framing documents for the world's first 'Earth Summit', it was a key part of generating the context in which calls for sustainability would be made.

1972–1992: the birth of the sustainability paradigm

Antecedents for the first 'Earth Summit'

The onset of popular environmentalism provided greater impetus to conservation efforts at the international level. The re-evaluation of the development trajectory of the richest parts of the world, in particular the influence of the view that it could end in catastrophe, is what led actors within the international conservation architecture to argue for the need to engage more with development debates. Yet these structures suffered a deficit of legitimacy, because they were not sufficiently representative of countries in the developing world (Adams, 2009; Boardman, 1981). One of the principal attempts to bridge this gap took the form of UNESCO's Man and the Biosphere Programme (MAB). The creation of MAB underlined a growing recognition on the part of conservationists that if they wanted to be taken seriously and to attract the kinds of funding required for the increasingly ambitious scope of their enquiries, they needed to be able to speak the 'language of development' (Adams, 2009). The idea for the Man and the Biosphere programme emerged from the 'Biosphere Conference' of 1968, held in Paris. Its main message, that conservation should be as much an objective as human profit in the use of natural resources, proved highly influential in the intergovernmental context (Batisse, 1982). The purpose for the MAB programme was therefore to establish a sound basis for the rational use of the biosphere's resources, which meant producing research on global environmental processes, but furbishing it in a format useful to resource managers. Its wide-ranging objectives were not, however, matched by funding or capacity (Batisse, 1975), meaning that its ambitions were from the outset unlikely to be realised. However, because of the links it made between the functioning of social and ecological systems, which were at the heart of MAB thinking, the programme can be seen as a

Figure 2.2 *Entrance to Pachmarhi Biosphere Reserve*

forerunner to the popular notion of 'sustainable development' that would follow (Adams, 2009).

One of the main outcomes of the MAB programme, in addition to its numerous research programmes and education outreach initiatives, were the biosphere reserves themselves. These remain the enduring legacy of MAB, and by the end of 2011, according to its website, 580 sites had been established in 114 countries, demonstrating their continued popularity. A biosphere reserve is essentially a form of protected area characterised by different zones, each with differing rules governing the use (or prohibition of use) of the environment. Though there is a diversity of types of biosphere reserve, it is quite common for them to consist of three zones:

1   at its core, a zone in which all uses of biodiversity are prohibited;
2   a buffer zone surrounding this core, in which some uses of biodiversity are permitted;
3   a zone in which uses of biodiversity deemed to integrate conservation and development objectives are encouraged.

The United Nations Conference on the Human Environment

The Man and the Biosphere Reserve Programme largely prefigured the themes of the meeting most commonly linked with the emergence of the concept of sustainability within the international conservation arena (McCormick 1989). Despite its reduced claims to originality, the United Nations Conference on the Human Environment, held in Stockholm in 1972, delivered in the broadest terms a twofold message which, it was hoped, would unite industrialised and developing countries in one cause. The message was as follows.

1   Global development had to take into account and reduce environmental impact if its own future was not to be compromised.
2   Environment and development objectives should not and need not be set up in opposition; rather they could be reconciled.

At this point, there was significantly less said on the matter of *how* to integrate environment and development objectives, which was perhaps in part owing to the novelty of the idea (Clarke & Timberlake, 1982). Yet there was likely also an element of expedience: even at this early stage, developing countries were dissatisfied with some of the concerns central to the growing environmental engagement with the development agenda (Clarke and Timberlake 1982). Suspicions abounded that pronouncements on the need to reduce population growth and reconfigure the use of natural resources was an attempt by industrialised nations to deny poorer countries the same development trajectories they themselves had enjoyed (UNEP, 1978). Indeed, such were the tensions and concerns over the UNCHE meeting that preparatory talks were held in Founex, Switzerland, to smooth the way. The message that came out of it was that greater environmental protection was not to be a brake on rapid industrialisation; it was, rather, to show how this objective could be achieved in ways which minimised its environmental impacts (McCormick 1989).

Even at the time of the conference which is credited with giving birth to the concept of sustainability, there were, then, clearly competing groups with divergent interests. Assertions that everyone's environment and development aspirations could all be accommodated merely papered over the early dividing lines over what sustainable development would look like, and *whose* vision of it would predominate. Just as much, then, as the message of reconciling conservation and development processes, the legacy of the Stockholm meeting was the inevitability of the trade-off. The important question even at that point was not so much what sustainability was, but by what political processes, power relations and sets of interests its meaning would be contested. Furthermore, it is somewhat anachronistic, not to mention simplistic, to attribute to this moment in the 1970s the emergence of the idea of sustainability. Chapter 1 has shown how the question of how to continue to use natural resources profitably in ways that do not damage the resource base, has been asked for centuries. It is more precise to suggest that what emerged was the question of whether the patterns of economic development could be sustained indefinitely with the environmental resources available; and if not, then how these might need to change. Nevertheless, Stockholm was key to establishing usage of 'sustainability' in the ways it is deployed in contemporary contexts.

The most tangible outcome of the conference, the formation of the UN Environment Programme (UNEP), was also rooted in the dynamics of the time. The UNEP was conceived as a means through which to coordinate all environmental activity within the UN and to ensure that environmental thinking and application infused the varied work of the various institutions which comprised the UN. In line with the increasingly internationalist ethos of the UN, Nairobi, Kenya, was chosen as the location for the UNEP headquarters. It was also given the unworkably broad remit that had surfaced in the UNCHE meeting – again, partly as a result of the necessity of incorporating a wide range of agendas and interests. When this was coupled with poor funding, a precedent for international 'action' on sustainable development was well and truly established, and the UNEP was left unable to fulfil its objectives (Boardman 1981).

Our Common Future: the Brundtland Report

For all of the fine talk about reconciling conservation and development that had surrounded the Stockholm meeting, guidance on how to translate these ideas into tangible outcomes was hard to come by. An early attempt to identify the practical measures that could be taken was spearheaded by the United Nations organisations working on conservation – the IUCN, UNEP, UNESCO and FAO – and took the form of a document known as the *World Conservation Strategy* (IUCN, 1980). Whilst it was a departure, in conservation terms, because it construed development as a means to achieve conservation objectives, it was not influential with the development constituency it sought to engage – largely because, suggests Holdgate (1999) it did not understand or address their priorities.

In large measure, the accolade of bringing the concept of sustainability more centrally into the international development fold went in the end to *Our Common Future* (WCED, 1987), the flagship publication of the World Commission on Environment and Development (WCED). It contains the most frequently cited (and contested) definition of sustainable development: 'development that meets the needs of the present without compromising the ability of future generations to meet their own needs' (WCED, 1987: 43).

There were two principal reasons for the influence Brundtland's report acquired. First, the WCED emerged within the UN General Assembly and was chaired by the then prime Minister of Norway, Gro Harlem Brundtland (hence the tendency to refer to *Our Common Future* as the Brundtland Report). In contrast, the starting point for the *World Conservation Strategy* was the IUCN and the UNEP, which had, tellingly, in the decades since their creation, remained in the backwaters of the UN system in spite of the strong impetus given to the environment-development nexus by the 1972 Stockholm conference. Second, *Our Common Future* much better spoke to the central concerns of the development community, arguing that environmental problems would remain unresolved in the absence of efforts to address 'the factors underlying world poverty and inequality' (ibid. :3). Whilst it shared, moreover, the common objective with the *World Conservation Strategy* of bringing development into line with environmental limits, its starting point was poverty reduction, which could only be resolved through more equitable forms of global economic growth. Environmental limits were not intrinsic to the environment itself, but were instead a question of finding the right technologies and modes of societal organisation (Adams 2009). If the *World Conservation Strategy* strayed frequently into 'doomster' territory, *Our Common Future* often seemed steeped in a cornucopian vision of things to come (Adams 2009). Moreover, if it was deemed to be considerably more politically aware, it was nonetheless characterised by a faith in post-war multilateralism and mutuality, coupled to reforms to open up the international economy as a means to deliver poverty reduction through redistribution. On either count this conviction looks hard to justify in the light of events of recent decades.

Nor did it really give a clear account of how its core vision, of accelerated 'world growth while respecting environmental constraints' (ibid.: 89) was to be brought about. It is for this reason that Adams concludes that *Our Common Future* 'hoped to have its cake and eat it' (2009: 78).

Rethinking environmentalism through (poor) people-centred framings

The 'push back' from Southern countries in the international environmental negotiations to leave space for development was one expression of mounting concern that conservation efforts on the ground, and indeed environmentalism more broadly, were missing important dynamics and perspectives through the prevailing focus on the harm that people were doing to the planet. A core motivation for Southern countries in the negotiations was the widely-held conviction that environmentalism was an exclusively Northern affair. As far back as the Stockholm conference in 1972, Indira Ghandi had argued that the South was, in the characterisation of Guha and Martinez Alier, 'too poor to be green' (1997: xvii). Northern commentators on the right and left argued something rather similar: the 'post-materialism' thesis of liberal political scientist Ronald Inglehart held that the environmentalist sentiment came about in a society only when material needs had been taken care of for the majority (1977). On the other side of the spectrum, the life-long Marxist historian Eric Hobsbawm deemed it 'no accident that the main support for ecological policies comes from the rich countries and from the comfortable rich and middle classes … the poor … wanted more 'development', not less' (1994: 570).

It is hard to argue with the proposition that global environmentalism is dominated by Northern resources, actors and organisations. Yet Guha and Martinez Alier (1997) have convincingly demonstrated that there are many shades of variety in environmental concern in the South both new and old, which have been forcing their way into the global environmental consciousness more noticeably since the 1980s. Some, though by no means all, of these stand at the opposite end of the spectrum from the 'full stomach' environmentalism of the North: they constitute the 'empty belly' environmentalism of the South (Guha and Martinez Alier, 1997: xxi). Throughout all the strands of Southern environmentalism, though, there is a tendency to be more concerned with the human dimensions of environmental problems. This can take the form of positing a link between poverty and degradation, documenting the adverse impacts of industrial resource extraction for poor people or their struggles to cling onto access to the natural resource bases – or even the habitats – on which their livelihoods depend. These socially, politically aware 'environmentalisms' find their Northern counterpoint in the 'environmental justice' movement, which developed throughout the 1980s. It emerged in response to the increasing number of studies demonstrating that, in the USA, 'ethnic minorities, indigenous persons, people of color, and low-income communities confront a higher burden of environmental exposure from air, water, and soil pollution from industrialisation, militarisation, and consumer practices' (Mohai, Pellow and Roberts, 2009: 406).

This tendency to reclaim the space for people within environmental discourses found echoes in changes to mainstream development thinking. Through the 1970s and into the 1990s, the idea that people were what was missing from the technical solutions posited, the economic calculations and the social cost benefit analyses became increasingly prevalent. It was argued with increasing frequency and urgency that fundamental decisions about the make-up and direction of development projects were being made by people both geographically and ideologically distant from the locations in which they were taking place (Ferguson, 1990; Escobar, 1995; Chambers, 1997, 1983; Gardner and Lewis, 1996). The value of local practices and knowledge was, in contrast, being ignored or underestimated because the superiority of data offered by 'hard science' was assumed, and sometimes erroneously (Chambers, 1997; Fairhead and Leach, 1998; Leach and Mearns, 1996). Another observation that dovetailed with this idea was that poor and marginalised groups were powerless, and that this was an issue which had to be addressed if their fortunes were to improve (Chambers, 1997; Escobar, 1995, 1991; Crush, 1995). Commentators varied on how it was to be addressed; Robert Chambers (1997) famously urged that 'expert' development practitioners should 'hand over the stick' to local people, and that it was crucial to 'put the first last', but without rejecting the development paradigm outright. Others looked to a 'post development' future (Escobar, 1992; Sachs, 1992) in which social movements in the South would build 'alternative visions of democracy, economy and society' (Escobar, 1992: 22), thereby throwing down the hegemony of Northern development in the mould of modernisation (ibid.).

The more radical visions of 'post development' thinkers seldom found their way into conservation initiatives, but a re-evaluation of the capacities of local actors to manage and conserve their environments grew exponentially from the late 1970s onwards, and which did draw more explicitly on proponents of participatory approaches to development such as Chambers (1983, 1997), Richards (1985), Brokensha et al. (1980) and others. Principally, the idea that wilderness was an uninhabited space, and that the best way to protect it was through enforcing a separation of it from humans, came very much under attack (Adams and McShane, 1992; Pimbert and Pretty, 1995). Much more attention was paid to the human costs of conservation, such as eviction and criminalisation of livelihood activities (Adams, Hulme and Murphree, 2001; Brockington, 2002; Neumann, 1997). Others questioned the sustainability of preservationist forms of conservation (Reij, 1991; Chambers, 1997; Fairhead and Leach, 1998) often by suggesting that they could in fact lead to environmental degradation by undermining livelihood security (Roy, Jackson and Kemf, 1993; Koch, 1994; Pimbert and Pretty, 1995). Anthropologists, ecologists and common-pool resource theorists made prolific efforts to document the ways in which a) much supposedly 'wild' land had in fact been inhabited (Adams and McShane 1992); b) much biodiversity with a conservation value existed because of the wise use of it made by local stewards (Berkes and Folke, 1998; Hviding and Baines, 1992; Murphree, 1997) and c) local forms of common-pool resource management provided sustainable alternatives to the state- and market-centric models of resource management that predominated (Ostrom, 1990; Ostrom, Gardner and Walker, 1994; Dolsak and Ostrom, 2003; Berkes, 2008; Holling, Berkes and Folke, 1998; Feeny, Berkes, McCay et al., 1990). These issues are explored in greater detail in Chapter 10.

Out of this process of reimagining development and conservation through a re-evaluation of local people, one which made links between social and ecological systems as opposed to trying to keep them separate (Berkes, Colding and Folke, 2008), came the notion of 'community conservation', explored in more detail in Chapter 10. The many different kinds of thought and praxis subsumed within this 'catch-all' term proved key in ensuring that local concerns and capacities were given prominence (at least rhetorically) at the international meetings and negotiations around sustainable development.

The rise and fall of sustainable development, 1992–2012

The Rio Earth Summit, 1992

Whatever the shortcomings of high-profile publications such as *Our Common Future* and the *World Conservation Strategy*, they did to some extent reignite the flame that had been kindled at Stockholm in 1972, but subsequently extinguished by the prevailing winds of economic change. Interest in the concept of sustainable development peaked twenty years later, with the holding of the UN Conference on Environment and Development, more commonly known by the friendlier term, 'Earth Summit'. Alternately described as a success (Grubb, Koch, Munson et al., 1993) and a failure (Redclift & Sage, 1995), the conference brought together 178 countries, including 120 heads of state and produced some landmark outcomes which would frame a global consensus around sustainable development and devise a plan of action for it which came to be known as Agenda 21 (see Box 2.1 for a breakdown of the 'outputs' from the Rio Earth Summit).

Figure 2.3 *Rio de Janeiro*

Source: Photographer Halley Pacheco de Oliveira; used under Creative Commons licence

Box 2.1

The progeny of Rio: the international conventions

The Rio Earth Summit produced five internationally negotiated outputs which reflected a global consensus on realising sustainable development, without always specifying what sustainable development was, and with varying requirements for compliance with stated aims. Three of the outputs were voluntary agreements:

1   The Rio Declaration on Environment and Development, which essentially established the principles for sustainable development

2   Agenda 21, a 600-page plan of action for sustainable development

3   The Agreement on Forest Principles, an expression of the principles for reducing deforestation and managing forests sustainably, originally intended to be a convention to go alongside the others produced at the conference

The other two outputs took the form of legally binding international treaties:

4   The Framework Convention on Climate Change, which aimed to reduce global carbon dioxide CO_2 emissions by 15% by the year 2012, taking 1990 levels as the baseline, as a means to tackle global warming

5   The Convention on Biological Diversity: an international, legally binding treaty on conserving biodiversity and promoting sustainable ecosystem and species use

Source: UNEP website – http://www.un.org/geninfo/bp/envirp2.html (accessed 03.03.2015)

The Rio Declaration comprised a list of 27 principles on which to base 'sustainable development'. These had been billed by the summit's chairman Maurice Strong as the basis for decisions 'that will literally change the face of the earth' (cited in Pearce, 1991: 24). However, others judged it too bland, watered down to the lowest common denominator in its contorted efforts to offer something to all in the rainbow coalition that had been entrusted with bringing about sustainable development (Holmberg, Thomson, Timberlake et al., 1993).

Beyond the hopes and the rhetoric, this framing sat very awkwardly with the diplomatic battleground that was the hallmark of the negotiations preceding and throughout the conference itself, and which, perhaps more than anything else, was the fundamental legacy of the international discussions around sustainability since 1972. In the twenty years since Stockholm, little had happened to suggest that environmental problems were being substantially addressed at the global level. Nor had the suspicions receded in many developing country negotiation teams that the environmental agenda would be used as a brake on their efforts to industrialise, or as an assault on their sovereign independence, which in many countries had been forged only recently.

Many commentators were impressed by the breadth of consensus that had permitted the production of these agreements (Grubb et al., 1993: 26). Others argued that they placed little in the way of binding legal commitment or obligation upon those party to the agreements (Grubb et al., 1993), a pattern that would be repeated in years to come, and as such source of regular and often bitter disappointment. Another complaint was that no concrete timetable for the actions envisaged under Agenda 21 had been agreed, nor had any specific mechanisms for their implementation been devised. Some concluded that despite the excitement and optimism surrounding the Summit, it amounted to little more than talk, and a failure therein (ibid.). Disillusion notwithstanding, the first Earth Summit had produced an international agenda for addressing environment and development concerns, predicated on the logic of sustainability.

Hindsight, unfortunately, bears out the views of those who were sceptical for the prospects of Agenda 21 and the Rio Principles. A review of the previous twenty years produced in the run-up to Rio+20 concluded that at best, limited and frequently no progress had been made towards the outcomes envisaged by Agenda 21 (UNDESA, 2012). Moreover, the 'spirit of partnership' that, internationally, would bring about the realisation of the Rio Principles had not been demonstrated by governments. On the contrary, when international agreements and targets for dealing with environmental externalities of economic growth were tabled at international meetings, 'good faith and partnership working seem to have been eschewed for individual goals and interests' (UNDESA, 2012: 14).

From Johannesburg (back) to Rio: sustainable development 2002–2012

Ten years on from Rio, there was little change in observed patterns of global consumption or the environmental externalities associated with them. Yet expectations had mounted for the Rio's 10-year update event, the World Summit on Sustainable Development, held in Johannesburg in 2002. An enormous gathering, it brought together 22,000 people, including many heads of state, 10,000 delegates, 8,000 representatives from civil society and the private sector and 4,000 journalists – one for every four-to-five of the Summit's attendees (UN, 2002a). Governments in attendance at the Summit tended to declare it a success, as typified by the optimistic response of Margaret Beckett, then UK Minister for Environment, Food and Rural Affairs (Beckett, 2002). The Plan of Implementation was lauded by the UN as a more focussed document than its predecessor, Agenda 21, advancing the cause of sustainable development through a clearer identification of the processes by which it could be achieved, and by bringing in a wider range of actors than had been involved in the Rio Summit of 1992 (UN, 2002b).

Other commentators, especially those within the civil society sector, expressed deep disappointment at the outcomes, perhaps most pithily captured by WWF's response to the UK Government's celebratory overtures: 'which summit did Beckett attend?' (cited in ENDS, 2002). Echoing one vein of response to the Rio Summit, the most frequent criticism that emerged was related to the perceived lack of commitment to setting timeframes for unambiguous targets, creating mechanisms for realising stated objectives and little or no mention of dedicated funding for priority activities (Seyfang, 2003; Von Frantzius, 2004). For some, it was even a 'step away' from sustainable development (Coates, 2002).

The greater disillusion that surrounded the Johannesburg meeting, relative to its predecessor 10 years before it, was not just a product of governments failing to translate the soaring, utopian rhetoric of sustainable development into concrete action. In many ways the sense of disappointment was indicative of the greater awareness of the difficulties and differences that lay behind the intuitive appeal and alliance-building power of the term 'sustainable development'. It had been adopted by a bewilderingly wide range of actors and organisations, but by 2002 it had become much clearer that they had disparate, often conflicting agendas. Far from being a pathway for reconciling global prosperity with environmental care, it was at risk of being used to describe irreconcilable objectives. Crucially, moreover, it was much clearer that, with different groups bringing conflicting definitions to the table, determining what the sustainable development agenda would comprise was not a question of semantics, but in fact a power struggle. Sustainability would depend partly on what was to be sustained, with differing implications for different interest groups. It would also depend, of course, on what use was made of the 'critical natural capital' (Pearce, 1991) on which much economic activity was based. This raises the political question of who owns such natural capital and who therefore has more power to determine how it is used and whose needs it most serves, recognising that there can be a trade-off entailed. As Redclift points out, 'Unless these processes are made more visible "sustainable development" discourses beg the question of whether, or how, environmental costs are passed on from one group of people to another, both within societies and between them' (Redclift, 2005: 545).

Twenty years after the Rio Earth Summit, more than 100 heads of state and tens of thousands of representatives from government, business, and civil society attended Rio+20, billed as the international meeting that would define for humankind 'The Future We Want'. If its organisers were hoping that a return to the same host city would allow the world to recapture the 'spirit of Rio' that had been remarked upon at the start of the 1990s, they may have been disappointed. By and large, the response to Rio+20 was not kind. The Bureau of Investigative Journalism deemed the event an 'epic failure'.[6] Environmentalist George Monbiot described the outcome document it produced as '283 paragraphs of fluff'[7] whilst Oxfam's chief executive, Barbara Stocking concluded that 'Rio will go down as the hoax summit'.[8]

The reason for such scathing critique is largely down to the lack of commitments and detail on how the fine sentiments about the importance of 'harmony with nature', or the 'deep' concerns about and continual recognition of environmental problems, would translate into actions – or at least, into legally binding ones. Rio+20's 'biggest idea', around the green economy, probably did not initially gain traction, as even supporters of it conceded;[9] but has developed a life beyond the summit and grown steadily in influence (see Box 2.2). For some commentators on the left of the political spectrum, this lack of specific commitments was to be expected even in advance of the conference. Salleh, for instance, argued that 'zero draft' of the text for 'The Future We Want' document confirmed 'Rio+20 as a quintessential neoliberal arrangement – "voluntary commitments" to be recorded'.[10] Indeed, the prominence of the private sector at the Rio +20 meeting – and what some saw as the 'privatisation' of the

Box 2.2

Deciphering the green economy

The 'green economy' concept has found a significant degree of traction, most notably at the Rio +20 Earth Summit and the Sustainable Development Goals (SDGs). Its prominence in these initiatives is at least in part indicative of its origins in the work of the United Nations Environment Programme (see especially UNEP, 2011), but it has since moved into the mainstream of politics and business discourse (Borel-Saladin & Turok, 2013). However, like other notions, such as sustainable development (or, increasingly, resilience), 'green economy' already means different things to different people. Such terms have a dual ability to convene and divide, often simultaneously. The 'green economy' enabled the Rio +20 process to reach out to private sector actors, even as it was being christened 'the new enemy' by some of the activists attending the summit (Terraviva, 2012). Some saw it as a way to deal with global financial crisis and climate change; but for others, it was another attempt to restrict the sovereign right of developing countries to use their natural resource base to leverage growth – a species of 'green protectionism' (Resnick, Tarp & Thurlow, 2012) feared by some developing countries since the 1972 Stockholm conference.

Faccer et al. (2014) have helpfully provided a means of starting to get to grips with the different meanings and positions that are emerging in relation to the concept of the green economy. They identify three broad 'stylised', overlapping discourses of green economy – incrementalist, reformist and transformative. 'Incrementalist' interpretations broadly accept the predominant economic system, with corrective, market-based measures which minimise or eliminate current environmental problems. Reformist discourses – of which the UNEP's is the most prominent – seek to refocus the economy to deliver equity and wellbeing benefits, but that hold also a green economy generates more growth than its 'brown' counterpart. The reformist paradigm requires not just a greening of 'brown' parts of the economy but a broader restructuring and new ways of innovating, generating and capturing value to drive green growth, reduce inequity and adapt to environmental change. Faccer et al. (2014) consider most mainstream discourse on green growth to be essentially reformist in character. Transformative discourses tend to be espoused more by civil society organisations and academics, than by multilateral organisations and governments, and critique rather than reject the green economy. Transformative perspectives call for a more fundamental organisation of society and economy, and one, typically, not based on economic growth. The 'steady state' economics of Herman Daly (2010) – in which the benefits offered by the economy grow, but not the physical size of the economy offering them – or Tim Jackson's 'prosperity without growth' thesis (Jackson, 2009; Victor & Jackson, 2012) could both be classified as transformative perspectives.

Some political ecologists/economists are either highly critical of or reject the concept of the green economy. Some objections focus on the trade-offs of transition. For instance, would mining workers in South Africa find themselves worse off if their industry were moth-balled? How would a transition ensure that they prospered (Newell & Mulvaney, 2013)? There is more, currently, in the way of heroic assumptions about these kinds of questions than there is understanding of how they might unfold. An even more fundamental critique has been offered by Edgardo Lander (2011), who argues that even well-intentioned visions such as the UNEP's do not recognise that existing political systems are very limited in their capacity to regulate or restrict the workings of free markets, and are therefore unlikely, underpowered vehicles for transition.

entire Rio process – was cause for concern. Writing in the Huffington Post, Eric Holt Gimenez captured the dilemma faced by environmentalists contemplating the efforts made at Rio+20 to engage the private sector to solve global environmental problems:

> While the Green Economy certainly has its bona fide believers, for some environmentalists, this is a last desperate effort to save the planet from the excesses of capitalism by mobilizing its most powerful institutions; the global oligopolies. For others, it is pouring fuel on the fire and risks causing a social, economic and environmental planetary meltdown.[11]

Certainly, the tendency to look to neoliberal policy agendas and instruments to reform (and perpetuate) a capitalist economic system fundamentally implicated in global environmental harm is opening up a rift amongst conservation actors (see Brockington, Duffy & Igoe, 2010 and also chapters 5 and 6).

Not everyone deemed Rio+20 a failure. UN General Secretary Ban Ki Moon was able to point to the US$513 billion in voluntary pledges and 692 commitments made to projects aimed at cutting fossil fuel use, boosting the contribution of renewable energy to the world's power grids, conserving water and reducing poverty.[12] Others argued that it managed ('just') to reaffirm commitment to the agreements that had been made at Rio in 1992, and that another of the 'big ideas' at Rio+20, the sustainable development goals, had been well positioned to influence the direction of development post 2015, the 'deadline' for the achievement of the millennium development goals.[13]

Whether anything of longer term significance comes of Rio+20 is a question that is perhaps better left until 2022, or even 2052. Yet what can plausibly be suggested more immediately is that it seemed to confirm that the multilateralism which was a necessary (if not sufficient) condition of the achievement of sustainable development advocated by its most influential expression, *Our Common Future*, was steadily losing ground. When 'keeping the talks alive' becomes a principal achievement of talks, and when there is a resounding lack of political will to sign up to the instruments of international collective action to translate the talks into activities, it is hard not to wonder what the future of multilateralism can be, at least in the environment-development arena. As a result, there is now a greater emphasis on looking beyond the multilateral negotiations, to the myriad efforts that are being made by local and regional governments, through, for instance, investments in 'green jobs'.

Rising tides and rising powers: climate change and North–South relations

The despondency surrounding the seeming failure of development to come to terms with environmental consequences was deepened by the persistently gloomy and ever worsening assessments of the state of the global environment (Hassan, Scholes and Ash, 2005; UNEP, 2012). Chapters 4 and 8 explore these in more detail. Of all of these concerns, the one that became most dominant by far in the first decade of the twenty-first century was climate change. Whilst it emerged as an issue within the context of the sustainable development and, as noted above, was established as an international convention at the Rio 1992 Earth Summit, it would go on to overshadow this broader agenda. In the words of a UN-sponsored reflection on the intervening years between *Our Common Future* and Rio+20, 'climate change has emerged as the de facto proxy for addressing sustainable development issues' (Drexhage and Murphy, 2010: 13). This was a mixed blessing for the environmental movement: on the one hand, the energy and political impact that the climate change agenda produced was astonishing. On the other, climate change was a much narrower agenda which diverted attention (and funding) from other equally or perhaps more pressing environmental problems.

In any case, the United Nations Framework Convention on Climate Change has been by quite some margin the most high profile of the international conventions that were launched at the Rio Earth Summit of 1992. Global media attention to the Conference of Parties meetings that happen yearly peaked in 2009 with COP 15, held at Copenhagen.

Yet even at Copenhagen, widely billed as the 'last chance' to agree legally binding carbon dioxide emissions reduction targets before the Kyoto Protocol expired, it did not prove possible to achieve this objective.

Whilst climate change was becoming the dominant environmental issue of the day, other changes were also happening. The rise to prominence of a clutch of countries in the South has changed the complexion of global order, and thrown into question the longevity of Anglo-American and European dominance within it (Khanna, 2009). Spearheaded by Brazil, South Africa, India and China – the so-called BASIC countries – but increasingly applicable to states like Mexico or Indonesia (who, along with Nigeria, and Turkey now have their own acronym in MINT), the notion of 'rising powers' has rapidly gained currency and is prompting much speculation over the prospects of a 'multi-polar world' (Hiro, 2012).

In many ways the 'rising powers' agenda is a reframing of the environment-development debate. On the one hand, the global economy is now held to be powered as much, if not more, by the dynamism of South Asia and some parts of Latin America as it is by industrialised, northern states. These are often cast as the new engines of development. On the other hand, this rapid industrialisation has quickly been characterised as a process likely only to accelerate and deepen the current environmental crisis (Anderson and Bows, 2011). Perhaps more than any other issue, climate change has served to reinforce this set of concerns around the development trajectories that the rising powers are forging.

The crux of these concerns is that higher rates of economic growth are leading to increases in the emissions of CO_2 and other greenhouse gases held to be driving climate change. For most of the climate change debate since 1992, climate change had been framed as a problem caused historically by the North and heaped upon the South, and one which Northern countries remained primarily responsible for. Yet increasingly, the contribution of economic powerhouses like China or India has come to be seen as a big, perhaps the biggest, part of the problem. Anderson and Bows speculate that 'Total cumulative emissions produced by nations that underwent industrialisation in the nineteenth century and first half of the twentieth century will be eclipsed if the five billion people currently resident in non-Annex 1 nations [i.e. currently without any legally binding obligations to reduce

Figure 2.4 *Aerial view of China*

their carbon emissions] remain or become locked into a fossil fuel economy' (2011: 21). Hurrell and Sengupta suggest that an increasingly common framing of the strong economic performance of the rising powers focuses on their 'incorporation into a far more complex and globalized capitalist order, much of whose dynamism is premised on ecologically unsustainable patterns of resource use' (2012: 466).

In this framing, what makes matters worse is that some BASIC countries are not willing to cooperate in the international negotiations on climate change. Some analyses – perhaps most famously Mark Lynas's article in the *Guardian*[14] – blamed China for obstructing and fatally weakening the agreed outcomes of the Copenhagen climate summit for which, they calculated, the USA would be held up as the culprits. For Lynas but also others (i.e. Roberts, 2011), what

Figure 2.5 *Chinese capacities in renewable energy production – wind farm in Xinjiang*

Source: Photographer Taylorandayumi; used under Creative Commons licence

Copenhagen demonstrated was the geopolitical shift that had taken place, one in which Europe was sidelined and China emerged as the ultimate 'winner'. This has led to the impression that the 'South' has become much stronger in the climate negotiations over time.

However, others have disputed this characterisation of the environmental politics of the rising powers and their increasing clout. Hurrel and Sengupta (2012) argue that if anything, the South was much more successful at getting its way at Rio 1992 than at subsequent meetings. In 1992, they extracted legal and financial commitments from the North and ensured that the North was held primarily responsible for dealing with global environmental problems, through the principle of 'common but differentiated responsibilities'. At the climate negotiations in Durban in 2011, in contrast, the principle was largely forsaken in a deal which 'saved' the future of international climate negotiations, but which for the first time committed some Southern countries, in principle, to legally binding emissions reduction targets of their own. The 'inability' of the Southern countries to ensure that they were not required to take on such commitments may, for Hurrel and Sengupta, have been down to the increasing fragmentation of negotiating blocs within the South relative to the coherence displayed at Rio in 1992. In one sense, this is an extraordinary outcome, as it represents a shift away from the South insisting on its right to 'develop' without 'unfair' environmental restrictions which make for an uneven playing field, and which denies the South the opportunities and avenues for wealth creation that the North has been able to take advantage of. Yet this debate is far from over: the principle of 'common but differentiated responsibilities' was re-enshrined in the outcome document of Rio+20; and there is a concerted effort to extend the principle to cover the remit of the Sustainable Development Goals (the replacement for the Millennium Development Goals, whose mandate was only set to run until 2015) (Constantine & Pontual, forthcoming).

Nor is it easy to characterise in clear-cut terms Southern power-houses like India, China or Brazil as environmental villains: all have voluntary commitments for quite drastic carbon emissions reduction; China is a world leader in renewable energy and Brazil has introduced very tough new laws on Amazonian deforestation and is, crucially, enforcing them more stringently and effectively. Moreover, in absolute terms, one paper has argued, Southern countries have pledged greater levels of emissions reduction than Northern counterparts (Kartha and Erickson, 2011). What is perhaps a much less contestable proposition is that in a multi-polar world with fragmented, conflicting interest blocs, the prospects for consensual multilateralism look bleaker than ever.

Summary

- In addition to the construction of an international conservation architecture, a popular environmental movement in the 'global North' arose in the 1960s, on the back of:

- increasing concern about the adverse environmental impacts of resulting from the post-Second World War economic boom, and;

- the emergence of parallel social movements, such as the civil right movement – even though, often, these two movements in particular had little in common and even less to do with each other.

- In the North, environmentalism was given impetus in part through the writings of commentators christened the 'prophets of doom', in reference to their bleak warnings about the potentially extreme consequences of human population growth and economic activity. In the 'global South', concern with the environment took the form of political struggle, whether over access to resources or unwanted exposure to environmental contamination – the 'environmentalism of the belly' rather than of the (scientifically trained) head.

- The different environmental problems, such as the contamination of the food chain with pesticides, documented by the 'doomsters' came to be seen as symptoms with a common, underlying cause: unfettered economic growth. In the 1970s, efforts to map the ecological limits to economic growth became central to Northern environmentalism, posing the question of what kind of development trajectories might be sustained within these limits, and how to effect the transition towards them.

- Increasingly, there were calls for development to be reconciled with environmental conservation in order for it to be sustainable. With the UN Conference on the Human Environment in 1972, the sustainable development agenda was born. However, the pains of labour were especially evident with fears from Southern countries that Northern ones would use environmental problems to prevent them from enjoying the same development opportunities they had benefited from.

- The publication of *Our Common Future* in 1987 enshrined the concept of sustainable development within international discourses and negotiations on development. Arguably, the notion reached the peak of its influence at the 1992 Rio Earth Summit, an event which gave rise to several important international conventions regulating the use of nature in different ways. However, the follow-up summits in Johannesburg in 2002 and Rio in 2012 were met with disappointment, as a result of a widely held perception that beyond the soaring rhetoric, little had actually changed in the ways that societies used nature. On the contrary, the concern was that the sustainable development agenda had been co-opted as a means of justifying and perpetuating a 'business as usual' economic model which was anything but sustainable. This concern continues to be the bedrock of critiques of terms such as the green economy, a more recent attempt to reconcile development trajectories driven by economic growth with the imperative of living within environmental means.

• The future of sustainable development is, perhaps more than anything, contingent upon the result of the ongoing reconfiguration of the global geopolitical 'order'. Countries like China and India – but also less obvious candidates like Indonesia, Turkey and Mexico – are becoming more powerful and more economically prosperous. As they do so, the structures of their economies, the extent to which they cultivate path dependency on fossil fuels and environmentally harmful practices, will determine how sustainable global development trajectories are going to be. China's position as the world's largest emitter of CO_2 and global leader in renewable energy technology aptly demonstrates the schizophrenia at the heart of the dynamics which will determine the future of sustainable development.

Discussion questions

1. How useful to the cause of conservation were the environmental prophets of doom?

2. What are the prospects for sustainable development?

3. Why has it been so difficult to get international agreement on efforts to reconcile conservation and development objectives to date?

4. What difference will the changes in the geopolitical balance of power between Northern and Southern countries make to the prospects for sustainable development?

Notes

1 For further information, see 'BP leak the world's worst accidental oil spill', Telegraph (03.08.2010, accessed 09.10.2012): http://www.telegraph.co.uk/finance/newsbysector/energy/oilandgas/7924009/BP-leak-the-worlds-worst-accidental-oil-spill.html (27.05.2010, accessed 09.10.2012); and "Obama, in Gulf, pledges to push on stopping leak". *USA Today*. Associated Press: http://usatoday30.usatoday.com/news/nation/2010-05-27-oil-spill-news_N.htm?csp=34news

2 For further information, go to http://www.iaea.org/newscenter/focus/fukushima/ (accessed 09.10.2012) or see "Explainer: What went wrong in Japan's nuclear reactors". *IEEE Spectrum* (4 April 2011, accessed 09.10.2012).

3 http://discovermagazine.com/2006/dec/25-greatest-science-books/article_view?
   b_start:int=2&page=2 (accessed 03.03.2015)

4 http://www.humanevents.com/2005/05/31/ten-most-harmful-books-of-the-19th-
   and-20th-centuries/ (accessed 03.03.2015)

5 http://www.nytimes.com/1990/12/02/magazine/betting-on-the-planet.
   html?pagewanted=all&src=pm (accessed 11.10.2012).

6 http://www.thebureauinvestigates.com/2012/06/22/analysis-rio-20-epic-fail/
   (accessed 2.11.2012)

7 http://www.monbiot.com/2012/06/22/how-%E2%80%9Csustainability%E2%80%
   9D-became-%E2%80%9Csustained-growth%E2%80%9D/ (accessed 2.11.2012)

8 Quoted in http://www.thebureauinvestigates.com/2012/06/22/analysis-rio-20-epic-fail/

9 http://www.iied.org/five-things-we-ve-learnt-rio20 (accessed 2.11.2012)

10 http://rio20.net/en/documentos/rio20-and-the-green-economy-technocrats-meta-
   industrials-wsf-and-occupy/ (accessed 05.03.2015)

11 http://www.huffingtonpost.com/eric-holt-gimenez/rio20-summit_b_1613733.
   html (accessed 05.03.2015)

12 http://www.bloomberg.com/news/2012-06-22/un-gets-sustainability-pledges-
   worth-513-billion-in-rio.html (accessed 2.11.2012)

13 http://www.southcentre.org/index.php?option=com_content&view=article&
   id=1791%3Asb64&catid=144%3Asouth-bulletin-individual-articles&
   Itemid=287&lang=en; http://sustainabledevelopment.un.org/; and http://www.
   guardian.co.uk/global-development/poverty-matters/2012/jun/27/rio20-reasons-
   cheerful (accessed 2.11.2012)

14 Mark Lynas, 'How do I know China wrecked the Copenhagen deal? I was in the
   room', *Guardian*, 22.12.2009. http://www.guardian.co.uk/environment/2009/
   dec/22/copenhagen-climate-change-mark-lynas

Further reading

Brundtland, G. H. 1987. *Our Common Future*, Oxford, Oxford University Press.

Carson, R. 1963. *Silent Spring*, London, Hamish Hamilton.

Ehrlich, P. R. 1975. *Population Bomb*, Rivercity, MA, Rivercity Press.

Hardin, G. 1968. The Tragedy of the Commons. *Science*, 162, 1243–1248.

Hurrell, A. and Sengupta, S., 2012. Emerging powers, North-South relations and
   global climate politics. *International Affairs*, 88, 463–84.

Jackson, T. 2009. *Prosperity without Growth: economics for a finite planet*, London,
   Earthscan

Meadows, D. H., Randers, J. and Meadows, D. L. 1972. *The Limits to Growth: a report
   for the Club of Rome's project on the predicament of mankind*, London, Earth Island
   Ltd; [London] (2 Fisher St., WC1R 4QA): [Distributed by Angus and Robertson].

Online resources

Declaration of the UN Conference on the Human Environment, Stockholm, 1972 – http://www.unep.org/Documents.Multilingual/Default.asp?documentid=97&articl eid=1503 (accessed 03.03.2015)

Garrett Hardin on the Tragedy of the Commons and Resources – https://www.youtube.com/watch?v=L8gAMFTAt2M (03.03.2015)

Rachel Carson wiki http://en.wikipedia.org/wiki/Rachel_Carson

Man and the Biosphere (MAB) – http://www.unesco.org/new/en/natural-sciences/environment/ecological-sciences/man-and-biosphere-programme/ (accessed 03.03.2015)

United Nations Environment Programme (UNEP) – http://www.unep.org/ (accessed 03.03.2015)

United Nations Framework Convention on Climate Change (UNFCCC) – http://unfccc.int/2860.php (accessed 03.03.2015)

The Future We Want (Rio+20) – http://www.un.org/en/sustainablefuture/ (accessed 03.03.2015)

Global Environmental Governance Rio+20 Roundup, Center for Governance and Sustainability, University of Massachusetts Boston – http://www.environmental-governance.org/rio20-responses/ (accessed 03.03.2015)

Sustainable Development Goals – https://sustainabledevelopment.un.org/topics/sustainabledevelopmentgoals (accessed 03.03.2015)

The green economy – http://www.unep.org/greeneconomy/ (accessed 03.03.2015)

PART 2
Conservation and development in the broader global context

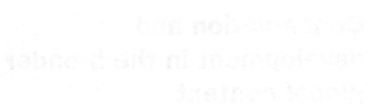

3 Conservation, development and the relationship between nature and society

- Nature and society: a tangled thicket
- Moving towards an understanding of social-ecological systems
- Constructing nature and society

Introduction

In short, if we want to gauge the prospects for reconciling conservation and development processes, we need first of all to understand the underlying relationship between nature and society. This is because what there is to conserve and what development trajectories we can pursue are both contingent upon the processes through which nature and society are co-produced. By way of cultivating a deeper understanding of the relationship between nature and society, we draw upon three key sets of conceptual resources: social-ecological systems theory; social constructivism applied to accounts of nature and society (as guided primarily by the work of sociologist of scientific knowledge, Barry Barnes); and political ecology. The first section starts this exploration with a consideration of why it is actually very difficult to separate nature from society. The second section reviews current thinking around social-ecological systems as a way of understanding nature and society together rather than separately, and introduces the concept of resilience. The third section explores how political ecology has been used to demonstrate the consequences for different people and environments of our taken-for-granted accounts of nature and society. At various points throughout the chapter, we consider the implications of these conceptual insights for how we think about and do conservation and development.

Nature and society: a tangled thicket

Nature and society as separate entities

The separation of nature from society is a common and quite taken for granted way of thinking in conservation and development circles; and, frequently, beyond them. This is in large measure because of how normal it is to see nature as that which is free from human influence, and society as a purely human phenomenon. We are simplifying, here, undoubtedly: sampling the work of Helena Siipi (2008), Table 3.1 gives a sample of just how many meanings the words 'natural' and 'unnatural' have taken on. Ben Ridder (2007) has likewise demonstrated the importance of distinguishing 'naturalness', in terms of a particular state of nature as measured by historical benchmarks, from 'wildness', that is, processes which occur autonomously of human action, even in landscapes which have been radically altered by the presence of humans. Conservation based on achieving 'naturalness', in trying to protect a particular state of biodiversity, would be a different kind of intervention from conservation seeking to leave space for 'wildness', which might not seek to restore a landscape to a former state of glory, but rather to let it change 'by itself', as it were. Nevertheless, the tendency to see nature as something where humans are not and society as somehow beyond nature remains remarkably prevalent.

Perhaps we should not be too surprised by this characterisation, which in one way or another can be traced back to the thinking of Descartes. His famous proposition, *cogito ergo sum* (I think therefore I am) epitomised a tendency to see humans as different from the rest of nature, precisely because thinking was something only humans – not animals, plants, rocks or soils – could do. This conceptual distinction underpins how we think and how we act in ways which are relatively easily traced in the context of conservation and development. Crudely put, the focus of conservation is commonly taken to be nature, whereas the focus of development is assumed to be society. As chapters 1 and 2 explore, despite the fact that both conservation and development imperatives arise from the same historical processes, they have often been organised to run along parallel tracks and to be kept separate from each other. Arguably, sometimes this is a deliberate effort by some conservationists to preserve some parts of nature from the environmentally destructive

Table 3.1 *Varieties of (un)naturalness (synthesised from Siipi, 2008)*

Kind/ Number	Definition	Example
History based (un)naturalness (something is or is not natural in binary terms):		
1	Naturalness as total independence from human beings	No contemporary example; historically, any environment free from human presence
2	Naturalness as independence from certain types of human activities	Domestic plants, restored ecosystems and GM crops
History based (un)naturalness (continuum)		
1	Entity X is more natural than entity Y; entity X exists/has its properties more independently of human beings than entity Y	Traditional breeding methods less unnatural than GM crops; some ecosystems have properties more independently of human beings than other ecosystems
4	(Un)naturalness as a degree of human-caused properties	Rural landscapes as opposed to conservation sites of minimal environmental restoration
Property-based (un)naturalness		
1	Naturalness as similarity to historically natural entities	Viewing homosexuality in humans as natural or unnatural because it is viewed as natural or unnatural amongst animals
2	Naturalness of actions as similarity to genetically and biologically based actions	Human predation of fish is more or less natural depending on how similar it is to the predation patterns of historically natural entities (i.e. seabirds or fish)
3	Naturalness as the absence of technology	Childbirth without technical surveillance, medical painkillers, or surgical operations
4	Naturalness as normality	A plant or insect functioning in its 'biologically normal state'; statistical normality, in which a thing is close to the mean or median
5	Naturalness as in accordance with human nature	Human behaviour compatible with healthy ecosystems; farming in ways adapted to local soil and climatic conditions; acts deemed to be in accordance with God's purpose

tendencies of human societies; arguably, it is sometimes simply a failure of coordination between people who work on conservation and those who work on development. The importance of the influence of Cartesian logic is, though, underlined rather than diminished by this state of affairs.

Not all attempts to comprehend the relationship between nature and society have found this difference between humans and the rest of nature to be important, causally. On the contrary, for much of human recorded history, it has been argued that the environment in which people live determines their social and cultural characteristics – a tradition of thought which can be traced back to early Greek, Roman and Arab empires (Moran, 2009). More recently, in the twentieth and twenty-first centuries, it has been argued that nature cannot be shown to be anything other than a social construct, whatever we imagine it to be – a position to which we shall return in the third section. Yet both of these viewpoints, which we might put at either end of a spectrum, end up trying to explain society in terms of nature, or nature in terms of society. The schools of thought which we explore in the second and third sections both argue that in fact, nature and society influence each other, that it is not possible to explain the one solely in terms of the other – a proposition with which we concur – and that therefore we have to look at them both together. This section seeks to demonstrate why it is so difficult to disentangle them.

In the twenty-first century it is quite normal for us to think about wilderness, a landscape free from human presence, as increasingly hard to find, but that historically, was more prevalent. More recently, it has been argued that the historical presence and reach of human societies, even before the age of imperial expansion and industrialisation, has been greatly underestimated. Denevan (1992) argued that, at the time of the arrival of Columbus to what are now called the Americas, the ostensibly 'pristine' forests of the Amazon played host to a human population of eight million people. Others have suggested that a feature of each of the world's tropical forest regions was human settlement (Wood, 1993). In a different field of enquiry, religious scholars have demonstrated the varying ways in which the world's predominant religions have set up humans as guardians of, or otherwise in relation to, nature (see Chapter 12). A burgeoning body of literature has led to the now much more commonly accepted view that most parts of the world's landmass have at some point been host to or embroiled in human activity of one form or another (Christensen et al., 1996; Katz, 1997; Vitousek et al., 1997; Vogel, 2003). Others, such as Bill McKibben (1989) have gone still further, arguing that if it is defined in terms of the absence of human intervention or presence, we have reached the 'end of nature' as a result of human pollution and, more recently, human-caused climate change. Humans have a global footprint.

Of course, it is only possible to make such grandiose statements by eliminating any sense of nuance. Most conservationists – in fact most people – would be open to a definition of something as 'natural' or 'unnatural' not as an either/or proposition, as McKibben puts it, but rather along the lines of a continuum, in which environments and the biodiversity they contain can be to a greater or lesser extent natural, depending upon the amount of human intervention. Nevertheless, what would have to be acknowledged if we do accept this continuum view is that there is not a landscape which is completely natural, in the sense of being free from all human intervention. Rather, the scale would run from unnatural to mostly natural.

Changing and surprisingly social ecologies

It is not just our thinking about nature as something beyond human influence which has shifted profoundly in recent decades; new understandings of how ecology functions, and in particular, how ecosystems change, have emerged. Three of the most important insights and implications that are evident in this new thinking can be characterised as follows:

1   Ecosystems are frequently not as predictable in their behaviour as we had supposed, and do not have just one 'natural state' to which they will return, or which will be maintained, in the absence of disturbance.
2   It is not always clear what causes changes in ecosystems, particularly those to which we sometimes refer as degradation and frequently attributed to human activity or behaviour.
3   Humans have not just occupied landscapes which previously we thought to be free from our presence; human presence is sometimes fundamentally implicated in the production of ecosystems which now have conservation value.

Historically, in the discipline of ecology, it has been common to conceive of a natural state to which an ecosystem or environment will return once it is no longer subjected to a particular form of disturbance – for instance human conversion of forest to agricultural land, or a heavy storm. However, this notion has been challenged by the increasing attention paid to the ways in which ecosystems change cyclically between different states, even in situations where there is little or no human intervention. The work of long-term ecologists has been particularly important in showing how dynamic ecosystems can be. For instance, Daniel Botkin (1990), analysing lakebed pollen records from the Boundary Waters Canoe Area of Minnesota and Ontario, argues that what is now a region of thick marsh and forest used to be, prior to the last glaciation, tundra (landscape characterised by shrubby vegetation, in which the growth of trees is limited by short growing seasons and low temperatures). The effects of glaciation, 10,000 years ago, were dramatic: the tundra gave way to spruce forest, itself replaced by jack and red pines 800 years later, suggesting a warmer and drier climate. White pine migrated into the area 7000 years ago, only to be accompanied, subsequently, by a return of spruce and jack pine, indicative of a cooler climate. Roughly every thousand years, the long-term ecological record bore witness to substantial change in the vegetation of the forest, driven by fluctuating temperatures and the return of species that had migrated further south during the ice age. This prompted a fundamental question for Botkin: 'which of these forests represented the natural state?' (1990: 58–59). In the absence of an answer to this question, it is difficult to fix upon a particular state which should be conserved, and against which to measure particular forms of disturbance, be they natural or human (Ridder, 2007).

Perhaps the most fundamental implication of conceiving ecosystems in terms of multiple states, as opposed to a single equilibrium, is the connotation of complexity inherent in this conceptualisation. Thinking about ecosystems as complex systems challenges notions of linear causality, and thereby predictability, central to earlier understandings in which ecosystems are seen as simple systems, i.e. ones that can be 'adequately captured using a single perspective and a standard analytical model, as in Newtonian mechanics and gas laws' (Berkes et al., 2008: 5). A complex system, in contrast, is characterised by nonlinearity, uncertainty, emergence, scale and self-organisation (ibid.). A complex system will organise around more than one state, which will remain relatively stable up to a point, but which can change quickly into another state if a threshold is reached. In the example given by Botkin (1990), a threshold was reached when tundra switched to spruce forest, and then again to pines associated with warmer and drier climates. From a complexity perspective, it is difficult to predict what will bring about such thresholds (Holling, 1986).

An important corollary of conceiving an ecosystem as a complex, rather than simple system, is that it may not return to a particular state, precisely because it has entered another 'stability domain' (Berkes et al., 2008), and may stay within that domain indefinitely. This process is known sometimes in ecology as hysteresis (Lockwood and Lockwood, 1993), and its consequences may be more or less desirable from a human standpoint. Perhaps the most common example given is that of a grassland rich in species which then becomes a grazing land. Conventional resource management would suggest that if the impacts from grazing become too heavy, the way in which to restore biodiversity is simply to remove the grazing for a while. However, the act of grazing can induce a state in which perennial grasses do not return and the landscape becomes character-ised by other woody species and annual herbs. It is not, nevertheless, always the case that ecosystems cannot or do not withstand the ways in which people interact with them. Moreover, there are examples of human interactions with ecosystems which are not as recklessly destructive as they are sometimes made out to be, and/or which are able to exploit the ways in which ecosystems change sustainably. See Box 3.1 for examples.

Moving towards an understanding of social-ecological systems

Changed understandings of change: the adaptive cycle

Examples like those given in the previous section have been drawn upon by a school of thought that maintains that because of these interconnections and complex relationships, the unit of analysis in fact needs to be social-ecological systems, as opposed to social- or ecosystems on their own. Thinking in terms of social-ecological sys-tems, from this standpoint, is the only way to capture the interactions, feedbacks and fundamental connections between society and nature. In fact, three of the scholars most heavily associated with social-ecological systems thinking, Fikret Berkes, Johan Colding and Carl Folke, maintain that 'the delineation between social and natural sys-tems is artificial and arbitrary' (2008: 5).

Thinking not in terms of ecosystems returning to stable states, but rather of multiple states and 'stability domains', requires us to think differently also about the nature of changes between different states, in particular to acknowledge the cyclical character of such changes. In ecology, a common model which captures these kinds of changes is known as the adaptive cycle. As well as conceptualising change as a cyclical activity, it derives from observations on how managing a single variable prompts changes in other variables, resulting in some instances in abrupt changes across the ecosystem and subsequent decline in function (Berkes et al., 2008). And although it has arisen

Box 3.1

Ecologies resilient to/produced by human activity

Hardy landscapes: A surprisingly common example is environmental degradation attributed to humans and then generalised across a landscape, without a full understanding of how some landscapes recover quickly, and how vegetative states can be induced as much – or more – by ecological processes as by human presence. An example can be taken from the work of Sian Sullivan (1999), who, in the context of north-west Namibia, contested the commonly held proposition that degradation was widespread. This 'narrative of environmental crisis' was not supported by ecological evidence, she argued. In her own study she found the predicted patterns of degradation only in smaller areas of settlement, but was not widespread. Moreover, she argues that the interpretation of available ecological evidence in initiatives such as Namibia's Programme to Combat Desertification did not take into account the implications of spatial and temporal scale. There was a tendency to attribute perceived degradation to the presence and impacts of people and livestock, without exploring 'the fundamental relationship between variable abiotic factors and primary productivity' (1999: 272). The work of other authors (i.e. Rohde, 1997) advanced a similar case to that made by Sullivan.

The work of Thembela Kepe and Scoones (1999) on the coastal grasslands of South Africa's Eastern Cape Province documents the role of local management in transitions between different ecological states. The ecology of the region makes important contributions to rural livelihoods, in terms of the economic, building and medicinal resources that it provides. It is also a diverse ecology, characterised by high quality grazing land (*Themeda triandra*), much less productive grasslands mostly comprised of *Aristida junciformis*, as well as lands rich in collection and gathering resources. Kepe and Scoones found that the environmental history of this area was one of transition between different environmental states, which were encouraged or discouraged by local level management techniques such as burning, grazing and selectively enriching or disturbing soils. So, for example, local livestock herders brought about transitions from unproductive grasslands (*Aristida junciformis*) to much better quality *Themeda triandra* pasture through careful application of seasonal burning and periodic rest. They could also be changed into *Cynodon dactylon* grasslands by soil enrichment. Perhaps most interesting of all is the way in which these transitions between different states come about as a result of different groups, pursuing distinct objectives, trying to shape the grassland in a variety of ways. Some of these transitions result in undesirable states which are difficult to reverse, for instance transition to low productivity, immature grasslands overrun by marginal species. However, transitions towards productive, perennial pasture, also feature, providing an instance of sustainable human–ecosystem interactions.

within the discipline of ecology, it is also applicable to social systems, and indeed is increasingly applied (see, for instance, Pelling, 2011).

Figure 3.1 is a visualisation of the adaptive cycle, taken from a canonical text within the study of adaptive governance, Gunderson and Holling's *Panarchy* (Gunderson and Holling, 2002; see also Holling, 1986). If we start our explanation of the diagram below at the point of 'exploitation' (r), we can see the cycle as charting the progression of a form of what we could call 'behaviour', be it in a plant species or in human society, that becomes widely adopted. In so doing, the behaviour stabilizes, changing only slowly over time, moving toward to the 'conservation' (K) point on the front loop. In the logic of the model, this period of stability sews the seeds of rapid change, as diversity and innovation are suppressed, in favour of maintaining the current system state. As a result, the system goes into the 'release' (Ω) phase of rapid change, in which alternative behaviours appear and processes of innovation lead to re-organisation (α) of the system. Once a new form of behaviour has become established, we go back into the exploitation phase as the loop repeats itself.

There are three commonly identified determinants of system behaviour in the four phases.

Figure 3.1 *The adaptive cycle*

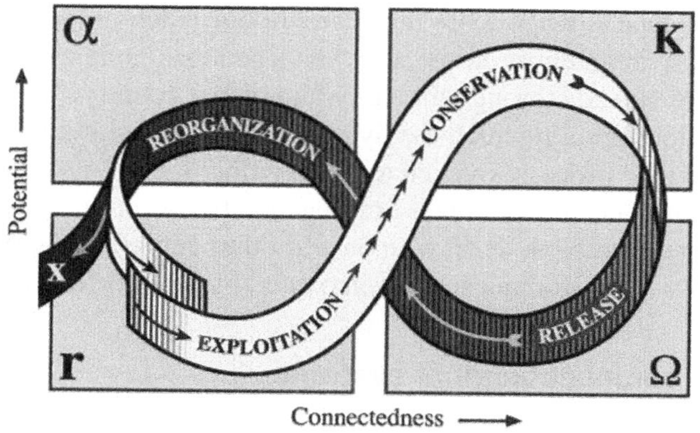

Source: *Panarchy*, 2002, p. 34.

(Source: Gunderson and Holling, 2002)

1   Potential, that is, the possibilities available for change and the limits of these.
2   Connectedness, that is, the extent to which system behaviour is self-regulating, as opposed to being driven by variables external to the system.
3   Resilience, that is, 'the amount of change a system can undergo and still retain the same function and structure while maintaining options to develop' (Nelson et al., 2007: 396).

In addition to this basic account of the reproductive cycle, there is another dimension to take into account: the scales across which systems span (i.e. geographical or jurisdictional) and the levels within these scales (local, national, global) (Cash et al., 2006). Local level ecosystems form part of bigger systems, such as catchments at the regional level. In the same way, municipalities at the local level are part of national government. The sum of these levels, links and interactions has been christened 'panarchy', in deference to Pan, the Greek god of nature, whose name carries connotations of enthusiasm, ecstasy and playfulness but also unpredictability and instability (Gunderson and Holling, 2002). Significantly, processes which happen at a lower level, say, within a geographical scale, can operate at different (quicker) speeds to those happening at a higher level (Berkes et al, 2008). Processes at one level can influence and be influenced by those at another. This is captured in figure 3.2.
Two other functions of adaptive cycles are also evident in this figure: revolt and remember (ibid.). A revolt is triggered when small, quickly occurring events prompt rapid change in the larger, slower cycles.
For instance, when a fire that starts in one patch of trees then spreads quickly to an entire forest; or when a protest in one neighbourhood or square spreads throughout a city and then a country. 'Remember' is a function through which the dynamics of larger, slower cycles affect those new processes which are happening in the reorganisation phase. For instance, after a forest fire, regeneration is possible because there is a seedbed bank to draw upon and other genetic resources which supply the disturbance or are activated by it (Berkes et al, 2008).

Resilience in social-ecological systems

If conservation and development objectives are to be reconciled, a social-ecological systems perspective would suggest that we need to be able to deal with transitions and change, rather than concentrating on maintaining and expanding a status quo. For this reason, research

Figure 3.2 *A panarchy of nested levels*

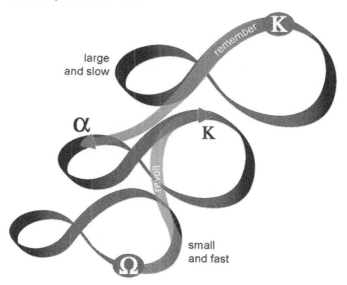

large
and slow

small
and fast

Source: Garmestani and Benson, 2013; reproduced with kind permission of the author

into the concept of resilience has gathered critical mass in recent decades. It has been used in different ways in diverse disciplines, but in ecology and across some parts of the social sciences, it has emerged as part of the broader reconceptualisation around social-ecological systems. In essence, in the face of uncertainty, surprise and disturbance, resilience becomes the fundamental characteristic towards which the management, indeed the governance, of social-ecological systems is bent. But if conservation and development objectives are to be understood against a background of social-ecological systems and oriented towards building and maintaining resilience, it is important to get a handle on the meaning of this term, especially given its ever increasing influence in both conservation and development policy and practice.

The theorisation of resilience has become increasingly sophisticated. It has been used in disciplines as diverse as psychology and engineering, but here, we focus on its intellectual history within the discipline of ecology and the ways in which social scientists working on natural resource management, common pool resources and development have drawn upon this particular body of work. Although we could go back considerably further, a common starting point for tracing the genealogy of contemporary resilience thinking is a seminal paper by

C. S. Holling, 'Resilience and stability of ecological systems' (1973). The reason why this particular paper is so important is because of the conceptual distinction it draws between different types of resilience. In the first type, which Holling labels 'engineering resilience', it is a characteristic which permits an ecosystem to return to a steady state following a perturbation or disturbance of some kind. This definition draws on ecological thinking which assumes that ecosystems will eventually return to a particular state; which, as we have seen, has been challenged. In the second type, resilience is defined as 'the magnitude of disturbance that can be absorbed before the system redefines its structure by changing the variables and processes that control behaviour' (Gunderson, 2008). This definition is more cognisant of the conditions and instabilities which can bring about a shift between one ecosystem state and another – as in the shift between tundra and forest that we saw in the work of Daniel Botkin (1990). Note that this definition of resilience does not dispense with the idea of stability. A change between states also implies a shift towards another 'stability domain', in which, once established, the conditions which determine a particular state will persist until such time as disturbances or perturbations give rise to a new set of pre-dominant conditions. Ecological resilience has, then, emerged from and been applied to work on the dynamics of freshwater systems (Fiering, 1982), forests (Clark et al., 1979) and interacting animal and plant populations (Dublin et al., 1990). Figure 3.3 visualises the difference between engineering and ecological resilience.

More recently, further refinements to the theorisation of ecological resilience have been made. Both engineering and ecological resilience as defined above presuppose the existence of ecosystems character-ised by different states (or stability domains) which themselves don't change. Conditions might flip an ecosystem between one or the other, but the states in themselves remain constant. More recently, it has been argued that the ecological processes that create these stability domains are in fact a series of slowly changing variables (Gunderson 2008). Species composition in semiarid rangelands (Carpenter et al., 1999), soil nutrient concentration in wetlands (Davis, 1994) or the spatial connectivity of old trees in spruce budworm forests (Ludwig et al., 1978) are all examples of such ecological processes. As well as flipping between stability domains, ecosystems also adapt to changes in the characteristics of stability domains themselves, in other words. This has been described as adaptive capacity (Peterson et al., 1998), though it also may be captured in the idea of transformation (Walker et al., 2004). See Figure 3.4.

Figure 3.3 *Engineering and ecological resilience*

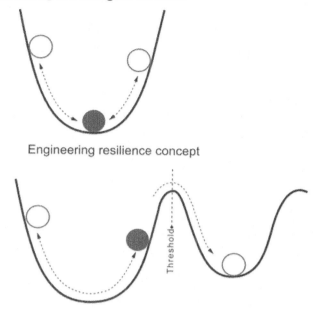

Engineering resilience concept

Ecological resilience concept

Source: Figure 2 from Liao, 2012

Figure 3.4 *Adaptations to shifting stability domains*

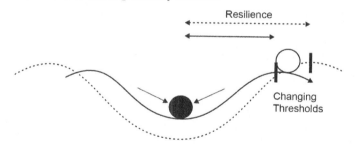

Implications of complex social-ecological systems thinking for conservation and development

If we accept the utility and importance of thinking in terms of social-ecological systems, then the implications for the relationship between conservation and development are profound. In essence, conservationists cannot afford to ignore the ways in which societies are inseparable from the nature that is the object of conservation efforts. Conservation, from this standpoint, cannot solely be about separating landscapes and protecting them from the destructive tendencies of humans, because of the pervasive influence of society upon global ecology. The ramifications of this kind of conservation for human wellbeing are revisited in Chapter 9, in the context of protected areas which, conventionally, have sought to demarcate and maintain 'wild' landscapes free from human presence; or as free as practicably possible. Nor can it be about trying to ensure that things stay the same, trying to maintain an ecosystem in one stable state. Partly, this is a question of values: what value is ascribed to different ecosystem states, and is one state privileged over another in terms of its conservation value? Which nature is worth conserving? Conservation cannot be based on an understanding of nature which is static and separate from the human beings. An often cited but very clear example of the limitations of such a strategy takes us back to our consideration of North American forests. Managing these in such a way as to maximise the levels of biodiversity within the forest is what can lead to the huge forest fires that often feature as international news stories, often on a yearly basis (for instance, in California). As we have seen, fire is a necessary mechanism for the reproduction of some species, but it is also an inevitable consequence of trying to maintain biodiversity at a climax state. If the object is to maintain the forest at this climax state, it is self-defeating. It is also extremely dangerous and often deadly for people living in or on the fringes of forests in this state.

Conversely, development cannot afford to unfold in such a way as to a) bring about shifts into ecosystem states which humans would then struggle to survive in and b) assume that we can manage nature in such a way as to maintain a permanent equilibrium of a particular resource and the ecosystem in which that resource is to be found. Current modes of economic activity attempt often to extract from the environment particular forms of benefits or services valued by society and are about making nature fit, as it were, the requirements of this economy. This is done by increasing 'control over resources through domestication and simplification of landscapes and seascapes to increase production, avoid fluctuations, and reduce uncertainty' (Folke et al., 2005: 442). Managing nature for stability is done with a view to producing the 'maximum sustainable yield' (MSY) of a particular resource that is deemed useful. We could take the example of fisheries to illustrate the issue at stake. In a landmark publication in the fisheries context, Larkin (1977) argued that maximum sustainable yield models simply do not come to terms with the complex food-web relations when predicting single species yields. Treating such resource units as discrete entities rather than as integrated parts of a much wider ecosystem is to fail to recognise that changing one part of it has repercussions for the rest of the system. Managing a resource by reducing its variability is, from a social-ecological systems perspective, not a good idea. Although successfully freezing it in one stage of natural change by blocking out environmental perturbations and feedback, makes it an easier resource from which to extract benefits in the short term, it may store up trouble for the longer term. In the words of Berkes and Folke: 'the accumulation of perturbations, [invites] larger and less predictable feedbacks at a level and scale that threaten the functioning performance of the whole ecosystem' (Berkes and Folke, 1998).

From the perspective of social-ecological systems thinking, in addition to thinking about conservation and development jointly rather than separately, it would make sense to try and reconcile them both through the concept of resilience. There is, moreover, a gathering momentum around using resilience for this purpose. Whilst, ultimately, there are some important issues and caveats with this project, which we identify in the conclusion, the notion of resilience in social-ecological systems is so clearly relevant to different dimensions of the relationship between conservation and development that we apply it at various junctures within the book. Table 3.2 catalogues the chapters in which it is applied, and the subject matter which it is used to analyse. We hope, through these applications, not just to show the utility of the concept of resilience itself, but the necessity of thinking about nature and society together, in order to improve our understanding of the prospects for reconciling conservation and development objectives.

Constructing nature and society

What does 'constructing nature and society' mean?[1]

If the idea of the irreducibility of nature and society seems complex and at times counter-intuitive, it is partly because of the difficulties of pinning down the meanings of nature or society. Such difficulties arise because of what veteran sociologist of knowledge, Barry Barnes, calls the problem of reference, which he characterises as

Table 3.2 *Application of resilience in social-ecological systems thinking across this book*

Chapter	Sphere of application
4	Environmental change – and specifically climate change – and the resilience of social-ecological systems
5	Vulnerability and resilience of the global economic system
6	Adaptive governance within social-ecological systems
8	Implications of overexploitation for the resilience of marine ecosystems
9	The contribution to social resilience of the provision of alternative livelihoods in the context of marine conservation
14	Resilience as a bridge between conservation and development?

'the relationship between our speech and that which is spoken of' (Barnes, 1983: 524). We may plausibly say that to know something entails classifying it as one thing, as opposed to something else (Barnes et al., 1996). Initially, it seems almost self-evident to say that once a thing's similarities and differences to other things have been established, we can safely refer to it as just that one thing; we know what it *is* and what it *is not*. However, how can we be sure that what we say about a thing corresponds to what that thing is? This is precisely the difficulty we have encountered with words such as wilderness, when applied to places supposedly beyond the reach of humanity, but which turn out after all to be influenced or even produced by some kind of human intervention or presence. Insofar as it is applied to a geographical location, we could argue that far from being a word which denotes the most natural of landscapes, 'wilderness' is actually a social construct. But what does this term actually mean? The answer may seem irrelevant at first glance to the achievement of real world problems related to conservation and development. And yet it is impossible to understand the relationship between conservation and development without answering this question.

The short answer is that social constructs are partially or fully self-referential knowledge claims; that is, they consist of references to other references to the same thing. Therefore, social constructs exist because we say they exist – although, crucially, also because we collectively agree that they exist. They might seem natural, to be statements referring to things which have their own existence independent of language. But when subjected to detailed scrutiny, their self-referential character is revealed.

That definition is probably too dense for all but the most intuitive of social philosophers to understand immediately, without further acquaintance with the thoughts and arguments that lie behind it. Again, the work of Barry Barnes (1983) serves here as a useful guide. For the purposes of analysis, Barnes proposes two types of thing to which we can and frequently do refer: 'S' terms and 'N' terms. 'N' terms are what Barnes calls natural kind terms, and proceed on the basis of seeing (or otherwise perceiving) an object and attaching a label to it by matching it to a pre-established pattern. Barnes takes the example of a leaf on a tree. If we come across a tree, we may well see small objects to which, where they are judged to be sufficiently similar, we attach the 'leaf' label. Where they are not considered sufficiently similar, by inference, we do not attach the 'leaf' label. Once designated, the designation is stable: it does not tend to change over time. Leaves today will still be leaves tomorrow. Crucially, with natural kind terms, the thing that is being referred to has, or is believed to have, an existence independent of references to it. A woman could see a leaf and teach her toddler to refer to it in the same way as she does during a trip to the park; but once they had returned home, the leaf would still exist (or at least most of us would be prepared to go along with this assumption).

In contrast, 'S' terms, or social kinds, are terms which are *not* applied to things on the basis of a consideration of their empirical properties. However, with established, routine usage they come to seem as real to us as any 'N' term. 'S' terms also proceed on the basis of recognition and pattern attachment, but they are often used to mark a distinction between one thing and another where no empirical difference is evident. For example, the difference between a queen and a slave is not marked by one having red hair and the other dark brown, by different eye colour, height, weight or shoe size. Rather, we recognise queens and slaves when they are so recognised by everyone else who applies the term. Whilst 'queen' tells us nothing in itself of the empirical characteristics of the person to whom it is applied, it indicates the sort of behaviour that is expected towards the person we designate 'queen'. One may choose not to be deferential on meeting Queen Elizabeth II, but only against the weight of considerable expectation. Crucially, moreover, we would not doubt, if we were one day called to Buckingham Palace, that she was a queen any more than we would doubt that she was sitting in front of us. She would be as real to us as the leaf on the tree. And yet the basis of us calling her queen would consist entirely in the taken-for-granted act of everyone else referring to her as a queen. She is the queen because, and indeed only because, we all say so. In consequence, in the case of some 'S' terms like this one, the stability of designation is almost as strong as it is with leaves: Elizabeth II is queen today and will continue to be so tomorrow (unless she dies, decides to abdicate or is overthrown by a revolution). But if we were to *deconstruct* this status, if we were to look into the history of how we came to refer to people as queens and kings, we would realise that it was a truth of our own making. This is the essence of the social construct.

However, there are lots of things which don't fit into these neat categories – as Barnes recognises and explores in detail in his book *The Nature of Power* (1988). There are terms which, we might say, have both 'N' and 'S' components in them, because they specify a relationship of something to something else (often what we might think of as natural kind term, with an existence independent of language), within a broader context of human activity. Taking the example of a summit, he contends that it is 'that part of the mountain which exists in a relationship with all its other parts'. If we want to verify that something is a summit, we will not be able to find it as an empirical property of the mountain itself, but rather in 'its relationship with its context' (ibid.: 47), which turns out to be 'a context of human activity' (ibid.: 49). A summit is only a summit because we treat it as such, because of our actions in relation to it. Of course, the mountain or hill would have to be present in order for us to be able to talk about summits in the first place, but that does not make a summit an inherent property of a mountain independent of its existence in language.
We could use other examples more pertinent to conservation and development: think, for instance, of a protected area or a health clinic. An example which is even more relevant to this discussion is the notion of a social-ecological system; although, importantly, because this is not a very common way of speaking about society and nature, it does not yet have the taken-for-granted status that health clinics or protected areas have acquired over time. The crucial thing to understand here is that many of the things which we take for granted, which we might put in the category of 'things out there in the world', with an existence independent of language, are only so because we take them for granted. They are constructs which describe our relationship with the world.

When ecology becomes political

We have seen now how we could understand objects and processes important to conservation and development through the lens of (one strand of) contemporary social theory. Yet, to borrow a turn of phrase from Paul Robbins, the question remains: why would we bother? Put simply, because how we refer to things determines the order into which they fall, the ways in which we do or do not use them, the kinds of societies we build upon the kinds of nature that we construct. And unless we all want the same thing and all benefit from it in the same way, this is frequently going to lead to situations in which there are winners and losers, to the potential for conflict, to considerations of equity and justice, to competitions over what we want nature and/or society to be. The relationship between society and nature and the subsequent relationship between conservation and development, seen from this standpoint, entails trade-offs between different interests and strategies, more than it entails win-win scenarios. Consider the example of a forest and the different, potentially incompatible relationships in which different people stand to it, as described in the words of Paul Robins, in the context of India.

On a day several months into the Indian dry season, a forester stands in a low alluvial plain, looking out at the stones in the stream bed and to the hills beyond. Pointing at a thick area of thorny trees on the opposite bank, he explains to me that the forest is returning in this area after years of abuse and neglect. Deep-rooted hardy tree species are securing the embankments and restoring greenery to the desert. The species responsible for this remarkable turnaround in the region is largely imported through global initiatives in scientific forestry.
The tree, prosopis juliflora, he explains, is salt-tolerant, nitrogen fixing, drought-resistant, and very productive. Along with several other introduced species – juliflora was brought from the Mexican/US Southwest a century ago – the tree has helped to triple forest productivity in the past 30 years.

For the forester, this increase in forest area represents not only an institutional victory, but also a personal triumph. Foresters in the crowded middle and lower ranks of the bureaucracy sometimes go decades without a promotion, watching projects develop with little or no success. To the degree that they are paid for their efforts, small bribes from local people and the occasional small "feast" from a local landlord are more common than official reward. The achievement of significant forest cover – the official goal not only of the Indian state but also of World Bank donors – is an important success for the local bureaucracy, one that will minimally assure the flow of already limited funding and support. In listening to the forester story, it seems that, far more abstractly, this achievement of forest cover is a deeply internal and aesthetic pleasure. Greenery is good.

*But standing again in the same stream bed a few months later with a local
herder, a member of the raika caste, an extended kin network of herders and
livestock breeders who have lived in the region for countless generations, I learn
that this tree cover represented something else entirely. The old man, leaning on
a tall gnarled staff and shaking his head at the cover of juliflora across the
rocks, explained that the tree was a hazard and a blemish, that it had no value,
and that it crowded out valuable grasses and forage. More to the point, he
insisted that the trees represented no kind of forest at all; on the contrary, they
had created Banjar, degraded wasteland. This wasn't junglat (forest), this was
simply angrezi, foreign English landscape.*

(Robbins, 2012: 122–3)

There are two important points to take from this vivid and compelling
description. First, as Robbins points out, once we stop taking the
forest for granted, we can see that it is not in itself a wholly 'natural'
phenomenon, even though it refers to something (an increasing
presence of *prosopis juliflora*) which exists independently of references to it. It is a term with, as Barnes might say, an 'S' component
and an 'N' component. Its status as a forest depends on the

Figure 3.5 *Prosopis juliflora* **flower**

Source: Photographer Bruno Navez; used under Creative Commons licence

relationship an observer has with it; and the forester and livestock herder have very different relationships with it indeed. This leads to the second point. The ways in which we would construct this forest – that is, the ways in which we would specify the relationship in which we stand with it – would in this example lead to a very different kind of forest ecology depending on our starting point as a herder or forester respectively. How the forest actually gets constructed is precisely where ecology becomes inherently political. If we just take a forest at face value, we fail to recognise that some forest constructs win out over others, that resources can be mobilised around them, that these resources will come to some people and not others, serve some purposes and not others. In consequence, we fail to interrogate how this has come to be. If we want to claim any level of legitimacy to the relationship between conservation and development that we envisage or aspire to, then these considerations of power and politics, which explain whose nature and society constructs count, have to be unpicked, rather than taken for granted.

This task of unpicking the taken-for-granted in social relationships with nature, and of looking at the (often unevenly distributed) consequences for people and environment is at the core of a body of work commonly referred to as political ecology. An increasingly diverse field, political ecology covers an expanding amount of ground, from a variety of different perspectives and starting points. Robbins (2012) identifies five broad research areas or theses within political ecology and summarises the things they attempt to explain – see Table 3.2. Because of this diversity, it is not easy to find a common thread which links all political ecologists, beyond a concern with bringing politics into a consideration of ecology, and ecology into a consideration of social and political processes. Frequently, however, political ecologists are concerned with the losers of contemporary relationships between nature and society, both in human and in environmental terms. Often, there is a focus on marginalised people and the environments they inhabit, with the injustices which result from forms of environmental degradation which result from people having constrained choices over how they use or interact with the environment. But equally, there is a concern with the social and environmental outcomes of global political and economic processes. As such, the work of political ecologists provide many entry points to a fuller understanding of the relationship between conservation and development, and is frequently drawn upon across various chapters in this book (as specified in Table 3.3).

Table 3.3 *Five broad theses within political ecology (after Robbins, 2012: 22)*

Thesis	What is explained?	Topic (and where to find coverage of it in this book)
Degradation and marginalisation	Environmental conditions (especially degradation) and the reasons for their change	How environmental degradation, blamed (sometimes erroneously) on marginal people, is a product of a larger political and economic context. (Coverage in this chapter, and **chapters 7 and 8**).
Conservation and control	Conservation outcomes (especially failures)	How efforts at environmental conservation often viewed as benign can have strongly adverse effects on people and environments, sometimes failing in consequence (coverage in **chapters 9, 10 and 11**).
Environmental conflict and exclusion	Access to the environment and conflict over exclusion from it	How environmental conflicts form part of larger gender, class and racial struggles and vice versa (coverage in **chapters 1, 7, 9, 10 and 14**).
Environmental subjects and identity	Construction of the individual and social identities	The way in which political identities and social struggles are related to livelihood and environmental activity (coverage in **chapter 6**).
Political objects and actors	The deep structures of socio-political conditions	How political and economic systems are affected by environmental processes (coverage in **chapters 1, 2, 5, 6 and 14**).

Varieties and strategies of constructivism

For people who want to engage with conservation and development, it is going to be necessary to assume that there is a world out there and that in some way, shape or form, the things that we say about it correspond or capture something about how it works, however imperfectly. Political ecologists probably don't have a definitive way of resolving the problem of reference, which has, after all, been with us in one way or another for thousands of years. But they do have some strategies, or at least perspectives, on it that can help us think through how to deconstruct the relationship between conservation and development, without giving up on our prospects of trying to change it for the better. It does mean being explicit about the values we bring to phrases like 'change it for the better', and it does require us to acknowledge that our accounts of this relationship are subject to, rather than derived from a definitive solution to, the problem of reference. Yet an inability to start from a position which can be shown to be unassailably correct may more easily facilitate processes of negotiation between different accounts of what the relationship between conservation and development should be.

One strategy is to give up trying to say anything about nature itself and look at how the ways in which it is socially constructed reflect the predominant distributions of power. From this perspective, the world is what we say it is, things are true because enough of us believe that they are. Coming back again to Robbins,

> Environmental conflicts are, therefore, struggles over ideas about nature, in which one group prevail not because they hold a better or more accurate account of a process – soil erosion, global warming, ozone depletion – but because they access and mobilise social power to create consensus on the truth (2012: 128).

Such an approach, which Robbins terms 'hard constructivism', may help us to avoid making claims about nature which we cannot absolutely justify, epistemologically speaking. And crucially, it leaves open space for alternate constructions or understandings of the world, like those of nomadic herders or religious philosophers, which can be at variance from those sanctioned by science. But by the same token, it leads us away from trying to describe nature at all, figuring out how it works and how it affects the societies we live in (which nomadic herders and scientists alike are keen to attempt). As much for the donor agency looking to fund better housing designs which can withstand tropical storms in developing countries, as for the biologist charting the novel ecosystems which arise in and around cities/post-industrial landscapes, simply explaining conflicting accounts of nature in terms of social beliefs about it is not enough. We have to make some claim about how nature is if we want to make any subsequent claims, either about the biodiversity character-istic of novel ecosystems or about the principles of earthquake-proof housing.

Another, more pragmatic standpoint is 'soft' constructivism, in which scientific knowledge can give us true accounts of nature – that is, what we say about things truly corresponds to how they are – but that the social character of science, the way in which it is done through communities of people and practices, means that bias is apt to slip in. This will get in the way of establishing a true account of nature, even if it can be eradicated over time and through more reflexive practices which encourage us to become more aware of the ways in which scientific knowledge is influenced and/or distorted through social processes (Sullivan, 2000). This line of thinking, then, leaves more space to acknowledge different knowledge claims – those of nomadic herders as well as those of scientists – but ultimately argues that these can be arbitrated by a scientific framework which engages rigorously with them both (Batterbury et al., 1997; Forsyth, 1996; Sullivan, 2000). This is quite a common approach, both within political ecology and outside of it. However, it gives scientific accounts of the world a privileged truth status which it denies to other, sometimes competing, beliefs describing that same world. It still allows for some forms of knowledge – for instance, belief in witchcraft – to be put down to distorted accounts of reality brought about by social or cultural beliefs which lead people to miss what is in front of their eyes. But it reserves for science believed to be true – for instance, internationally-recognised soil classification systems – a description of a form of reality free from any of the social processes which govern the formation of other beliefs.

Robbins finds both of these approaches ultimately unsatisfying, the one because it doesn't allow us to 'get at' nature itself, looking only at our beliefs about it, the other because it still only sees a role for social explanations in scientific knowledge deemed to be false. What he proposes instead is a view in which both soils and soil scientists are 'co-produced' in a dialectical understanding of the relationship between how nature influences society and how society influences nature:

> *In this view, the landscape is produced from the very ideas through which it is apprehended, even while is those ideas are rooted in the material activities and changes of the landscape. Constructions of the environment are not solely pernicious politicised "inaccurate" accounts but are instead the scaffolding of knowledge that allows us to understand the material context of that knowledge.*

This proposition is intuitively attractive, not just in terms of providing us with the basis for a theory of knowledge, but also for providing us with a helpful way to understand the relationship between nature and society. An important part of the metaphor of 'co-production' for Robbins is about the kinds of environments that get produced, which can be very undesirable – for instance, parts of the Niger Delta contaminated by oil spills – or which can be desirable. See Box 3.2 for an example of a desirable co-produced landscape. From a philosophical point of view, arguably, the idea of co-production does not

Box 3.2

The 'co-production' of Amazonian Dark Earths

In the context of the Amazonian basin, the term 'dark earths' refers to a range of soils with different colours, including *Terra Preta do Índio* (black earth of the Indians) and *Terra Mulatta* (brown earth). These soils are at variance from the more predominant oxisols and utilsols, because they are fertile, retaining nutrients and moisture better than oxi- or utisols. They are full of organic matter, which is a result of the presence of charcoal in the soil, and may make up as much as 10% of the Amazon. There remain questions over quite how the soils were formed and the extent to which their facility is a result of deliberate human action in their creation. Some scholars (i.e. Fairhead and Leach, 2009; Leach et al., 2012; Woods and Denevan, 2009) argue that they are the product of human occupation and action: by the use of maidens to discard organic materials in, in combination with periodic light burning, the productivity of the soils has been enhanced. The soils hold out the prospect of Amazon regeneration in the face of deforestation, the possibility of recovery where previously it was presumed that forest clearance was irreversible. Paul Robbins (2012) argues that Amazonian dark earths present us with the possibilities of human interaction with environments to be characterised not just in terms of landscape destruction but also of the production of desirable ones. The rangeland ecologies documented by Kepe and Scoones (1999 – see Box 3.1) constitute another example

resolve the fundamental problem of reference which we have discussed at some length in this chapter. There is disagreement over this by different proponents working in the field of science and technology studies. Bruno Latour, on which Robbins draws in his account of co-production, has been for some decades now writing about the theory of knowledge which, for him, gets beyond setting up the argument in terms of the difference between things in the world and the knowledge claims we make about those things (see, for instance Latour, 1993). But his account of co-production is controversial, occasionally borders on the incomprehensible and has been rejected outright by some scholars in the field (see Bloor, 1999 for a highly critical account of Latour's theory of co-production).

Nevertheless, regardless of these differences, we might be able to agree that, whatever our philosophical starting point, the idea of a dialectical relationship in which nature affects what we believe about it and we affect nature through our beliefs about it, leads to something like a co-production of nature and society and a contingent requirement to understand these things together, rather than to try and separate them. Indeed, this is not an altogether dissimilar proposition from the idea of the social-ecological system. This comparison may make Robbins and other political ecologists a little uncomfortable, because whilst it has the same kind of intuitive appeal, proponents of social-ecological systems thinking arguably take both their constructs of nature and society more for granted than any political ecologist would be happy to sanction, and are not renowned for paying attention to the power-politics dimensions around which kinds of nature and society are co-produced (see Chapter 14). Nevertheless, albeit via rather different pathways, these different groups of thinkers share at least some common ground when it comes to the dialectical way in which they set up the relationship between nature and society.

We do not need to privilege science or any other belief system as uniquely true in order to use this metaphor as a working hypothesis. Instead of trying to adjudicate between different belief systems, moreover, it may be more useful to set up processes of knowledge exchange between groups with differing accounts of the relationship between society and nature, with a view to achieving a consensus where possible around the kinds of society and nature to co-produce, and of mediating conflicts where the chances of consensus are low.

What does this all mean for the relationship between conservation and development? On one level, it means understanding the social, political and economic dimensions of conservation, and making room for accounts of the relationship between nature and society that are not grounded purely in science. But at another level, it is about who makes decisions in either conservation or development projects, or those which attempt to do both simultaneously. This is one way of not taking nature for granted, but rather, coming to understand how it is differently conceived by different groups, with a view to having an inclusive debate and process for determining the kinds of societies and natures that we want conservation and development to work towards.

Summary

If we want to understand the prospects for reconciling conservation and development, we need first to attend to the relationship between nature and society. In the nineteenth and twentieth centuries we have tended to treat nature and society as separate entities. Yet these taken-for-granted ideas (constructs) are unhelpful because they do not help us to appreciate the ways in which society and nature appear to shape each other fundamentally. Drawing upon social-ecological systems thinking and political ecology, we can generate an account of the co-production of nature and society which more richly describes the relationship between the two. Understanding nature and society as being co-produced has two profound implications for conservation and development:

1 Managing nature, and particularly doing so in ways which attempt to keep an ecosystem in equilibrium, in order to support a particular form of resource use, may bring about changes in the ecosystem which make it less useful, or even useless, in human terms. Therefore, even some of the 'wise use' varieties of conservation, at the heart of the idea of sustainable development, may not in fact be sustainable across a medium or long term timescale. The resilience of social-ecological systems to shocks and stresses also requires adaptive capacity.

2 Managing nature is fundamentally political. What we do to nature depends on our definitions of it, and whose definitions and uses predominate. If we take nature as a given, we fail to acknowledge the extent to which social, political and economic activity is organised not so much around nature itself, but how we understand (construct) nature – and society. Privileging one understanding over another creates situations where there are winners and losers.

Discussion questions

1. Is it possible to define conservation in a value-free way?

2. How convinced are you of the need to consider nature and society in conjunction, rather than separately?

3. What would efforts to achieve conservation and development objectives look like, if systematically grounded in the notion of a (complex) social-ecological system?

4. Why bother to deconstruct our taken-for-granted notions of both conservation and development?

Note

1 Much of this section draws heavily on Newsham (2007).

Further reading

Barnes, B., 1983. Social-life as bootstrapped induction. *Sociology* 17, 524–545.

Berkes, F., Colding, J. and Folke, C., 2008. *Navigating Social-ecological Systems: building resilience for complexity and change.* Cambridge University Press, Cambridge; New York.

Gunderson, L.H. and Holling, C.S., 2002. *Panarchy: understanding transformations in human and natural systems.* Island Press, Washington, D.C.; London

Robbins, P., 2012. *Political Ecology: a critical introduction.* John Wiley & Sons.

Walker, B., Holling, C.S., Carpenter, S.R. and Kinzig, A., 2004. Resilience, adaptability and transformability in social-ecological systems. *Ecology and Society* 9, 5.

Online resources

Resilience Alliance – http://www.resalliance.org/ (accessed 14.04.2015)

Ecology and Society (open access journal) – http://www.ecologyandsociety.org/ (accessed 14.04.2015)

Centre for Political Ecology – http://www.centerforpoliticalecology.org/ (accessed 14.04.2015)

Cultural and Political Ecology Specialty Group of the Association of American Geographers – https://capeaag.wordpress.com/ (accessed 14.04.2015)

Journal of Political Ecology –http://jpe.library.arizona.edu/ (accessed 14.04.2015)

STS (Science and Technology Studies) wiki – http://www.stswiki.org/index.php?title=Main_Page (accessed 14.04.2015)

ENTITLE (European Network of Political Ecology) – http://www.politicalecology.eu/ (accessed 14.04.2015)

Members of ENTITLE explaining political ecology – https://www.youtube.com/watch?v=HLVE69QZt5w (accessed 14.04.2015)

④ Conservation and development in an era of global change

- **Global change**
- **Global change and resilience of social-ecological systems**
- **Climate change, conservation and development**

Introduction

In chapters 1 and 2 we take a historical overview of conservation and development. These chapters set conservation and development within the context of global change. Anthropocene, 'the age of humans', has changed the parameters of how conservation and development should operate. The widespread human footprint on the planet has implications for the sustainable use and management of natural resources. These implications have been portrayed as a 'perfect storm' of a wide range of issues that humanity faces in the future. 'Planetary boundaries' has been put forward as a framework that strives to define the limits within which humanity should live. Such conceptualisation of boundaries, however, also raises the question of who should set such boundaries and whom are they for? Despite increasing economic prosperity, a large section of the human population on Earth still lives below the poverty line. This section of the population is perhaps most vulnerable to the global change. Enhancing their adaptive capacity and making them more resilient is one of the challenges that requires careful thinking about the conservation–development nexus. In the meantime, climate change has become the most prominent environmental issue of our time and building resilience to climate change has become somewhat of an obsession. This has implications for conservation and development because climate change is closely tied to a wide variety of other issues.

But can climate change provide the overarching *raison d'être* for joined-up conservation and development? While climate change does address a wide variety of issues about sustainability, it perhaps does not have the same moral grounding as the concept of sustainability. Should conservation and development align their moral agendas with that of sustainability, living within our means? This chapters describes a range of issues around global change that directly relate to conservation and development and it sets the scene for the rest of this book. The issues raised here are revisited in the following chapters in greater detail.

Global change

Unprecedented change in ecological, economic and social spheres

There is now a growing recognition that human activity on the planet over the last 200 years has propelled us into the Anthropocene. Paul Crutzen, the Nobel Laureate, in an article published in the year 2000 argues that human activities have grown to become significant geological forces, such as through land use changes, deforestation and fossil fuel burning and therefore it is justified to call this period over the last two centuries by the name 'Anthropocene' (Crutzen, 2000). Over the Holocene epoch (last 11,700 years) human civilisation has developed on Earth in a remarkably stable climate. The beginning and expansion of agriculture has been a landmark development over this period, meaning that our societies have become much more settled in comparison with our hunter-gatherer predecessors. This relative stability and resource abundance with improvements and innovation in agriculture has caused a growth in the human population. In the recent past, however, the pace of this growth has increased remarkably – particularly after the Industrial Revolution of the late eighteenth century. The speed of transformation over the last 200 years, therefore, has required the renaming of this most recent period as Anthropocene, 'the human age'.

While the rapid growth in world population, projected to reach 9 billion by 2050, is a defining factor of the Anthropocene, changes in other spheres are also acknowledged – mainly as a consequence of human population growth. Such growth is putting increased demand on natural resources, and in some instances runs the risk of causing degradation of many ecosystems beyond 'tipping points' (although, as Chapter 7 demonstrates, the relationship between population growth and environmental degradation is not straightforward). Tipping point is a metaphor used to describe change in the state of ecosystems to such an extent that crossing a certain threshold may lead to its rapid collapse. Ecosystem degradation is leading to the loss of some species and gain of others, changing the composition of these systems. This has implications for natural-resource dependent societies, who are having to adapt to the changing resource base. Population growth, coupled with increased urbanisation, is changing the fabric of society. In 2007, the projected human population estimates suggested that in 2008 for the first time, half of the human population will live in cities (UNFPA 2007). This proportion is set to increase with rural to urban migration. Cities are known to have large 'ecological footprints', meaning that rapid urbanisation also has an effect on natural ecosystems far beyond the boundaries of these cities (Wackernagel et al. 2006). Alongside changes in the organisation of human society, the growing economic wealth is also putting increasing pressure on natural resources because the creation of wealth depends directly on the exploitation of natural resources. These changes in economic, social and ecological spheres amount collectively to global change, which has implications for resource use and management. It is not clear that efforts to achieve conservation and development simultaneously respond substantively to these changes, a point we will return to in the conclusion.

Implications for 'sustainability' of current resource use and management

A transition from agrarian to industrialised society during the nineteenth and twentieth centuries led to economic growth in many Western countries. This industrialisation was largely driven by fossil fuels – beginning with coal and then oil. Fossil fuels enabled mechanisation in many instances where manual labour was used previously. Agriculture is one sector where such mechanisation led to dramatic increase in food production, followed by the introduction of technologies to improve seeds and more recently genetic modification. Intensification of agriculture, however, did not mean improvement in the quality of natural ecosystems. The degradation of ecosystems continued simultaneously owing to the increasing footprint of human activities other than agriculture. These predominantly included manufacturing industry and transport which relied on fossil fuels. Fossil fuel-driven industrial and economic growth contributed to environmental degradation through the release of effluents and other pollutants in the environment.

This paradigm of industrialisation and economic growth came under scrutiny in the 1970s and 80s, when environmental activists started to raise concerns about the 'sustainability' of this economic growth and began pointing to the fact that growth based on purely economic criteria does not pay attention to social and ecological issues – see Chapter 2 for more detail. This aspiration for sustainability is reinforced, discursively at least, through global programmes such as the United Nations' Millennium Development Goals (UN, 2000) or Aichi Biodiversity Targets (CBD, 2010) and their local implementation. Despite a move towards sustainable development, however, the exploitation of natural resources has continued unabated alongside emissions from fossil fuel use. The discussion on its effects came into the mainstream during the 1990s and 2000s culminating in climate change being identified as the most important environmental concern (IPCC, 2007). The discussion on climate change reinvigorated various environmental issues, including food security, biodiversity conservation, water and energy sustainability.

The 'perfect storm' and 'planetary boundaries'

In 2009, the UK government chief scientist Sir John Beddington warned of a 'perfect storm' of food, energy and water shortages by 2030 (BBC, 2009). He attributed these to the growing world population and the effects of climate change. He also suggested that this may lead to public unrest, cross-border conflicts and mass migration. In 2009, a group of scientists led by Johan Rockström of Stockholm Resilience Centre, Sweden, identified a number of 'planetary boundaries' and defined a safe operating space for humanity with respect to the Earth system – the planet's biophysical subsystems or processes (Rockström et al., 2009). These scientists argued that we have already crossed three planetary boundaries, namely that of biodiversity loss, climate change and nitrogen cycle. Drawing on this analysis Kate Raworth of Oxfam, UK, proposed a safe and just space for humanity, acknowledging that many of the world's poor still live below acceptable standards of living and therefore the safe operating space for

Figure 4.1 (a)   Planetary boundaries: safe operating space for humanity

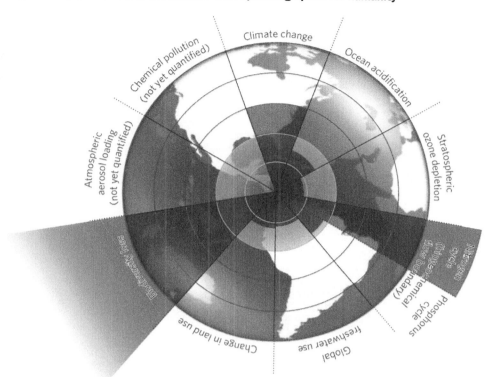

Source: www.nature.com/news/specials/planetaryboundaries/index.html; permission kindly granted by the Nature Publishing Group

humanity needs to be defined in such a way that the planetary boundaries are also socially just (Raworth, 2012).

The discussion on planetary boundaries and the need for a safe and just space for humanity poses a variety of challenges. There is no doubt that enhancing standards of living of the world's poor is of utmost importance from a moral standpoint, particularly as Western countries have enjoyed significantly higher standards of living over the last half century. But to alleviate poverty of a large number of the world's poor, what is an acceptable standard of living? They have the right to aspire to Western standards of living, but will that goal exceed the planetary boundaries far beyond what the Earth system can cope with, leading to various tipping points? Also, who decides what an acceptable standard of living is? Should these decisions be

Figure 4.1 (b)   Oxfam doughnut: safe and just operating space for humanity

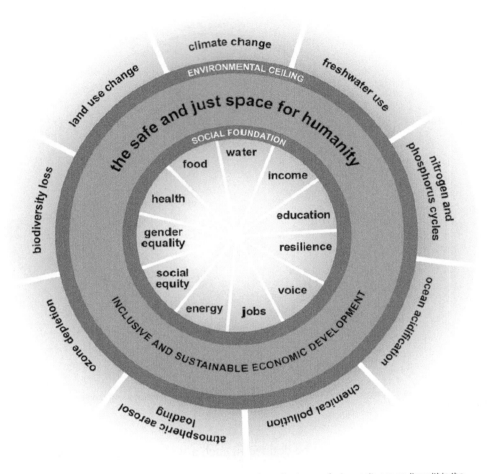

Source: http://policy-practice.oxfam.org.uk/publications/a-safe-and-just-space-for-humanity-can-we-live-within-the-doughnut-210490); permission kindly granted by Oxfam

made from a moral standpoint, from a standpoint of social justice and human rights, or from a standpoint of planetary boundaries? These are difficult questions and they pose difficult challenges for policy makers, planners and managers. What then is the way forward?

The thresholds in Planetary Boundaries or Oxfam doughnut frameworks are determined using a wide variety of quantitative global datasets. Not crossing these thresholds would seem to be the best way forward, but what are the means to achieve this? Two different solutions have been proposed: one requires a greater austerity in resource use by the Western world and the other requires technological solutions to maintain Western standards of living, whilst also bringing up the standards of living of the world's poor. Raworth (2012) argues that the biggest source of planetary-boundary stress is excessive resource consumption by the wealthiest 10% of the world's population. On the other hand, Beddington suggests that a major technological push is needed to boost crop yields, better utilising existing water supplies and developing renewable energy supplies. Rockström et al. (2010) acknowledge the need for not crossing thresholds, but suggest that provided this is achieved, 'humanity has the freedom to pursue long-term social and economic development' (p. 475).

But how can such long-term social and economic development be brought about in a globalised world marked by control on resources by large corporations? The improvement in crop yields, according to Beddington's suggestion, for example, has in the past been achieved through the Green Revolution and more recently through genetic modification of crops. However, experience of both of these has highlighted that commercial control over crop varieties has not benefited farmers in developing countries and therefore increases in crop yield have been somewhat elusive. For example, Vandana Shiva in her book *The Violence of the Green Revolution* reports the destruction of genetic diversity and soil fertility – counterproductive to yield increases – in parts of India that were at the helm of the Green Revolution in the 1970s (Shiva, 1991). A wide range of natural resources are increasingly controlled by large private corporations, posing hurdles in the social and economic development of some of the poorest people in the world. Existing water supplies, for example, have often been excessively used for commercial as opposed to domestic consumption – leading to water shortages in rural areas. The development of renewable energy technologies is similarly hampered by the lack of will on the part of established corporations in the energy sector, notably oil and gas corporations. In addition, on-the-ground operations, particularly those of extractive resource industries pollute the environment and put pressure on natural resources collectively contributing to the transgression of planetary boundaries. This transgression of planetary boundaries compromises the ambition to provide a safe *and just* space for humanity.

Global change and resilience of social-ecological systems

As outlined in Chapter 3, Scholars such as Buzz Holling, Lance Gunderson and Carl Folke (Gunderson and Holling 2002; Folke et al. 2010) have suggested that humanity must aim to build resilient social-ecological systems particularly if we are to weather Beddington's 'perfect storm' and to remain within Rockström et al.'s 'planetary boundaries'. Resilience in an ecological system is its capacity to 'tolerate disturbance without collapsing into a qualitatively different state that is controlled by a different set of processes', and resilience in a social system has the 'added capacity of humans to anticipate and plan for the future' (see the Resilience Alliance website). In this chapter, it suffices to view a resilient social-ecological system as one that can withstand shocks and rebuild itself when necessary as well as adapting to changes that are inevitable. The ability of social-ecological systems to cope with change is discussed at length in the following section.

Climate change and unexpected changes in ecological systems

The resilience of social-ecological systems depends largely on underlying environmental factors. There are numerous examples, where climatic changes have been responsible for the collapse of civilisations – due to undesirable effects on soil fertility and the consequent breakdown of agricultural systems, social structures or human values. These examples suggest that changes in environmental conditions can bring ecosystems up to a threshold beyond which they can no longer function in their current state, inducing often rapid collapse. These changes in the ecological systems were followed by detrimental changes in the structure and organisation of society. The lack of ability to adapt was the primary cause of societal collapse in many of these cases, meaning that the societies had to migrate somewhere else and start anew. In cases where there was sharp decline in population due to resource crunch, starvation or hunger, the societies simply collapsed.

The example of Easter Island, which is the best historical example of a societal collapse in isolation, is illustrative (Diamond, 2005). Diamond suggests that environmental degradation and extreme deforestation destabilized the island's ecosystem. With all large trees lost, the Easter Islanders were no longer able to build seaworthy vessels, significantly diminishing their ability for fishing, a major source of livelihood and dietary protein. Other important sources of dietary protein such as land and sea birds and large sea snails also disappeared following the loss of forest. Furthermore, deforestation enhanced soil erosion and as a result crop yields declined substantially. Diamond's thesis of the linkages between environmental degradation and societal collapse has been contested, however, and alternative explanations for the collapse have been put forward. Peiser (2005) argues that the diseases brought by the Europeans were responsible for the collapse of Easter Islanders while Hunt and Lipo (2006) argue that the domesticated animals brought by the Europeans caused the collapse of native flora and fauna. Collectively, these various factors led to the collapse of the social-ecological system on the island and by the eighteenth century, the remaining Easter islanders are known to have become marginal farmers, who reared chicken as their primary source of protein.

Scientists are now convinced that modern-day climate change is driven by humans, that it started with industrialisation, and that the levels of increasing atmospheric carbon-dioxide are changing the temperature of the Earth's atmosphere (IPCC, 2007). There are concerns over the speed of this change and the evidence put together by IPCC indicates that the pace of climate change overshadows the pace of any of the past climate-change episodes. Furthermore, the effects of climate change are likely to be unpredictable, although there is generally agreement among scientists working on the IPCC report that the frequency of extreme events will increase in the future. Climate change is predicted to have an effect on ecological systems, altering their structure and function, but also changing their species composition. Such climatic changes will therefore give rise to novel ecosystems, characterised by species assemblages that have no analogue in the past, and marked by abundance of weeds and other invasive species.

Research has suggested, however, that such species-impoverished systems are the ones that might be more resilient than those with complete assemblages of species. A study by Côté and Darling found that coral reef communities that are partially degraded are more resilient to the effects of climate change (Côté and Darling, 2010). Such strands of evidence are also emerging from tropical forests. For example, a 7,500-year-old record from the Western Ghats of India indicates that a disturbed and degraded landscape is more resilient to the effects of anthropogenic activities such as land clearing and burning as well as climatic factors such as fluctuations in monsoon intensity (Bhagwat et al., 2012). Why might disturbed and degraded ecological communities be more resilient to the effects of anthropogenic or climatic changes? One explanation is that such systems have already lost their habitat-specialist species, which are most sensitive to disturbance while the assemblage of species remaining is able to withstand disturbance. Due to uncertainties in the effects of climatic and anthropogenic changes on ecological systems, scientists are suggesting that even degraded or disturbed ecosystems might have some conservation value in the future because they are more resilient to disturbance and can provide continuity to the ecosystem through opportunities for recovery when favourable conditions arrive. Adaptive capacity in ecological systems might therefore be related to heterogeneity of landscape mosaics and the diversity that such landscapes provide at genetic and species levels (Bengtsson et al., 2003). Much as anthropogenic factors alter the ecology, people are also instrumental in developing landscapes in ways that support human needs and often those of other species.

In social systems, research has asked a different set of questions to understand what makes these systems resilient, but researchers are unanimous in their conclusion that the adaptive capacity of these systems is the key to their resilience (Berkes et al., 2003). What factors enhance adaptive capacity in social systems? Researchers have concluded that the existence of institutions and networks that learn and share knowledge and experience, create flexibility in problem solving and balance power among interest groups play an important role in adaptive capacity (Berkes et al., 2003). Folke et al. (2002) identify four critical factors that are necessary for adapting to the changing provision of natural resources during periods of change and reorganisation. These include (1) learning to live with change and uncertainty; (2) nurturing diversity for resilience; (3) combining different types of knowledge for learning; and (4) creating opportunity for self-organisation towards social-ecological sustainability. Examples of such systems can be found in terrestrial and marine environments (Berkes et al., 2003). To promote such properties within a system, the fabric of the social system in question is very important; and this fabric would be determined by the nature of institutions and the quality of governance within that system.

Societies that are dependent on natural resources are the ones that will bear the brunt of the effects of climate change on ecological systems. Such changes are likely to affect the fabric of social-ecological systems that have developed over the history of human presence on Earth. Would many societies face the same fate as the Easter islanders did? Based on the past evidence of collapse and reorganisation of social-ecological systems following extreme climatic events, suggestions have been made that modern-day climate change will have similar consequences for human society (e.g. Lynas, 2006). Building resilience into these social-ecological systems and increasing their adaptive capacity is therefore considered the primary objective in policy circles. This is based on the foundation of systems thinking: 'Systems with high adaptive capacity are able to re-configure themselves without significant declines in crucial functions in relation to primary productivity, hydrological cycles, social relations and economic prosperity' (Resilience Alliance, 2002: 1). Adaptive capacity of social-ecological systems is therefore characterised by their ability to survive in the face of change and uncertainty.

What do human societies require to adapt to changes imposed on them by extreme climatic events? This adaptation may take three different forms characterised in colloquial terms by (1) 'hanging in', (2) 'stepping out', (3) 'jumping up' (Dorward, 2014). Hanging in would require societies to withstand extreme changes whilst modifying their existing livelihood activity to suit these changes. For example, agrarian societies might have to change the length of their crop-rotation or modify planting cycles. Stepping out would require societies to change their livelihood activity, or migrate to a location that provides them with more suitable conditions to continue the same livelihood activity. For example, mountain-dwelling hunter-gatherer communities might have to take up agriculture if low-altitude agro-climatic zones migrate upslope. Jumping up would require societies to diversity their livelihood activities in order to adapt to changing conditions. For example, loss of agricultural productivity in drylands due to further drying of climate might force communities to take up cattle rearing.

Climate change, conservation and development

Climate change, however, is only one factor in an era of changes in ecological, social and economic spheres. Responding to global change therefore requires combined conservation and development efforts and begs for a greater understanding of social-ecological systems, their adaptive capacity and resilience. Dealing with climate change cannot only be about reducing carbon emissions if we do not maintain the 'ecosystem services' which regulate climate, and which underwrite current levels of prosperity enjoyed by the rich world, and aspired to by poorer countries. Therefore, drivers of ecological, social and economic changes and their effect on the provision of these services also need to be addressed simultaneously. In spite of the perceived failure of integrated conservation and development efforts, we are still left with the same imperative of squaring conservation and development now for the continued provision of 'ecosystem services'. To what extent is this shift in discourse from sustainability to climate change to ecosystem services beneficial for integrated conservation and development?

Ecosystem services within the discourse on climate change

The discourse on climate change has dominated the discussion on the environment in the twenty-first century, since the scientific consensus on the link between anthropogenic activities and carbon emissions is becoming stronger (IPCC, 2007). This has prompted efforts to reduce carbon-dioxide emissions in various ways including curtailing entities that emit carbon-dioxide (e.g. infrastructure and transport) and promoting entities that absorb carbon dioxide (e.g. tropical rainforests). As such efforts are directed towards understanding quantitatively the emission and absorption of carbon-dioxide due to various human activities as well as natural processes. The importance of natural ecosystems such as tropical forests in fixing atmospheric carbon, coupled with the growing discussion on the need to conserve biodiversity of tropical forests has brought to the forefront the concept of 'ecosystem services' – benefits that humans derive from nature. Deforestation and the degradation of tropical forests contribute to a large proportion of carbon emissions (Ramankutty et al., 2007). Reducing those emissions by preventing deforestation and degradation, therefore, is considered an effective way of mitigating climate change (Miles and Kapos, 2008). This has shifted the focus from conservation of natural ecosystems to conservation of natural ecosystems for human wellbeing, providing a new opportunity to integrate the conservation of global biodiversity and poverty alleviation.

Tropical deforestation has been a persistent environmental issue for the last three decades and measures to reduce deforestation have met with mixed success. One of the key reasons for the lack of success is that economic development has been at the forefront of policies in countries at the tropical forest frontier. This means that although there is awareness that deforestation is harmful to the environment, the economic needs of the countries in question compete with the environmental concerns at stake. The international community sees tremendous potential in tropical forests to mitigate the effects of climate change because of their value for fixing atmospheric carbon. But the only way to reduce deforestation in countries at the tropical forest frontier is to provide them with incentives to do so. This is where ecosystem services and the proposals for the payment for ecosystem services come into picture. If payments are made to these countries to protect their forests, then in theory they can continue their economic development trajectory whilst protecting the forests for which payments are made. This measure to mitigate climate change can potentially provide a win-win solution for conservation and development. But does this win-win solution address key environmental problems?

Tropical deforestation is a complex environmental issue. Much of the recent deforestation has happened due to the expanding agricultural frontier such as large-scale soya farms in Brazil, or expanding oil palm plantations in south-east Asia (e.g. Pacheco, 2012). Logging industries have also long been responsible for deforestation and illegal logging in remote tropical forests in Africa, South-east Asia, and South America (Bowles et al., 1998). While payments for ecosystem services (PES) as an instrument is targeted towards the local communities in the hope that they will protect their forests, it is often the case that these communities are at loggerheads with much larger and much more powerful economic forces. Therefore, while paying for ecosystem services, in theory, can address the problem of tropical deforestation, there are often challenges that such payments may not be able to tackle. Soya, palm oil and timber are commodities that are much sought after and the industries which control their production are large and multi-national. It is doubtful whether payments for ecosystem services can provide a large enough incentive for these companies not to convert tropical forests into land that produces agricultural commodities. These industries might, however, be willing to pay for ecosystem services to offset the negative impact of their activities under the guise of 'corporate social responsibility'. With climate change becoming a sensitive issue, there might well be growing interest from these industries to make a difference so that they are seen as 'good guys' in the public eye.

But are such gestures likely to address the dual challenge of conservation and development? Critiques have suggested that this may not necessarily be the case (e.g. Redford and Adams, 2009). An ecosystem service only has a value when it has a user and this means only those 'ecosystem functions' that generate services beneficial to the user are likely to receive attention from the industry. For example clean water is a service that has value for people, but nutrient cycling *per se* does not have an obvious value. This means that while the industry is willing to pay for clean water, it is unlikely to be interested in paying for nutrient cycling. Yet, nutrient cycling is an important element in maintaining the tropical forest ecosystem. This juxtaposition of services with an obvious use and those without an obvious use means that ecosystem service payment schemes are likely to be selective. Should tropical forests then be conserved for their own sake rather than for the services they provide? While the climate change argument supports this proposition, the ecosystem services argument does not. Should we then take a step back from ecosystem services and look at climate change in the wider context? The next section explores climate change as it applies to a wide variety of other environmental issues.

Climate change discourse and sustainability

Climate change discourse touches upon many other environmental issues and these issues are now widely discussed within the context of climate change. There is growing concern over the effects of climate change on agriculture and food production (Lobell and Burke, 2010). In Africa and Latin America, for example, climate change is likely to reduce crop yields by 10–20% and threaten food security by the mid-twenty-first century (Jones and Thornton, 2003). Simultaneously, there is concern over the effects of climate change on biodiversity (Lovejoy and Hannah, 2005). Some estimates suggest that climate change will lead to large-scale migrations of some species (e.g. Fitzpatrick et al., 2008), and where suitable climate is not available large number of species extinctions (e.g. Isaac, 2009).

The suitability of the existing network of inviolate protected areas to provide habitat for species in the future has therefore been questioned and suggestions have been made for conservation beyond reserves in order to provide habitat to migrating species (Willis and Bhagwat, 2009). Climate change is projected to exacerbate desertification of arid environments in the future (Verstraete et al., 2008) and concerns have been raised with respect to livelihoods of people who inhabit these areas (Stringer et al., 2009). Similarly, climate change is suggested to have an effect on already depleting fish stocks, possibly due to change in the physical environment of oceans and consequent biodiversity loss (Worm et al., 2006).

Figure 4.2   *Climate change discourse connects with various facets of sustainability*

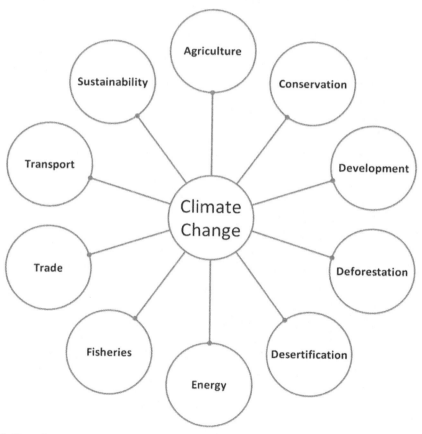

Source: S. Bhagwat

Climate change-induced resource degradation combined with the lack of food security is likely to raise the issue of economic development in less developed countries, particularly in sub-Saharan Africa (Brainard et al., 2009). Concerns have been raised over poverty of 'the bottom billion' (Collier, 2008) who may be further impoverished due to climate change and its consequences for agriculture, fisheries and the natural environment. Resource degradation due to climate change is likely to mean that poor countries have even fewer tradable commodities, are less powerful in global markets and therefore vulnerable to exploitation (Collier, 2010). The discourse on climate change links issues of energy including fossil fuels, non-renewable and renewable sources of energy. For a long time, scientists have been concerned about 'peak oil', the decline in non-renewable fossil fuel energy sources (e.g. Duncan, 2003). While concerns have also been raised about carbon emissions caused by fossil fuel use, and therefore 'cleaner' sources of energy such as solar power or wind have long been explored as alternatives, climate change has provided a strong reason to move away from non-renewable and towards renewable sources of energy (Reis and Miguel, 2009). The discourse on climate change has also influenced discussion on transport, particularly a move towards strengthening public transport networks in large cities so that future fossil fuel energy use is minimised (e.g. Robert et al. 2007). The key goal in energy and transport is to reduce carbon emissions as a contribution to preventing climate change.

To some extent, these issues link into the wider discourse on sustainability – meeting current needs without jeopardising needs of the future generations (Brundtland, 1987) – but the motivation for these is to prevent or mitigate or adapt to climate change. Sustainability, on the other hand, has a moral standpoint – one that is motivated by the concern for the wellbeing of the future generations. Although climate change is often used as a 'joker in the pack' in policy circles (Redford and Adams, 2009), arguably, it does not have the same moral basis as sustainability. Conservation and development both have moral agendas (Adams et al., 2004): conservation is founded on the moral responsibility of humans towards other species; and development is founded on the moral responsibility of humans towards other human beings. Should we then cast aside climate change or any other 'capture-all' twist on environmental concerns and return to the concept of sustainability, almost three decades after it was first proposed and after two decades since it was embraced by the world leaders?

Summary

- This chapter started by outlining the unprecedented ecological, economic and social changes that have prompted the discussion on global change. It walked you through that discussion and picked up two contemporary themes that have come forward. First, we explored 'the perfect storm', a metaphor used to define the synergistic effects of the unprecedented ecological, economic and social changes on the environment. Second, we unpicked 'planetary boundaries', a concept that put forward the limits within which humanity should live in order to avert catastrophic environmental consequences of human actions. Both these themes bring up a wide range of issues relevant to conservation and development.

- Joined-up conservation and development have always had to be justified in the context of a changing *raison d'être* – be it climate change, ecosystem services or planetary boundaries. These (and indeed the concept of resilience – see Chapter 14) have arguably superseded 'sustainability' in terms of their discursive importance. But the character and magnitude of global environmental problems appears to suggest that sustainability, living within our means, remains fundamentally relevant to how we think about the relationship between conservation and development.

- A return to emphasising sustainability, as a way of understanding the prospects for reconciling conservation and development, may be particularly important as conservation and development are having to operate in a globalised world, where free-market capitalism has dramatically changed humanity's relationship with the natural world – as the next chapter explores.

Discussion questions

1. What are the implications of the widespread human footprint on the planet captured in the concept of the Anthropocene?

2. 'Planetary boundaries' have come forward as the most recent conceptualisation of the need for humanity to live within its own means. Who should define planetary boundaries and whom they are for?

3. What environmental issues does climate change connect and how is the discussion on climate change relevant to conservation and development?

Further reading

Adams, W.M., Aveling, R., Brockington, D., Dickson, B., Elliott, J, Hutton J., Roe, D., Vira, B., and Wolmer, W. (2004) Biodiversity conservation and the eradication of poverty. *Science*; 306:1146–1149.

Crutzen, P. (2000) The anthropocene. *Global Change Newsletter* 41(1): 17–18, Springer

Rockström, J., W. Steffen, K. Noone, Å. Persson, F.S. Chapin, III, E.F. Lambin, T.M. Lenton, M. Scheffer, C. Folke, H.J. Schellnhuber, B. Nykvist, C.A. de Wit, T. Hughes, S. van der Leeuw, H. Rodhe, S. Sörlin, P.K. Snyder, R. Costanza, U. Svedin, M. Falkenmark, L. Karlberg, R.W. Corell, V.J. Fabry, J. Hansen, B. Walker, D. Liverman, K. Richardson, P. Crutzen, and J.A. Foley (2009) A safe operating space for humanity. *Nature*, 461, 472–475.

Berkes, F., J. Colding, and C. Folke (eds). (2008) *Navigating Social–ecological Systems: building resilience for complexity and change*. Cambridge University Press, Cambridge, UK.

Gunderson, L. H., and C. S. Holling, editors. (2002) *Panarchy: understanding transformations in human and natural systems*. Island Press, Washington, D.C.

Raworth, Kate (2012) A safe and just space for humanity: can we live within the doughnut? Discussion paper, Oxfam International.

Online resources

BBC (2009) Averting a perfect storm of shortages http://news.bbc.co.uk/1/hi/8213884.stm (accessed 20 January 2015)

Folke, C., S. R. Carpenter, B. Walker, M. Scheffer, T. Chapin, and J. Rockström. 2010. Resilience thinking: integrating resilience, adaptability and transformability. Ecology and Society 15(4): 20. [online] URL: http://www.ecologyandsociety.org/vol15/iss4/art2/

UN (2000), The Millennium Development Goals http://www.unmillenniumproject.org/goals/ (accessed 20 January 2015)

5 Conservation and development in the context of globalisation and neoliberalism

- The beginnings of globalisation
- Globalisation, neoliberalism and conservation
- Ecosystem services and market-based mechanisms
- Markets within a global economy
- Vulnerability and resilience of the global economic system

Introduction

In Chapter 2, we saw that the conservation movement was consolidated after the end of the Second World War. Since the end of the Second World War, the world's economies have also intensified the extent and global reach of the production of goods, trade of these commodities and financial investments. These trends can be traced back to imperial expansion from the 1700s onwards and, following that, the industrial revolutions. In this chapter, we will look at conservation and development in the context of post-Second World War globalisation and neoliberalism. In this context, the market-based mechanisms that have emerged since the 2000s for operationalising conservation and development initiative are notable. Nature is increasingly seen as a tradable commodity by valuing nature's services. Such economic justifications for the conservation of nature, however, have their limitations particularly because the markets on which these mechanisms are reliant upon are volatile and at the mercy of global economic forces. The chapter goes on to explore the governance of common pool resources beyond the state and the markets. It concludes by examining the vulnerability and resilience of global economic system.

The beginnings of globalisation

While the trade links across the world have existed for millennia (e.g. Boivin and Fuller, 2009), the post-Second World War trade links are marked by their acceleration and pace, facilitated by rapid transport links across the world. Furthermore, these changes have been driven by the liberalisation of trade and finance, leading subsequently to the free-market economy, one which is driven by demand and supply of goods and commodities and is not regulated by the state. Neoliberalism promotes the privatisation of state-owned enterprises and infrastructure and deregulation of markets, enhancing the importance of the private sector in the economic order (Brenner and Theodore, 2002). Globalisation and neoliberalism have progressed hand-in-hand. For instance, since the 1950s the volume of trade between countries has grown much faster than the global economy. Global GDP increased 10 times between 1950 and 2005, while trade increased by more than 100 times (WTO 2011). In many countries, therefore trade has been the engine of economic growth – examples include countries such as Japan, South Korea and China, who have aimed at export-oriented growth. The flows of money around the world have also increased since the 1950s, with some of the banks owning assets worth trillions of dollars overseas. This was facilitated since the 1980s following greater liberalisation of world capital markets until the Global Financial Crisis, starting in 2008. Other changes that have taken place in a globalised, free-market economy since the 1980s include the increased market share by multinationals, who have benefited from decreasing costs of transport and communication as well as cheap labour in developing countries.

The most obvious benefit of globalisation is the increase in prosperity for some in the last 50 years – initially in the United States and Europe and later in developing economies such as Brazil, China and India. However, over this same time period, the world distribution of wealth has also become more unequal (Wilkinson and Pickett, 2010). For example, the richest 10% of households in the world have as much yearly income as the bottom 90%. The total assets of the richest and the poorest are even more unequal – the richest 1% who are concentrated in the USA, Europe and Japan own 40% of the world's wealth. On the other hand, poverty is widespread across the developing countries, where more than 80% of the world's population lives (UNU, 2008). There are serious concerns that economic growth is unequitable. Although some sections of the society are benefiting from the economic growth, there is increase in disparity in the distribution of this growth. A 2015 report by Oxfam 'Wealth: Having it all and wanting more' argues, for example, that in 2014 the richest 85 billionaires in the world had the same wealth as the bottom half of the world's population (Hardoon, 2015). According to Oxfam's estimates, this disparity will only grow.

The disparity in the distribution of wealth has prompted some thinkers to call into question the merits of neoliberal capitalism (Jessop, 2002; Harvey, 2005), and this has particularly come into a sharp focus in the wake of the financial crisis of 2008. Saad-Filho (2010), for example, argues from a Marxist perspective that the financialisation of five entities, namely, (1) capital accumulation, (2) the global economy, (3) the state, (4) the ideology, and (5) the reproduction of the working class has been responsible for the current crisis in neoliberal capitalism. With the increased integration of the global financial sector, the proportion of global capital contributed by the finance sector has increased dramatically. For example, the import of cheap goods and commodities from the East Asian countries to the Western countries, particularly the USA, in the 1990s has 'subsidised' Western economies considerably. This, combined with low interest rates, gave a false sense of economic buoyancy since the early 2000s until the financial crisis of 2008. During these series of financial crises the state often acted as 'guarantor' stabilising the financial sector at the cost of public funds. Simultaneously, neoliberal capitalism has promoted an ideology of economic competitiveness legitimising the accumulation of wealth by individuals within society. Furthermore, salaries, pension funds, education, health and most areas of social provision are now increasingly linked with the finance sector. This increased integration of society with the finance sector has relied heavily on credit from households and the state, which was necessary for financial institutions to produce and increase their financial assets. Furthermore, minimal regulation of the control of financial institutions by the state is said to have laid the foundation for the financial crisis. There is a feeling, therefore, that free-market capitalism without regulation has had an adverse impact on society.

Globalisation, neoliberalism and conservation

The crisis in free-market capitalism that started in 2008 has had 'ripple effects' in the environmental sector, shaking the faith in market-based solutions to environmental problems promoted by many businesses and corporations. Whilst conservation and capitalism have existed side-by-side for a long time (Adams, 2004; Brockington, Duffy and Igoe, 2008), concerns have been raised about the increasing and direct links between conservation and corporate capitalism (Brockington and Duffy, 2010; Buscher et al., 2012). There is acknowledgement, however, that conservation throughout history has been driven by elites within society. This section illustrates the linkages between capitalism and conservation in the context of globalisation and neoliberalism.

Neoliberalism can be defined as: 'a theory of political economic practices that proposes that human well-being can best be advanced by liberating individual entrepreneurial freedoms and skills within an institutional framework characterised by strong private property rights, free markets, and free trade' (Harvey, 2007: 2). Scholars such as Brockington et al. (2010), Castree (2008), Büscher and Dressler (2012) suggest that conservation has bought into the idea that neoliberal capitalism can be the means to achieve its objectives. The linkages between capitalism and conservation in the context of globalisation and neoliberalism are illustrated by a number of useful examples from a 2010 Special Issue of the journal *Antipode* (Brockington and Duffy, 2010). In this issue Holmes (2010), for example, talks about the influence of powerful elites on conservation in the Dominican Republic; meaning that while some of the conservation initiatives have been successful, they have often been detrimental to the interest of the local communities – leading to their eviction from strictly protected areas. This has been brought about by the powerful elites' privileged access to political power, their access to and links with corporate capitalism and their ability to shape public and policy discourses on the environment (Holmes, 2010).

Similarly, Spierenburg and Wels (2010) explore linkages between conservation elites and philanthropists in trans-boundary conservation in southern Africa. They argue that nature conservation in southern Africa is heavily influenced by capitalist ideologies promoted by business philanthropy (Spierenburg and Wels, 2010). Conservation NGOs are also arguably tied to capitalist ideologies. Brockington & Scholfield (2010) argue, for example, based on their study of conservation NGOs in Africa, that these NGOs are creating the conditions that are conducive to the expansion of capitalism. They further demonstrate that these organisations are instrumental in turning the financial wealth from the rich countries into new commodities, and of creating new markets for some industries related to conservation, e.g. tourism (Brockington and Scholfield, 2010).

Within the conservation sector, there is some scepticism for the increased linkages between conservation and corporate capitalism. MacDonald (2010) argues that conservation interests are now increasingly aligned with the interest of businesses, and business has become a major actor in shaping contemporary biodiversity conservation under the guise of 'corporate social responsibility'. According to him, the organisational characteristics of 'modernist conservation' might be responsible for the subordinate position that conservation takes against larger societal and political projects such as neoliberal capitalism (MacDonald, 2010). The expansion of protected areas, seen as one of the key instruments for conservation, might in fact be closely linked to the expansion of neoliberal ideologies. Igoe et al. (2010), for example, observe that proliferation of protected areas earmarked for conservation has gone hand in hand with neoliberal market expansion. They argue that mainstream conservation, dominated by large non-governmental organisations, has promoted ideas and agendas of their corporate donors and imposed them on the conservation sector, which inherently has wide-ranging values and interests (Igoe et al., 2010). The close links between conservation and neoliberal capitalism have led to disenfranchisement of many local communities because these links have often failed to address poverty, a pressing issue in many developing countries. Corson (2010), critiquing the US Agency for International Development's biodiversity conservation agenda, argues that neoliberal conservation has reinforced the separation of environmental concerns from their political and economic drivers. In doing so, this form of conservation has enabled conceptualisation of environmental goals without changing the existing distribution of economic power or resource flows; instead reinforcing the political institutions that are responsible for disparities in economic distribution of resource flows (Adams, 1995; McAfee, 1999). Sachedina (2010) gives the example of the African Wildlife Foundation in Tanzania, whose agenda, which is influenced by neoliberal capitalism, meant that the local people were routinely disenfranchised from conservation initiatives promoted by African NGOs. In many such cases NGOs put nature before people because neoliberal capitalism can claim to save nature from people through a variety of market-based instruments.

Increased links between environmental conservation and capitalism are also demonstrated by initiatives to promote ethical consumption.

Carrier (2010) argues that initiatives such as Fairtrade identify ethical consumption as the most effective way to promote the protection of natural environment, thereby closely linking markets with environmental conservation. West (2010), with reference to coffee production in Papua New Guinea argues that the speciality coffee sector is made up of consumers who are socially conscious, roasters who are well-meaning individuals or small companies that are politically engaged. Furthermore, the poor but ecologically conscious producers in the coffee sector aspire to live in close harmony with the natural world (through, for example, ecologically friendly methods of coffee production) whilst being integrated into the flows of global capital. Tourism is another sector, which offers opportunity to consumers to make ethical choices, but is often closely tied to capitalist principles and practices. Neves (2010) argues from a Marxist perspective that whale hunting and whale watching, two activities that are contrasting on the face of it – the former against conservation and the latter for it – are in fact very similar in their operationalisation, marked by 'fetishized commoditisation of cetaceans and the creation of a metabolic rift in human–cetacean relations' (p. 719). While whale hunting is to do with the use of ocean resources for consumption far away from the source of extraction, whale watching views and commoditises the ocean as a provider of services to be consumed and enjoyed at the place of their origin. In both contexts, however, nature is viewed from a capitalist perspective (Neves, 2010). Duffy and Moore (2010), using examples from Botswana and Thailand argue that tourism redesigns and repackages nature for global consumption such that it imposes new values and uses of nature for international markets, but without completely obliterating existing ways of valuing, using, owning and approaching nature (also see McAfee, 1999; Fletcher, 2009). There are also cases where 'hybrid' forms of organisation are observed. For example, Wilshusen (2010) argues that in the community forestry sector in Mexico people have embraced neoliberal policies and programmes creating a novel set of property regimes, forms of organisation and modes of exchange.

Ecosystem services and market-based mechanisms

In the globalised world that we live in today 'hybrid' forms of organisation are inevitable. Does that mean then that working with large corporations, rather than in antagonism with them, is a rational way forward? Large corporations and markets are now closely tied to each other; and both in turn are closely linked to the financial sector and the global economy. A more careful accounting of the negative impacts of activities of large corporations is, however, necessary if the mighty corporations are to offset the impacts of their activities on natural ecosystems. The concept of 'ecosystem services' – benefits that people derive from natural ecosystems – has been proposed as a framework to enable this (MEA, 2005) – see also Chapter 4. This chapter revisits the concept of ecosystem services in the context of market-based mechanisms for conservation. If all negative impacts of business activities by large corporations are accounted for, then these corporations can be asked to pay for mitigating those impacts. Such payments can be useful to maintain or enhance the ecosystem services from nature. This is an attractive idea in theory, but does it work in practice? What are the opportunities for working within the ecosystem services framework and what are the challenges? The rest of this section addresses these questions.

A market is a place where the sellers and buyers can exchange goods or services. Such exchangeable goods and services can be bought with a currency. Similarly, ecosystem services can be considered as tradable commodities. Although such services are produced in nature, their supply depends on how people manage nature. For example, in order to derive ecosystem services from a forest, people living in the vicinity of this forest need to keep that forest as forest and not cut it down. Such services can then be sold in the market in exchange of money. The buyers in this case are people who do not produce ecosystem services, but consume them. For example, a city-dweller who benefits from the clear air and clean water from the forest should pay people who maintain that forest and ensure supply of these services.

Such market-based instruments apply economic principles to the supply and demand of goods and services derived from nature, namely, habitat, water quality, air quality, biodiversity, etc. Therefore, in theory, market-based instruments can be potentially useful to harness market forces. It has been suggested that in some cases this will improve conservation outcomes as well as alleviating poverty (e.g. van Noordwijk and Leimona, 2010). By paying for ecosystem services, the communities are compensated for guarding a habitat and its services because they forego benefits that would otherwise be derived from the conversion of this habitat to a more productive land use. If such arrangements are contractual between the provider and the recipient of an ecosystem services, then this would enhance conservation outcome as well as address poverty of the guardian communities. In theory, this would mean that by lifting people out of poverty, economic growth would continue whilst ensuring conservation of some of the last remaining patches of wild nature.

Market-based mechanisms, however, are not without their critics. Redford and Adams (2009), for example, identify a number of problems with the ecosystems services approach – problems that are representative of other similar market-based instruments. The first problem is that there are markets for only a limited set of services. While clean water may have a well-defined consumer market, this is not necessarily true for soil nutrients; and there is no identifiable market for biodiversity. The second problem that stems from the limitations of markets is that markets are inherently competitive and therefore will compete to gain control over a service that is market-able. Such private control over an ecosystem service may not always be in the interest of people or nature, meaning that such control may not always lead to positive outcomes for biodiversity conservation and poverty alleviation. These problems are further compounded by uncertainties in the provision of ecosystem services in the future, possibly due to climate change, making some of these services scarce. An imbalance between supply and demand of a service is likely to make it more expensive; and if markets have a vested interest in profiteering, then scarcity of a service would be in the markets' interest, but neither in the interest of ecosystems nor people.

Another set of problems with ecosystem services pertains to the anthropogenic bias in defining and/or creating a service. For example, a service such as food production is essential for humans, but this is not necessarily in the interest of other living beings that depend on a habitat that gets converted to agricultural field for food production. Furthermore, in an anthropogenic world, it is possible to create a service artificially without supporting the natural ecosystem processes and functions that are necessary for the creation of such service. For example, a technologically sophisticated water filtration system can provide clean water without the necessity of maintaining a tree-covered habitat. Such technological solutions to ecosystem provision may not necessarily be in favour of ecosystems. Finally, ecosystem services are not always congruent with biodiversity as many studies at various scales have demonstrated (e.g. Bennett et al., 2009; Eigenbrod et al., 2009; Strassburg et al., 2010), meaning that a focus of conservation on the provision of ecosystem services is not always beneficial for biodiversity conservation.

Furthermore, the ecosystems services approach undermines the non-economic justification for biodiversity conservation and for poverty alleviation; and assumes that decisions on conservation or development are tied to economic choices that individuals, communities or societies make. This overlooks the intangible value that people ascribe to nature and its components. Similarly, the alleviation of poverty is often justified on moral grounds more than economic. The ecosystem services framework does not explicitly address these kinds of values of nature. However, the argument goes that while intangible values are important to people, most decisions over management of 'natural resources' are made at a higher level, i.e. by states or corporations. These agents are driven primarily by economic motivation and use those criteria in making decisions rather than criteria based on intangible values that individuals may ascribe to nature. The 'ecological pragmatists' therefore argue that in order to appeal to these agents, the ecosystem services framework provides a robust means to the dual goals of conservation and development.

Markets within a global economy

Ecosystem services is an approach which is important in many circles for its attempts to make links between markets, conservation and development. Despite some scepticism for ecosystem services, such an approach is considered a pragmatic solution to conservation and development by many 'ecological pragmatists' (Spash 2000), who argue that conservation and development both have to operate within the existing global economic paradigm. This has led to a variety of market-based instruments which have been founded on the principle of valuing, accounting and paying for ecosystem services. The current global economy, however, is influenced by large multinational corporations, who have substantial power over how decisions are made on the use and conservation of natural resources. Many of these corporations have a greater turnover than the GDP of most countries. For instance, in the year 2000, of the 100 largest economies in the world, 51 were corporations and 49 were countries, and these corporations had sales figures between US$51 billion and US$247 billion (Anderson and Cavanagh, 2000). Since then, corporations have gone from strength to strength in increasing their collective wealth.

These include multi-national giants such as Wal-Mart, Coca-Cola and Shell. The ecological pragmatists argue that conservation and development have to operate within the existing economic paradigm until that paradigm changes. As such, conservationists have to work with the transnational corporations rather than against them. This is one reason why ecological pragmatists are enthusiastic about market-based mechanisms for conservation and development because these, in theory, can allow conservationists to influence the behaviour of transnational corporations, hopefully to the benefit of nature and people.

The enthusiasm for market-based mechanisms also stems partly from the assumption that markets are competitive and they will also compete with each other for good behaviour – benefitting the dual goal of biodiversity conservation and poverty alleviation. However, recent research shows that the networks of these corporations operate in a 'rich gets richer' fashion because a large portion of control flows to a small tightly-knit core of financial institutions (Vitali et al., 2011). This research is based on ownership pathways of 43,060 trans-national corporations identified according to the Organization for Economic Co-operation and Development (OECD) definition. The research demonstrates that many of these corporations have ownership ties, facilitating the formation of blocks, which hamper market competition. As such, these corporations drive markets in their own favour overriding the assumption that markets are competitive. This has implications for the attitude of large corporations to biodiversity conservation and poverty alleviation, which, without competition may not have the incentive for 'good behaviour'.

A further caveat of market-based mechanisms is that fluctuations in markets have substantial influence on natural resource use and management. When markets are strong and profits healthy, there is sufficient incentive for corporations to channel some of these profits into supporting 'good behaviour' under the banner of 'corporate social responsibility'. However, when markets are weak and profits low, corporations are less willing to channel their profits into biodiversity conservation or poverty alleviation. Such 'boom and bust' cycles in markets therefore directly influence the willingness of corporations to engage in environmentally and socially responsible behaviour. Markets in commodity crops such as coffee and cocoa have historically suffered from such boom and bust cycles (e.g. Garcia et al., 2010; Tschnartke et al., 2011). Presence of shade trees is an important feature of coffee and cocoa plantations because shade trees are important for creating conditions necessary for flowering and fruit bearing of certain varieties of these commodity crops. Shade trees are often present in early stages of plantation, when the forest has been cleared recently for planting. However, this tree cover gradually depletes if the farmers are unwilling to replace shade trees. The willingness of farmers to maintain tree cover is influenced largely by markets. When coffee or cocoa markets are buoyant, the industry is willing to maintain such shade trees, but during lean periods shade trees are removed to make extra space and to compensate for the lost income by selling timber trees from plantations. This boom and bust cycle is detrimental to maintaining tree cover in the landscape and to creating condition necessary for the conservation of forest-dwelling fauna and flora beyond protected areas.

Vulnerability and resilience of the global economic system

The vulnerability of global markets leads to further concerns about the resilience of market-based mechanisms in conservation and development. Resilience of a social-ecological system in a globalised world is highly dependent on the resilience of an economic system, which, in the context of neoliberalism is defined by free-market economy. Local and regional economies are often at the mercy of the prevalent economic system and therefore shocks in the larger economy often have severe consequences for local and regional economies. These latter economies are also influenced by global markets and unexpected shocks within those markets can lead to negative consequences for local economies. The concept of 'economic resilience' therefore encompasses the ability of a local or regional economy to bounce back from the negative effects of external shocks. Briguglio et al. (2009) classify economic resilience as shock counteraction (ability of an economy to recover quickly following adverse shocks) and shock absorption (ability of an economy to withstand shocks). Simmie and Martin (2010), however, argue that such 'equilibrist' versions of economic resilience, where the system is expected to remain in the same state or return to the previous state after encountering shocks, are unhelpful. Instead, drawing on the panarchy theory (Gunderson and Holling, 2002) they propose an 'adaptive cycle' model (which posits a four-phase process of continual adjustment in ecological, social and environmental systems) as a framework for analysing regional economic resilience.

Figure 5.1 *A four-phase adaptive cycle model of regional economic resilience*

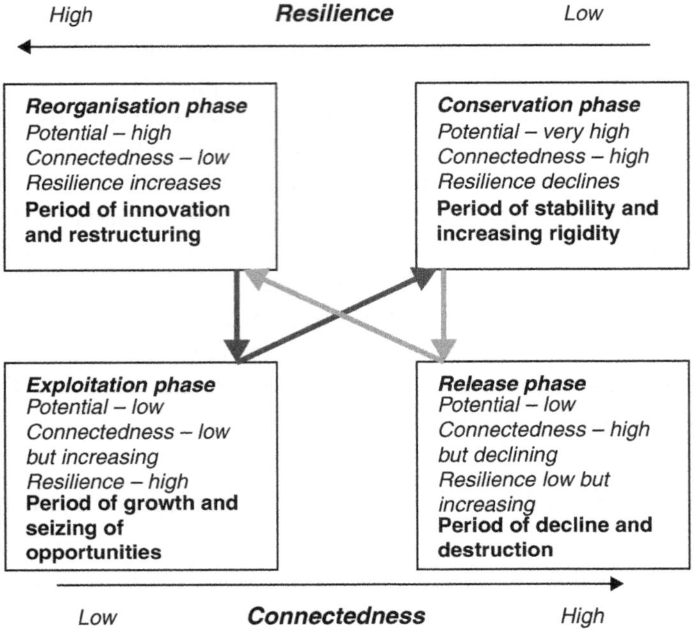

Source: Simmie and Martin, 2010; permission kindly granted by Oxford University Press

It can be argued that the global economy is also at the mercy of similar ups and downs. Although the political discourse on economy is that of perpetual growth, it is widely acknowledged that the global economy has been dynamic over the last 50 years, with periods of growth followed by periods of recession. Since the end of the Second World War, however, many Western countries introduced fiscal and monetary policies that have restrained the extremes of economic growth or slump to a large extent. However, flying in the face of such measures, the global economy was hit in 2008 by a severe downturn, demonstrating that the global economic system has become far too complex to make accurate predictions and that the global financial systems have become more interdependent, raising the possibility of the so-called 'domino effect'. Hamel and Välikangas (2003) predicted this crisis and argued that a greater recognition of the dynamic nature of business environment and the ability to adapt to sudden and large changes to that environment are key factors in building resilience at the level of organisations and, when scaled up, at the level of the global economy. Following the 2008 crisis in global economy, there is increasing interest in a more resilient economic system. Recognising the vulnerability of the global economic system, initiatives to build a more resilient economic system, however, started even before the global economic crisis. Around 2005–2006 Rob Hopkins founded the first Transition Towns in Kinsale, Ireland and in Totnes, Devon as an experiment in 'localising' the economy. There are now over 400 Transition Town around the world affiliated to a UK-based charity called the Transition Network founded in 2007 (Hopkins, 2008). Among other things, the Transition Towns movement has introduced local currencies to ensure that wealth stays in the community where it is used in a more conscientious way with the hope of building a more durable local economy that is somewhat sheltered from the ups and downs in global markets. While local currency can be an important component of a larger vision of social change and local currencies can be used as an alternative form of local protection from capital flight, local currencies may not necessarily improve the resilience of the wider global economic system, which is driven by factors beyond a local control. Local currencies do, however, encourage the purchase of locally-produced goods and services, which means that the local community receives a greater share of the benefit and less of it drains out to other parts of the global economic system. This might make local communities more resilient to ups and downs in the global economy.

Summary

- Since the end of the Second World War, the world has been increasingly connected in its economic and political systems. Conservation and development have to operate in a world that is increasingly reliant on free market capitalism. As such, different ways of connecting markets with conservation and development have been put forward.

- Viewing nature as a tradable commodity has given rise to market-based mechanisms for conservation and development. However, the global economic crisis that started in 2008 has shown that markets are volatile, raising concerns over purely economic justifications for the conservation of nature.

- The level of such management of nature often has to be 'local' as opposed to 'global' and this raises the question of whether 'localisation' of economy might be a way forward to make economic and ecological systems more resilient.

Discussion questions

1. What are the limitations of viewing nature as a tradable commodity?
2. How does the volatility of economic markets impact on the purely economic justification for the conservation of nature?
3. Can the governance of natural resources at a local level safeguard natural resources from the vulnerabilities in the global economic system?

Further reading

Brockington, D. and Duffy, R. 2010. 'Conservation and capitalism: an introduction.' *Antipode* 42(3): 469–484.

Buscher, B., Sullivan, Sian, Neves, K., Igoe, J. and Brockington, D. (2012) Towards a synthesized critique of neoliberal biodiversity conservation. *Capitalism Nature Socialism* 23(2): 4–30.

Redford, K.H., Adams, W.M. (2009). Payment for ecosystem services and the challenge of saving nature. *Conservation Biology* 23: 785–787.

Hardoon, Deborah (2015) Wealth: having it all and wanting more. Research report, Oxfam International. ISBN 978-1-78077-795-5

McAfee, K, 1999, "Selling nature to save it? Biodiversity and green developmentalism" *Environment and Planning D: Society and Space* 17(2): 133–154.

Saad Filho, Alfredo (2010) Neoliberalism in crisis: a Marxist analysis. *Marxism*, 21. pp. 235–257.

Online resources

WTO (2011) World Trade Organisation International Trade Statistics 2011 URL: http://www.wto.org/english/res_e/statis_e/its2011_e/its11_charts_e.htm. [accessed 26 January 2015].

UNU (2008) World Income Inequality Database V2.0C May 2008 URL: http://www.wider.unu.edu/research/Database/en_GB/database/ [accessed 26 January 2015].

Anderson, Sarah and John Cavanagh (2000) Top 200: The Rise of Global Corporate Power. Global Policy Forum. https://www.globalpolicy.org/component/content/article/221/47211.html [accessed 03 February 2015]

6 Global environmental governance

- From government to governance
- Definitions of environmental governance
- New actors and institutions in environmental governance
- Varieties of environmental governance

Introduction

Environmental problems, which have become increasingly global in reach, are the focus for efforts to understand and bring about environmental governance. Examples of global environmental problems include climate change, biodiversity degradation and species loss, land cover and land use change, acidification of the oceans and the global fisheries crisis. All of these have, in their different ways, adverse implications for both conservation and development, and all of them arise directly as a result of the interactions between conservation and development objectives. This is because they are problems of collective action: they relate to the functions and services of ecosystems whose usage is essential to the workings of the global economy and also the seemingly insatiable pursuit of ever-increasing growth levels. Therefore these patterns of usage, good or bad, have implications not just for the user but for much broader groups of people and, potentially, for all humanity. Broadly speaking, the rules, people, processes and mechanisms through which decisions are made to frame and respond to environmental problems are the subject of environmental governance.

This chapter seeks to provide an introduction to work done on environmental governance by geographers, political scientists, ecologists and others. It begins by considering the question of how it is that we got from government to governance. It then proceeds to define governance and, more concretely, environmental governance, before introducing new sets of actors who are heavily involved in governance, and to whom much more attention has recently been paid. Following this, it outlines five archetypal varieties of governance: hierarchy, network, market, adaptive and environmentality.

From government to governance

Moving away from state-centric understandings of how things are governed

The acceleration of the process of globalisation has had profound implications for how we conceive of governing, be it in relation to conservation or development. For hundreds of years government, in the form of the nation-state, has been the core focus of thinking around how goals are set and big decisions are made within societies. Since their inception from the sixteenth century onwards, states have been concerned with more than the 'high politics' of war, peace, diplomacy and constitutional change; they have become increasingly focused on the 'low politics' of administering to the everyday needs and affairs of the population (Evans, 2012). Indeed, when the modern nation-state administration emerges, 'not only does the idea of a measurable and manageable population come into existence, but so also does the notion of the environment as the sum of the physical resources on which populations depend' (Rutherford, 1999, p. 39). As the state continued to expand in the 19th and 20th centuries, it came to be seen as the appropriate mechanism for environmental protection (Agrawal and Lemos, 2006). Even as late as the 1970s and 1980s, some actors continued to advocate a higher degree of centralised control as the best way to deal with the increasing magnitude of environmental problems. Yet the importance of the state in matters of environmental governance is by no means new. As chapter 1 explores, environmental historians have documented its centrality in the very processes of the commodification of nature, brought about by the industrial revolutions and rapidly expanding networks of global trade that characterised colonial expansion, which gave rise to global environmental problems (Beinart and Hughes, 2008; Grove, 1995). Even where the state was seen as part of the problem, many felt that the solution could only come through its intervention.

Above all, the concept of governance challenges state-centric approaches precisely because in most (though not all) of its guises, it broadens the scope of enquiry to include non-state actors, processes and mechanisms which are just as frequently – and sometimes more – central to processes of governance (Jordan, 2008; Newell, 2008a; van Kersbergen and van Waarden, 2004). Given the dizzying variety of arrangements through which environmental governance now unfolds, it is hard not to conclude that a focus solely on the state is at best woefully incomplete, bordering on myopic. For a number of reasons, governments have thereby lost ground when it comes to governance – and especially in the global context. The first reason is that what needs to be governed frequently spills over national borders. Because the scale of environmental problems and economic processes extends so far beyond the reach of a single nation, an international governance architecture has emerged (Krasner, 1983; Young, 1989). Substantial parts of this architecture remain the preserve of interactions between nation-states, but in the practice of international (and lower-level) governance, a much greater array of actors comes in to play (Biermann and Pattberg, 2012; Newell, 2008a).

Second, greater space for the private sector, civil society, social movements and other actors is partly a result of the greater connectedness between societies and economies that comes with globalisation. Yet it is also partly in response to the 'hollowing out' of the state in response to the dominance of neo-liberal thinking on the global stage (Armitage et al., 2012; Evans, 2012; Lemos and Agrawal, 2006) – see Chapter 5 for more details. This has created a wide variety of mechanisms and processes for global governance, which are no longer dominated by one type of actor.

Third, and linked to this idea of the hollowing out of the state, processes of the decentralisation of decision-making processes have taken away power from the centre, distributing it between lower level actors and non-state actors. Partly this reflects a loss of faith in the state as a competent manager of the environment, and indeed of the economy (Corbridge, 1991). It is also to do with the decreasing resources available to states to manage their environments (Lemos and Agrawal, 2006). Accompanying this weakening of the state has been a discourse which posits decentralisation as a more effective way of resolving environmental problems for three principal reasons:

1 Greater competition between subnational units leads to greater efficiencies.
2 It brings decision making much closer to, or directly into the hands of, people affected by those decisions, thereby increasing the legitimacy of the decision-making process.
3 It helps decision-makers ground their decisions in better, more relevant local information pertaining to environmental management questions (Lemos and Agrawal, 2006).

Definitions of environmental governance

Definitions of governance vary according to the level at which they are applied, but they replicate the division between state-centric and 'beyond the state' approaches. The notion of governance first emerged in the national context to describe new kinds of regulation at variance with long-standing forms of hierarchical state activity (van Kersbergen and van Waarden, 2004). When defining global governance, some scholars tend to restrict its application to multilateralism and interstate cooperation, such as Young's definition of global governance as 'the combined efforts of international and trans-national regimes' (Young, 1999: 11). Others take a broader view, as in James Rosenau's definition: 'the sum of the world's formal and informal systems at all levels of community amounts to what can properly be called global governance' (2002: 4). Biermann and Pattberg (2012) note that the more restricted definitions simply miss many of the "socio-political transformations" that are occurring within global governance, and which occur outside of the realm of traditional political analysis. Conversely, the broader definitions are so extended, they suggest, that they become synonymous with politics, and therefore analytically useless, owing to their indistinctness.

Environmental governance is a subset of the broader global governance literature. Whilst it therefore shares conceptual underpinnings, environmental problems and their intersection with development priorities have particular features, which need to be understood on their own terms (Newell, 2008b; Biermann and Pattberg, 2012). Nevertheless, its definitions are quite similar to those of global governance. The following three (to which many more could be added) give an idea of the different emphases and nuances that appear within the literature:

1   'The set of regulatory processes, mechanisms and organisations through which political actors influence environmental actions and outcomes' (Lemos and Agrawal, 2006: 298).
2   'Environmental governance should be understood broadly so as to include all institutional solutions for resolving conflicts over environmental resources' (Paavola, 2007: 97).
3   'The interrelated and increasingly integrated system of formal and informal rules, rule-making systems, and actor-networks at all levels of human society (from local to global) that are set up to steer societies toward preventing, mitigating, and adapting to global and local environmental change and, in particular, earth system transformation, within the normative context of sustainable development' (Biermann et al., 2009: 3).

Because non-state actors have gained in importance so rapidly in recent decades, it is less common to speak of a state-centric world. At the start of the twenty-first century, hierarchical governce is accompanied by a polycentric (Galaz, Crona, Österblom et al., 2012) or multi-centric order (Newell, Pattberg and Schroeder, 2012). These terms suggest not just one 'centre' (the state or the market) with everything clustered around it, but numerous networks of actors joined together, with multiple centres within them. In other words, the traditional system of regulatory frameworks negotiated between governments co-exists with states of affairs in which there are multiple centres of power, decision-making structures and processes of

Figure 6.1   *From state-centric to polycentric orders*

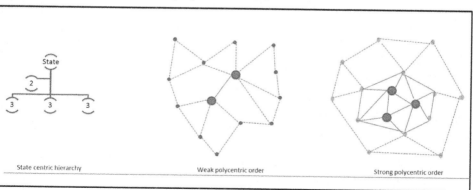

State centric hierarchy   Weak polycentric order   Strong polycentric order

Source: adapted from Galaz et al., 2012

contestation, through which conservation and development are governed (see Figure 6.1). This has led to "the dynamic mixing of the public and private, with state based public power being exercised by state institutions alongside and along with the exercise of private power by markets and civil society institutions" (Thynne, 2008, p. 329).

The state-centric hierarchy is characterised by strong formal ties, with the state being the authority at the heart of the decision-making process and other actors lower down in influence and importance. Polycentric orders are less hierarchical and less rigid than state hierarchies and may involve a wide variety of actors. The weak polycentric order is characterised by key actors at the centre with informal, weak ties which permit communication between different actors in the network but not much coordination. The actors at the centre of the order do not have as much authority as the state. The strong polycentric order is characterised by formal ties and relationships between key actors, which permit greater levels of coordination and collaboration. There are also looser, weaker ties with a range of other actors with which there is communication but less scope for coordination and collaboration. Authority in the order is spread across the key actors more evenly than in the state-centric hierarchy (see also the section below on network governance).

New actors and institutions in environmental governance

As mentioned previously, one of the key features of 'governance beyond government' is a focus on the array of non-state actors that are now involved in arrangements for governing the environment. This section details the principal actors that have been studied beyond the state:

- private-sector
- civil society.

Private sector

If it is not quite accurate to refer to the private sector as a new set of actors, the ways in which they have become involved in processes of environmental governance has changed. There are perhaps two principal 'identities' commonly attributed to the private sector in relation to the environment. Perhaps the dominant one is as perpetrator of environmental problems, whose typical response to environmental protection tends to range from dismissal to hostility (Tienhaara et al., 2012).

Corporations devour a large percentage of global energy, deplete natural resources and are heavily implicated in the production of environmental pollution (Elliott, 2004). For example, Heede (2013) calculates that, between 1751 and 2010, 63% of cumulative worldwide emissions of industrial CO2 and methane were generated by 90 "carbon major" entities, 50 of which were investor-owned producers of coal, oil and natural gas. (The other entities were state-owned or nation-state producers of these commodities.) Moreover, the promotion of ever-greater levels of consumption is directly in the interests of many private companies; indeed it may accurately be described as their *raison d'être*.

A more recent identity is as a set of actors whose involvement and wherewithal is crucial to the success of efforts to resolve environmental problems. In this framing, private sector actors can be seen as a source of research capacity and as drivers of the technological change required to make the global economy more environmentally friendly (Falkner, 2008; Levy and Newell, 2005). This identity has become more prominent, for instance, at the Rio +20 Earth Summit and the discursive framings of the 'green economy' associated with it (see Chapter 2). Corporate responses to efforts to curb environmental harm stemming from economic activity have also, for some commentators, changed. Efforts by multinationals to fund research dedicated to undermining the credibility of mainstream science on climate change demonstrate that this strategy endures.[1] Others, however, espouse a commitment to more environmentally responsible behaviour as a stronger business case and a more certain route to profitability (Braithwaite and Drahos, 2000).

Some commentators, nevertheless, remain sceptical, arguing that changing the rhetoric is one thing, whilst demonstrating that corporations have changed their behaviour is another entirely (Sklair 2001). This perception of private-sector actors relates to a bigger set of questions around the global political and economic status quo, which is heavily influenced by neoliberal thinking, and one which serves the interests of powerful factors – such as the private sector – more than other actors. The question for those concerned with environmental governance, and indeed with the goal of sustainable development, is whether this global system can be reformed, so as to produce better environmental governance outcomes, or whether the system itself is the problem, with the implication that forms of environmental governance rooted in it will cause, rather than solve, global environmental problems (see chapters 2, 4 and 10 for more discussion of these points).

In this view, the power relations underpinning current political and economic system are set against thoroughgoing environmental governance; an underlying condition which, for some commentators, mainstream thinking on environmental governance fails to interrogate (Newell, 2008a).

Civil society

Civil society has taken its place as an influential and organised actor in governance processes relating to conservation to development. One clear role for civil society in this regard has been environmental activism and advocacy. Environmentalist advocacy groups were amongst the first non-state actors to be analysed extensively (i.e. Conca, 1995; Wapner, 1996). These activists serve a number of purposes: research and policy advice; monitoring and evaluating the progress (or lack therein) states make towards environmental objectives to which they have made a public commitment; and offering feedback to diplomats at international meetings (Betsill and Corell, 2001). Chapter 2 chronicles some of the early, attention-grabbing exploits of environmental activists to influence the foreign policy of powerful nation states, and this form of intervention remains popular; and is indeed often carried out by the same actors. A recent example would be the protests that Greenpeace activists made against the first attempt by Gazprom, the Russian state owned oil company, to initiate drilling activities in the Arctic.[2]

As well as seeking to influence the processes of international meetings and agreements, such as those of the Earth Summits, civil society seeks to give a voice at the global level to marginalised peoples, for instance the indigenous peoples forum in Cochabamba, 2010 (Newell et al., 2012), as well as to provide space for greater stakeholder participation in policy and practice from the local to the global level (Bäckstrand, 2006; Newell et al., 2012; Tarrow, 2005).

Because of their success in widening participation in environmental governance processes, civil society actors find themselves partnering with governments in the production of public environmental regulations, or being included in state delegations that are sent to international negotiations, thereby having a hand in the making of international law (Betsill and Corell, 2008). Beyond these forms of engagement, they are also very active in the formation of civil regulation and efforts to monitor, regulate or otherwise influence the behaviour of the private sector (Bendell, 2000; Hemmati, 2002).

Figure 6.2 *South African protest against the arrest of Greenpeace activists*

Source: www.greenpeace.org/africa/en/News/Blog/watch-thousands-of-people-in-nearly-50-countr/blog/46930/ (accessed 9 March 2015); permission kindly granted by Greenpeace

Social movements are very important actors in environmental governance, and perhaps especially in South America. In Brazil, in the years from 1985 to 2007, there was a tenfold expansion in the size of the protected area network, from less than 250,000 km² to over 2,500,000 km² (Zimmerer, 2011). Indigenous groups have long been fundamental to the actual, every day conservation of vast swathes of the Amazon rainforest (ibid.), but the social movements that they have created were also critical to ensuring that this impressive expansion was possible (Hochstetler and Keck, 2007). In particular, they were instrumental in the establishment of extractive reserves and indigenous conservation areas as official and common protected area categories in Brazil (Salisbury and Schmink, 2007; Schwartzman and Zimmerman, 2005). However, it would be simplistic, bordering on romantic, to equate indigenous groups automatically with conservation values; some indigenous people, along with rubber tappers, have become increasingly interested in cattle-ranching, which is heavily implicated in Brazil's still-high deforestation rates (Zimmerer 2011).

In Peru, a similar expansion in the size of the protected area network has unfolded in recent decades. Bankrolled to the tune of tens of millions of dollars by the World Bank and the Inter-American Development Bank since the 1990s (Stocks, 2005), and amplified by the efforts of international conservation NGOs such as The Nature Conservancy, Peru has played host to a fourfold increase in the total area under conservation (Zimmerer, 2011). As in Brazil, much of the expansion in protected areas involved the recognition of indigenous land titles and territorial claims, via the re-categorisation of land as communal reserve (Newing and Wahl, 2004; Stocks, 2005).

Another important role civil society actors often try to play is to provide a bulwark against the so-called "democratic deficit" in the governance arrangements for an increasingly globalised world, in which decision-making processes with profound ramifications for environmental governance take place outside the boundaries of democratic processes (Biermann and Gupta, 2011). By way of following some of these decisions, which may be more important for determining both conservation and development outcomes, some civil society organisations have shifted their priorities away from sites of intergovernmental processes. Instead, some have tracked the activities of multilateral and regional development banks, bilateral donors and the financial sector, given the importance of their day-to-day decision making for a wide range of environment and development issues (Newell, 2008a). However, civil society actors themselves are not immune to claims of a democratic deficit in the way that they operate. Bernauer and Betzold (2012) point out that many civil society organisations claim to speak on behalf of the general public, but often make decisions which reflect their own priorities and for which they cannot always be held accountable.

Varieties of environmental governance regimes

In this widened field of governance actors, different regimes of interaction and decision-making can be discerned. Partly following and partly expanding upon Evans (2012), here we present five ideal types which serve as useful analytical categories to highlight particular differences in approaches to environmental governance:

1 hierarchy
2 network
3 market
4 adaptive
5 environmentality.

In the practice of governance, of course, things are much messier and there is a great deal of overlap between these ideal types. The most frequent kind of arrangement for environmental governance is probably the hybrid, which combines different aspects and actors from across these varieties (Armitage et al., 2012; Lemos and Agrawal, 2006; Plummer et al., 2013). Table 6.1 summarises the principal attributes of different modes of governance.

Hierarchy

Hierarchy is conventionally the most common form of governance and is characterised by a chain of command structure in which

Table 6.1 *Varieties of environmental governance*

Property/ type	Hierarchy	Network	Market	Adaptive	Environmen- tality
Relationship	Authority	Complementarity & trust	Contractual & competitive	Complementarity & trust	Authority embedded in social and political relations
Means of interaction	Routine & procedure	Relational, deliberative	Pricing & contract stipulation	Deliberative, iterative, provisional	Through conformity and resistance of social norms
Tools for governing	Regulation & enforcement capacity	Collaboration, negotiation & persuasion	Economic incentives, contracts	(Social) learning, experimental, monitoring	Social norms which create environmental subjects
Dispute resolution	Procedure, formal diplomacy	Reciprocity, negotiation, informality	Uneven bargaining	Reciprocity, negotiation, informality,	Recourse to social norms
Flexibility	Low	Medium-to-high	Medium-to-high	High	Low
Commitment to regime	High	Medium	Low	Provisionally high	High
Ethos	Formal structures, efficiency & clarity, reach	Mutual benefits, collaboration, collective	Competition & expedient cooperation	Mutual benefits; cooperative, evolutionary, experimental	Conformity and control
Decision- making process	Dependent, chain of command	Inclusive, interdependent	Independent	Inclusive, interdependent, iterative	Interdependent, embedded in social norms
State role	Provides & enforces rules	Encourages & persuades	Enforces rules & offers incentives	Encourages & supports	Creates environmental subjects

Adapted and expanded from Evans, 2012

authority rests at the top and decisions are executed by actors lower down. Formal rules, obligations and routines tie actors together within a hierarchy, and aim to commit those involved to work together towards 'a clear route to a desired outcome' (Evans, 2012: 35). Both state bureaucracies and private companies are typically characterised by a hierarchical arrangement. Because a hierarchy is so conventional, some scholars would seek to exclude it as a mode of governance: it is the operation of existing governing structures as opposed to the newer, more loosely structured alternatives which have emerged more recently (Bierman and Pattberg, 2012). As such, a well-functioning hierarchy can engender a coherent and systematic approach towards collective action, but is less likely to offer flexibility or be a source of innovation in practice (Evans, 2012).

The state is a paradigmatic example of hierarchy. Departments and ministries have long devised command and control hierarchal structures through which to address the problems, environmental or otherwise, of industrial society (Bäckstrand and Lövbrand, 2006; Evans, 2012). Experts lead bureaucratic units which set the rules through which public goods – such as air quality or clean water – are to be protected and properly managed. Often, these rules prohibit or heavily restrict usage of such goods, and are captured in the form of environmental regulation; they are thereby enforceable by law. A hierarchical structure links all of the people who are involved in the production of such legislation, from the technical experts who attempt to determine the parameters of what is to be regulated, to those that draft the legislation, to those that are responsible for enforcing it.

The command and control model has helped to solve some environmental problems that have arisen from development processes – and perhaps especially those at the national and subnational levels; see Chapter 2 for examples. Yet they have also led to the side-lining of the environment as a political issue, better left in the hands of technical experts (Lowe and Ward, 1998). Moreover, command and control approaches have led to 'disjointed incrementalism' (Lindblom, 1979); that is, a piecemeal approach to regulation which treats environmental problems separately, without identifying underlying causes or situating particular pieces of legislation within an overarching structure of environmental planning and coordination. Given these difficulties at the national level, it is easy to see why a command and control approach would be even harder to achieve at the international level. Not least, this is because of the added complication of states having conflicting interests in relation to a public good, the regulation of which can hinder one government's objectives even as it helps another's.

Network

A network assembles a range of stakeholders with complementary abilities, which may or may not include government, that seek to achieve commonly held objectives that are of mutual benefit, but which would not be achievable without collaboration (Rydin, 2010). Because of the way in which actors voluntarily come together to form a network, they are held to be self-organising and, thereby, self-governing. As such, in its ideal form at least, the network is set up in opposition to the hierarchy, as it is based on cooperation rather than coercion. Networks can be inclusive and innovative but may be less cohesive, coherent or robust because there are fewer requirements for the actors to remain within and commit over the long term to the network.

Flexibility and agility are two key attributes of networks, which can arise in response to the much slower processes of governance of nation-states, which are bound up in and slowed down by regulation, protocol, hierarchy and bureaucratic inertia (Klijn and Skelcher, 2007). Networks can gain or shed constituent organisations or actors and pool resources in ways which allow them to respond more quickly to problems and opportunities (Klijn and Skelcher, 2007; Evans 2012). For this reason, networks may be seen as important counterparts to government-led transnational processes, given how slowly these can unfold (see Chapter 2 for an example of the climate change negotiations). Indeed, they often arise out of frustration with the sloth-like pace of inter-governmental interaction.

The study of networks is different from that of individual actors. The analysis of networks often focuses on the structural characteristics of the network which are a function of how connections are formed between individuals, as opposed to the nature of a particular connection between two individuals. In other words, it is about the sum, not the parts. This is evident in the work of Marco Janssen and colleagues (Janssen, 2007; Janssen et al., 2006a, 2006b), for instance, on social networks, who use the concepts of connectivity and centrality to explore the interactions that characterise a social network. Connectivity refers to the qualities of density (how many connections one 'node' in a network has to other nodes) and reachability (how easy or difficult it is for all the nodes in a network to connect to each other). Centrality (akin to polycentrism) refers to whether there are central nodes in a network which bear more weight in coordinating between all the network nodes. High levels of connectivity can help with the dissemination of new information and with eliciting a response from all network nodes, but can also facilitate the spread of misinformation or spread bad practices very quickly (think of how quickly information can be spread on Facebook or Twitter). High levels of centralisation can engender more co-ordinated efforts, but can become more rigid and hierarchical, ending up mimicking the structures that networks are often supposed to circumvent (see also Figure 6.1).

Others take this idea of relational ontology – that is, what things are, not in isolation but in relation to each other – a step further. Actor Network Theory (ANT), perhaps best represented by the work of Michel Callon and Bruno Latour (Callon, 1986; Callon et al., 2009, 1992; Latour, 2007, 1996, 1987), seeks to understand the world as one related network of people and 'things' (i.e. nature), acting upon and being changed by each other constantly. Controversially, and for some, counter-intuitively (Bloor, 1999), this body of theory gives 'things' an agency to act which is not too dissimilar to human agency – that is, of acting with intention, or something like it (though ANT scholars argue that their definition of agency does not presuppose agency; cf. Latour, 1996). It is difficult to characterise the underlying philosophy of ANT's main proponents, because of the dense way in which it is articulated and the extent to which it remains contested (see Bloor 1999). Yet ANT's attempt to understand nature and the way in which it acts upon humans, as well as vice versa, in terms of 'co-production' (see Chapter 3), is one that scholars of conservation and development would do well to take seriously. It also resonates with some of the thinking in adaptive governance (see below). As applied to the relationship between conservation and development, Actor Network Theory might be about social-ecological networks, not just social networks (see Chapter 3 for a conceptual introduction to social-ecological systems).

Whilst networks, therefore, offer up new avenues for environmental governance, they are not without their problems and detractors. According to Ansell and Gash (2008), the advocacy of network governance is determined, enabled and constrained by five factors:

1 history of conflict and cooperation between network members;
2 incentives structures for stakeholder participation;

3 resource access and power differentials between network members;
4 leadership;
5 institutional design.

Even if a network successfully navigates these considerations, the question as to how influential they really are persists. Networks lack the political authority of a nation state (Evans 2012). Moreover, whilst they are often vaunted as governance spaces beyond the state, the literature on how networks function is frequently at odds with the idea of a perceived shift from a "sovereign" to a "post sovereign world" (Bäckstrand, 2008). Some hold that transnational networks function only in "the shadow of hierarchy" (Evans 2012:121), because the power to set rules, or to delegate this function to partnerships and networks, remains at the discretion of the state (Berry and Rondinelli, 1998).

Other commentators raise the concern of the democratic deficit: private actors, NGOs, quasi governmental organisations and others can exert considerable influence without ever being elected or accountable to the public (Weber and Christophersen, 2002). Networks are often also quite closed spaces or, as Cornwall and colleagues call them, invited spaces, raising the question of who is and who is not invited to participate in the space (Cornwall 2002, 2004; Brock et al., 2004). Others have argued that networks allow a wider range of stakeholders a voice in decision-making processes than does governance by parliamentary democracy (Sørensen and Torfing, 2008). Whatever the merits or failings of network governance, it is difficult to see it as a wholly separate phenomenon from the state. Nor is it applicable solely to well-financed, northern institutional networks (see Box 6.1 for an example of network thinking applied to a Southern context).

Market

Markets seek to foster individual and collective behaviour by offering incentives to act in one way and not another. In relation to the environment, this is done through attempting to minimise the costs and maximise the benefits of forms of behaviour which yield environmentally positive outcomes (Cashore, 2002; Lemos and Agrawal, 2006). Participants stand in a relationship of supplier and consumer of resources, products or services, regulated through property rights and

Box 6.1

The role of civil society networks in amplified voice and preference of marginalised people in the Gran Chaco, South America

Alcorn et al. (2010) discuss the importance of civil society in the governance arrangements of the Gran Chaco ecosystem extending across Bolivia and Argentina. In both countries, efforts had been made by local civil society organisations to connect local inhabitants with local and national government, and to try to ensure that local knowledge is one of the key sources of information fed into ecosystems governance. Using a watershed management approach, NGOs have focused on mobilising social networks and civic science to achieve conservation and development objectives simultaneously.

Differing experiences with network governance can be seen in work in Bolivia and Argentina, as documented by Janice Alcorn et al. (2010). In the Upper Parapeti Watershed, Bolivia, a small local NGO, Yangareko, sought to establish a system of watershed management that could draw on existing local practice and knowledge, to ensure that it contributed to conservation and poverty reduction objectives. Alcorn et al. claim that the zoning plans and institution building done by Yangareko opened a space for consultation with local people with respect to government policy changes. The work with the local inhabitants was also held to facilitate the collection of biodiversity specimens. Meanwhile, in the Pilcomayo basin, Argentina, the Management and Regional Research Foundation (FUNGIR) sought to involve local indigenous and *criollo**populations in the Pilcomayo Master Plan Project, a government project funded by the European Union. FUNGIR trained local environmental monitors who could liaise with the project staff to ensure that the project made use of indigenous knowledge in environmental decision-making, and garner support for ecologically appropriate development activities such as the production of *algarrobo* groves. Unfortunately, FUNGIR apparently made little headway with these objectives, and the people it represented remained on the margins of decision-making within the project.

To a greater or lesser extent, in both cases, the preference of both governments for the commercialisation of the natural resources in these areas may hinder the prospects for these kinds of efforts to realise conservation and development. In both Argentina and Bolivia, substantial change is occurring as a result of the Integration of the Regional Infrastructure in South America initiative (IIRSA), an international investment plan to improve the transport infrastructure links across South America. Greater links to the rest of the country and the continent make these areas more attractive to investors seeking to extract forest resources, oil and gas or for expansion of the agricultural frontier especially soy plantations. These may be more highly prioritised than local efforts, even where these might offer better prospects for sustainable development. This is just one of many examples which might lead us to wonder about the efficacy and power of civil society to influence the big government decisions.

Criollo refers to Latin American people who self identify as wholly or predominantly Spanish in origin, as opposed to claiming a pre-hispanic identity.

contracts (Evans, 2012). Engagement in markets is contingent on the value and cost which attach to the goods and services available; both value and cost are expressed and mediated through the price given to products and services. As such, there is greater flexibility to participate or not in market transactions. Concomitantly there can be less commitment to the forms of behaviour markets may try to bring about if conditions change in such a way that costs are perceived to outweigh benefits. Examples of market governments include energy taxes, tradable permits, voluntary agreements, eco-labelling and certification of particular products, such as fish or kitchen roll, to show that they have been produced in ways which are not environmentally unfriendly. Box 6.2 gives the example of the work of the Forest Stewardship Council. See chapters 4, 6 and 13 for further coverage of the role of markets in efforts to achieve conservation and development simultaneously.

Box 6.2

Forest certification as market governance: examples from Brazil and Argentina of market governance

One well-known example of an 'eco-consumerist' certification scheme which private actors commit to on a voluntary basis is the Forest Stewardship Council (FSC). In the UK, the FSC brand is prominently displayed across a variety of commonly available products, from toilet rolls to tables. The FSC is an international network of civil society organisations interested in environmental and development issues, principally equity. Founded in 1993, the FSC's aim from the outset was to provide a single global label for existing labels that already denoted the use of sustainable forestry practices in the making of wooden products. Through offering both 'Forest' and 'Chain of Custody' certifications, the FSC covers both the source of the product and the supply chain it goes through to reach the consumer. Its international headquarters in Berlin defines the global principles, standards, compliance, auditing and enforcement procedures, in conjunction with regional working groups coordinated by national chapters. Through these mechanisms, the FSC aims to establish a credible guarantee for consumers that a forest product has been made and supplied in an environmentally responsible manner.

Comparing the experiences of Argentina and Brazil suggests that the FSA model works better in some places than others and not always for the reasons commonly cited. Despite their similar economic histories and important forestry industries, and despite the fact that the same environmental NGOs promoted the establishment of FSC processes in both places, the Brazil chapter thrives whilst its Argentine counterpart languishes. Often, it is assumed that market benefits and the presence of transnational intermediary actors role are the two key factors which explain the level of commitment to and success of FSC certification schemes. However, the profit and market access benefits conferred by FSC certification are very similar in Argentina and Brazil. Crucially, for the majority of producers they have been low. In Argentina, the lack of market benefits has limited the interest that producers have in seeking FSC-certified status, even though there is interest in gaining access to European markets. Local civil society has been engaged but mainly through the Argentine equivalent of WWF. Therefore, the biggest producer companies have not signed up to the FSC certification process. Moreover, the reaction from government has been mixed: whilst the middle Ministry of Environment supports the establishment of the scheme, the Ministry of Agriculture and National Institute for Standardisation have been involved in establishing rival schemes with more flexible system standards. Industry observers doubt that the schemes could be credible with consumers because they enjoy significant support from the country's largest forestry companies and are suspected to be unduly influenced by these players.

In Brazil, many more producers have signed up to the FSC certification process and 65 forests have been declared compliant, in comparison with Argentina's nine. Like Argentina, market returns from access to European markets have not yet delivered great benefits to FSC certified companies, which tends to have forest plantations. Only a very small amount of native Amazon forest has yet been declared FSC certified, but the logic is that in the future the extremely high returns made on this kind of tropical wood will be sufficient to ensure much more extensive certification. In spite of the lack of market benefits, companies which are involved in legal forestry operations seek the legitimacy of the FSC certified status because the legacy of failed forest management in Brazil brings them under intense public scrutiny. Additionally, the FSC Brazil chapter, whilst in many ways dominated by civil society actors, also engaged industry leaders at an early stage, isolated opponents and established a link between forest certification and successful, well managed businesses. However, whilst there is evidence to suggest that environmental benefits are being generated, the FSC process has been criticised in the Brazil context for not being very equitable, favouring large forestry companies over community-based enterprises.

Sources: Espach (2006), Guedes Pinto and McDermott (2013)

Adaptive

In contrast to thinking on other modes of governance, adaptive governance stems much more directly from a consideration of the relationship between environment and society. It is rooted in an understanding of this relationship as a complex, adaptive system whose social and ecological components are linked and indeed interdependent (Berkes and Folke, 1998; Gallopin et al., 2001; Turner et al., 2003; Waltner-Toews et al., 2003). What this effectively means is that governance arrangements need to able to take account of how the way ecosystems work impinges on the forms of society that are possible and, conversely, how the forms of society that we have shape the way ecosystems work. The underlying premises of this approach therefore differ fundamentally from conventional understandings of the role of the environment in producing and supporting society. Current modes of economic activity attempt often to extract from the environment particular forms of benefits or services valued by society and are about making nature fit the requirements of this economy. This is done by increasing 'control over resources through domestication and simplification of landscapes and seascapes to increase production, avoid fluctuations, and reduce uncertainty' (Folke et al., 2005: 442). Traditionally, one of the goals of natural resource management science has been to maximise the use of a specific resource unit, for instance fish or timber, and to manage it in such a way as to stabilise the yield that can be extracted from year to year (Berkes and Folke, 1998). With adaptive management, the essential goal is to maintain social-ecological resilience in the face of disturbance, surprise and other forms of change which are frequently difficult to anticipate (Folke et al., 2005). See Chapter 3 for more detailed treatment of resilience in social-ecological systems.

Adaptive governance starts from the premise that the functioning of ecological and social processes are uncertain, difficult to predict and frequently in a state of flux. In fact, from this perspective, efforts to try to fix the services and functions provided by nature, with a view to being able to control a stable supply of ecosystem functions and

Table 6.2 Determinants of social-ecological resilience (after Seixas and Berkes, 2008)

Factors that weaken social-ecological resilience	Factors that strengthen social-ecological resilience
Breakdown of 'traditional' institutions and authority system	Strong institutions, i.e. robust local institutions with strong enforcement and strong, credible leaders
Rapid technological changes leading to more efficient resource exploitation	Good communication between local, regional and national levels, allowing for co-management, drawing on local and scientific knowledge
Rapid changes in the local socio-economic system	Equity in resource access
Institutional instability at a higher political level	Use of memory and knowledge as a source of innovation and novelty
	Political space for experimentation

services, will in fact generate turbulent forms of change that can be difficult to deal with. Herein, from a complex adaptive systems perspective, lies a fundamental flaw of natural resource management science in treating such resource units as discrete entities rather than as integrated parts of a much wider ecosystem: to change one part of it has repercussions for the rest of the system. Managing a resource by reducing its variability may make it easier for us to manipulate it to our best advantage, but variability is central to the functioning of ecosystems. To take the example of forests, managing them in such a way as to attempt to eliminate or minimise forest fires has led to generating the conditions for ever greater and more devastating fires, of the kinds experienced in places such as California and parts of Australia, that tend to make headlines around the world. In this case, there are adverse effects for both the social and ecological components of the system. Box 6.3 offers an example of adaptive governance, by way of contrast, which also demonstrates that efforts to manage continual change remain imperfect, and from which can be drawn some lessons about the factors that are likely to strengthen or weaken social-ecological resilience (see Table 6.2).

Box 6.3

Adaptive governance in Ibaquera Lagoon, Brazil

One example, which gives a flavour of the constantly shifting changes and rearrangements entailed by adaptive co-management, comes from the work of Seixas and Berkes (2008) on the Ibiraquera Lagoon in Santa Caterina State, Southern Brazil. In the 1960s the lagoon played host principally to a community of fishers who also practised subsistence agriculture. With the construction of roads, the greater access to market and the arrival of the state, the socio-economic dynamics changed, for example with the greater trade in fish and seafood with larger cities and with the establishment of tourism activities in the area. The lagoon is seasonally connected to the Atlantic Ocean: for much of the year there is a sandbar between the lagoon and the ocean. However, when the water level rises throughout the rainy season, the water pressure builds up and forges an ephemeral channel through the sandbar. This allows young fish and shrimp to enter into the Lagoon, where they grow, returning to the ocean as adults when the channel next opens. Fishers have used a variety of strategies to prevent the return of fish and shrimp to the ocean, in order to harvest them. As a result of the changes to the uses of the lake and the rules for fishing, the state of the resource base rose and fell over a time period of approximately 40 years.

From the 1960s to 2000, Seixas and Berkes (2003) chart four periods of resource decline and renewal. These mirror changes to the management structures and resource use practices found in the Ibiraquera Lagoon. In the 1960s, 'traditional' management practices, with informal but effective enforcement of rules embodied in respect for elder fishers, produced high levels of social and ecological resilience. In the 1970s, even though fishing practices and technologies did not change drastically, socio-economic changes adversely affected ecosystem resilience. The opening of a road, which permitted greater access to the lagoon, led to incipient opportunities to sell fish in roadside outlets and in cities. This encouraged fishers using the larger gillnets to increase their catch. As a result, inequality grew between gillnet fishers and those who used smaller cast nets, prompting social disquiet.

In the 1980s, the users of the smaller nets regained control of the fishing organisation and established better relations with local government. The use of gillnets was banned and the rules governing fishing in the lagoon were strictly enforced. However, the way the lagoon was used was changing with population increases and with the arrival of tourism. Building on the shores increased erosion and siltation, whilst the rising popu-larity of water sports threatened to scare away returning fish. In the early 1990s the local fishers organisation requested help from a local university and local government to introduce fish and shrimp breeding into the lagoon. This innovation appeared to resolve the problem; although by this point, many fishers were operating only on a part-time basis, having diversified their activities by seeking employment in the tour-ism and construction sectors. After 1994, however, the state support that had been received was withdrawn, whilst the local fishers' organisation had become 'brittle', resulting in overfishing and a decline in the fish and shrimp resources. By 2000, never-theless, a community council was resurrected and institutional renewal seemed again a distinct possibility.

Environmentality – governance technologies and environmental subjects

One critique of 'mainstream' thinking on environmental governance is that it does not necessarily get to grips with the power structures inherent in the global political and economic system which forms the background to efforts to achieve more effective environmental gov-ernance. We shall explore these critiques further in the next section, but another way of understanding the operation of power in matters of environmental governance is via the concept of 'environmentality', coined by veteran scholar of the commons and political ecologist, Arun Agrawal, in his book (and summary paper) of the same name (Agrawal, 2005a, 2005b). Agrawal's primary concern is with explain-ing how, in the Kumaon province of India, a people who, in the 1920s effectively resisted colonial efforts to impose forest use restrictions upon them, came to manage their forests in ways the government approved of and indeed 'far more systematically and carefully than the forest department could' (2005b: 8). He identifies a number of different factors, revolving principally around new forms of knowledge, institutions and political relationships. Above all, he sets out to explain how it is that people come to care about the environment, and how this changes collective behaviour towards the forest quite dramatically.

Agrawal draws on political ecology and environmental feminism literatures quite extensively, but leans most heavily upon the work of Michel Foucault on governmentality. This idea posits that governments have expanded power over 'their' populations not (solely) through coercion, but through 'technologies of government' that produce subjects who are sympathetic to the government's priorities and interests. 'Subject' in the sense used here refers to actors or

Figure 6.3 *Dwarahat, a town in the Kumaon hills, India*

Source: Photographer Rautela wiki, used under Creative Commons licence

agents with a degree of autonomy in their actions but who are nevertheless subject to a source of authority (central government in this case). With 'environmentality' Agrawal recasts the idea of governmentality as a more 'specific optic' (2005b: 226) for exploring the contestations and collaborations of environmental politics. In the case of Kumaon, it is to explore how people become environmental subjects that are prepared to use forest resources available to them in ways which are compatible with government objectives around forest management.

When, in 1929, the British colonial administration realised how powerless it was to stop the forest fires and other modes of resistance shown by Kumaonis, it changed tack and, much in the vein of indirect rule found across many other parts of the British Empire, devolved responsibility for forest management to the local level. It did this through the creation of forest councils, with locally elected leaders who were able to take decisions about how to manage the forest. Agrawal refers to this change as an instance of creating "governmentalized localities", in which forest councils accept and take on the regulatory functions of government, including monitoring, dealing with infractions, adjudication, etc. Above all, these institutions engender the participation of individuals in the regulatory regime as a means of ensuring compliance with it. On the basis of longitudinal study and a wealth of statistical data, Agrawal builds a large evidence base for his claim that before these councils were introduced, people had negative attitudes towards the forest, whilst after their introduction, their attitudes and behaviour changed. He goes on to argue that it is precisely people's participation in this new regulatory process that has turned them into 'environmental subjects'. This self-disciplining behaviour represents a power relationship: it is not a power of the government *over* the people but a power manifested *through* the (compliant) behaviour of the people in line with government priorities.

In many ways, the workings of the forest councils in Kumaon can be seen as a success story for community-based conservation grounded in decentralisation and devolution of power, and indeed Agrawal's work is refreshing in that it permits a sense of optimism around the possibility of using nature wisely, in contrast with the doom and gloom surrounding so much of what is written on environmental governance and environmentalism more broadly. However, some commentators have noted that whilst Agrawal does acknowledge environmentality as a phenomenon which has brought about a 'good' state of affairs, there is a certain ambivalence on his part towards it: following the internal logic of Foucault's notion of governmentality to its conclusion would, after all, lead us to identify forest councils as a tool of co-optation and control. In other words, environmentality can also be read as 'a clever way for authorities to pacify local resource users while obtaining what they want' (Raymond, 2006: 264). This more sceptical framing inevitably raises the question of whom the arrangements in Kumaon best serve and whether the fruits of local labour remain at the local level or are vulnerable to exploitation by state and external private actors (Robbins, 2006).

Critical perspectives on environmental governance

The 'neoliberalisation' of environmental governance

An entry point into critical perspectives on environmental governance comes via what we might call the 'elephant in the room syndrome' (see, for instance, Sullivan, 2006). In essence, this is the idea that conventional accounts of environmental governance do not get to grips with the deeper political implications of the extent to which governance is structured by the thinking, mechanisms and processes of neoliberalism, or the extent to which both conservation and development have been 'neoliberalised'. (See Chapter 5 for a definition of neoliberalism.) This is the prevailing status quo but its effects, for critics, have not been fully acknowledged within the environmental governance literature (Newell, 2008a, Buscher and Arsel, 2012, McCarthy, 2012). What makes the silence, from this standpoint, on these underlying conditions even more bizarre is the patent influence of neoliberal thinking on the tools and mechanisms through which governance objectives are to be achieved. In short, market solutions, grounded in an ever more clearly delineated and strictly enforced global private property regime, and accompanied by a substantial curtailment of the power of the state, are the answer – even though the patterns of global consumption, fuelled by market mechanisms, are still the source of conservation and development problems, at the global level.

This standpoint runs counter to the longstanding idea that conservation is a struggle against capitalist overexploitation of nature in the name of progress or development. In Chapter 1, we discussed the idea that conservation and development both emerge as a result of the commodification of nature. That is, conservation emerged because of the tremendous damage that was being done by turning large parts of the natural world into tradable objects, spurred by industrial revolution and ever greater waves of global trade, underpinned by imperial expansion. Conservation in that vein of thought emerges as the 'conscience' of capitalism, perhaps, a plea for the better governance of nature in the face of the inexorable march of human progress. However, as Brockington et al. (2010) point out, this attitude may remain prevalent amongst some conservation organisations and actors, but it no longer characterises the mainstream, international conservation organisations. What these actors increasingly do is work hand-in-hand with actors who are fundamentally implicated in fuelling ever-increasing patterns of global consumption (ibid.). What Brockington et al. and other scholars (i.e. Castree, 2008, Büscher and Dressler, 2012) suggest is that this kind of conservation has been co-opted. It has instead bought into the idea that neoliberal capitalism can be the means for the better governance of nature.

If this analysis seems a little far-fetched, bordering perhaps on a conspiracy theory, then consider the example of the Qit Minerals/ Rio Tinto ilmenite[3] mine in Fort Dauphin, Madagascar, which looks on the surface very much like a case of network governance with the potential to deliver real conservation and development benefits. One objective is to contribute to biodiversity conservation by turning parts of the mining concession land into protected areas, which will increase the size of the country's formal protected area network. Another is to contribute to local development by providing employment, development projects to boost livelihoods and through the construction of a port which is accessible to cruise liners, thereby bringing tourism to this hitherto inaccessible part of Madagascar (Rio Tinto, 2012).[4] The project was brought about through collaborations with a network of local and global actors, including international conservation NGOs such as Earthwatch and the Royal Botanical Gardens at Kew, the Malagasy government and the World Bank (Duffy, 2008; Brockington et al., 2010). The common objective that this network was assembled to address was the mitigation of social and environmental impacts caused as a result of the opening of the mine.

The ilmenite mine has gone on to become one of the most important examples of foreign direct investments in Madagascar and was valued early in 2013 at US$940 million.[5] However, the initiative has come in for criticism on a number of fronts. A Panos report commissioned by Friends of the Earth recounts that not enough was done to prepare and explain to the people affected by the opening of the mine – and in particular for the 1000 people who were displaced as a result – what the impacts were going to be or across what time frame (Harbinson, 2007). Perhaps more crucially, there have been reports that the benefits that were promised to the residents of Port Dauphin have not materialised: the compensation people were paid was insufficient owing to rises in the price of land; and the promise of employment, according to local leaders, has fallen through[6]; though Rio Tinto rejects this claim and claims instead that 76% of employees in the Fort Dauphin mine were local Malagasies. The Friends of the Earth report warned that in the continued absence of a credible discussion and negotiation space, which all stakeholders involved could accept and trust not to be dominated by Rio Tinto, the potential for conflict would persist (Harbinson, 2007). This insight proved to be prescient as five days of local protests against the mine occurred in January 2013, which closed the road leading to the mine before the crowds were eventually dispersed with the use of teargas[7].

As Brockington et al. (2010) point out, integral to the governance arrangements for the mine is the act of re-classifying the landscape through a convergence of neoliberal capitalist and conservation agendas. This gave rise to processes of commodifying nature through establishing new connections to markets, and fundamentally altering the relationship between humans and the environment in this part of the world. There is also a pattern here in which the benefits are unevenly distributed with the poorest missing out and most benefits going to wealthy local elites and/or international investors (Buscher and Arsel, 2012). It could be argued that the consequences of this form of network governance contribute to patterns of global inequity, which has only become more pronounced in recent decades.

This pattern is evident in other studies. For instance, Dressler and Roth (2011) argue that in the Philippines, neoliberal policies have been used to achieve conservation objectives, under the guise of community conservation. In essence, there have been sustained efforts to discourage farmers in Palawan from shifting cultivation, a practice which conservation agencies and local politicians maintain is implicated in environmental degradation. Instead, farmers are being encouraged to adopt more intensive forms of conservation which use less land and, and therefore require less forest clearance both within and adjacent to the Puerto Princesa Subterranean River National Park. The idea is that this kind of agriculture will yield higher benefits, through being aimed at and more integrated into local and regional markets.

Alongside this push has been an effort to establish formal, private land titles, ostensibly with a view to giving farmers more tenure security and therefore better rights to derive economic gain from the land. However, according to the authors, such reforms serve better to consolidate wealth and control over land in the hands of a property owning elite, but do not improve the lot of the broader populace. It is these people who can capitalise on some of the more lucrative investments, such as restaurants and other higher and tourism activities. But poorer farmers have found themselves unable to practise shifting cultivation and worse off in terms of their new livelihood activities. This form of poverty reproduction is a common outcome of an agrarian transition which has been documented across South Asia and more widely (Rigg, 2006). Dressler and Roth (2011) suggest that conservation practice in this context is contributing to this impoverishing process of transition. They also make similar claims for conservation practice in Thailand.

The reason this pattern is manifest in so many places, according to this account, is because of a transnational class which is reinforcing its interests (Sklair, 2001). The global environmental elite is part of this class, and its 'job' is to 'distance global capitalism from the sources of environmental problems' (Newella, 2008: 516, after Sklair 2001). Mild criticism of consumerism and globalisation is permissible but not radical critique. From this viewpoint, analysis of the Rio Earth Summit in 1992 points to the success of this bloc in obscuring its role in processes that it actually has powerful influence over (ibid.); a trick that Hildyard (1993) has likened to putting the foxes in charge of the chickens. One powerful manifestation of this at the global level is the way in which trade conventions and regulation so frequently trump environmental counterparts whenever these are perceived to be a threat (Newell, 2008). A quick look at the numbers sums this up: currently, we are spending US$10bn on measures to deal with climate change, whilst we need to spend US$100bn. In contrast, we spend closer to US$1trn per year on fossil fuel subsidies and other activities which contribute massively to putting more carbon into the atmosphere (Bierbaum et al., 2014). These figures are testament to the extent to which patterns of trade remain unrestrained by environmental considerations. Following Newell (2008a) it is hard not to conclude that even as the international negotiations on climate change have failed year after year to establish a regime capable of reducing carbon emissions to within safe limits, the World Trade Organization continues much more successfully to negotiate increases in trade.

Summary

- The field of environmental governance is shaped by an attempt to go beyond government in order to understand who really makes decisions about how the environment is used and treated. If we want to understand the relationship between conservation and development, it is necessary first to get to grips with the implications of environmental governance.

- Environmental governance is characterised by the emergence of new actors, institutions and networks. In particular the private sector and civil society have become much more visible actors in processes of governing the environment. Hybrid actors, which cannot easily be categorised as state, private or civil society but which incorporate elements of all three, have also become increasingly prevalent and influential.

- In addition to the more conventional governance structures of hierarchy and markets, network and adaptive governance arrangements have also emerged. These are characterised by more fluid and flexible approaches; although their efficacy and influence have yet to be comprehended fully.

- What is increasingly clear from this pattern is that the mechanisms of the market, in tandem with the ideology of neoliberal capitalism, have gained much ground, in terms of their use as tools for environmental governance. Some important conservation actors appeared to have embraced (or at least accommodated) aspects of neoliberal capitalism as a means through which to achieve their objectives. However, a number of commentators maintain that these forms of political and economic activity continue to be the root cause of environmental degradation and uneven, deeply inequitable forms of development.

Discussion questions

1. Which of the definitions presented do you think best captures the broad sweep of the varieties of environmental governance covered in this chapter?

2. Why is it important to understand the role of non-state actors in environmental governance?

3. How fair is the critique that 'mainstream' accounts of environmental governance do not get to grips with the underlying power dynamics which determine environmental governance outcomes?

4. To what extent are we all 'environmental subjects'?

Notes

1. See, for instance, http://www.theguardian.com/science/2013/sep/20/big-business-funding-climate-change-sceptics accessed 02.01.2014.

2. Video footage for attempts by Greenpeace activists to scale a Russian oil rig can be seen at http://www.youtube.com/watch?v=AI4UeVHQEOo

3. Ilmenite is a mineral used in the production of titanium dioxide, a white pigment often found in toothpaste, make up and paint.

4. http://www.riotinto.com/ourbusiness/qit-madagascar-minerals-4645.aspx, accessed 18.12.2013

5. http://world.time.com/2013/02/08/the-white-stuff-mining-giant-rio-tinto-unearths-unrest-in-madagascar/ accessed 18.12.2013.

6. http://www.foe.co.uk/sites/default/files/downloads/mining_madagascar.pdf accessed 18.12.2013.

7. http://world.time.com/2013/02/08/the-white-stuff-mining-giant-rio-tinto-unearths-unrest-in-madagascar/ accessed 18.12.2013.

Further reading

Agrawal, A. (2005) *Environmentality: Technologies Of Government and The Making Of Subjects*, Duke University Press.

Berkes, F., Colding, J., Folke, C. (2008) *Navigating Social-ecological Systems: Building Resilience for Complexity and Change*, Cambridge University Press: New York.

Biermann, F., Pattberg, P.H. (2012) *Global Environmental Governance Reconsidered*, MIT Press.

Brockington, D., Duffy, R., and Igoe, J. (2008) *Nature Unbound: Conservation, Capitalism and the Future of Protected Areas*, London: Earthscan.

Evans, J. (2012) *Environmental Governance*, Abingdon: Routledge.

Gunderson Lance, H., and Holling, C.S. (2002) *Panarchy: Understanding Transformations in Human and Natural Systems*, Washington, D.C.; London: Island Press.

Harvey, D. (2007) *A Brief History of Neoliberalism*, Oxford: Oxford University Press.

Newell, P., Pattberg, P., and Schroeder, H. (2012) 'Multiactor Governance and the Environment', in Gadgil, A. and Liverman, D.M., eds., *Annual Review of Environment and Resources, Vol. 37*, Annual Reviews: Palo Alto, 365–387.

Ostrom, E. (1990) *Governing the Commons: The Evolution of Institutions for Collective Action*, Cambridge: Cambridge University Press.

Online resources

Earth Systems Governance Project – international social science research network on global environmental governance and change: http://www.earthsystemgovernance.org/ (accessed 11.02.2015)

Global Governance Project – collaboration of European universities investigating the new actors, mechanisms and institutions associated with global governance – http://www.glogov.org/

International Human Dimensions Programme – overarching initiative looking at the human dimensions of global environmental change, which houses the Earth Systems Governance Project – http://en.wikipedia.org/wiki/International_Human_Dimensions_Programme

World Trade Organisation – one of the key institutions in a globalised world – http://www.wto.org/english/res_e/statis_e/its2011_e/its11_charts_e.htm

UNU wider initiative – repository for information on globalisation (http://www.wider.unu.edu/research/Database/en_GB/database/)

Diversitas - international organisation focusing on halting biodiversity loss with a focus on governance and linked with other organisations which focus in some of their work on global environmental governance – http://www.diversitas-international.org/activities

7 Population, consumption, conflict and environment

- **Historical and contemporary debates around population and environment**
- **Disentangling the effects of consumption from population growth**
- **Population, environment and conflict**

Introduction

The most recent UN publications on population growth present three scenarios, each based on differing assumptions about fertility rates in the future (UNDESA, 2013).[1] In the low scenario, population increases to 8.3 billion by mid-century and declines to 6.8 billion by the end of the century, from the starting point of a 7.2 billion in July 2013. In the medium scenario, population increases to 9.6 billion by 2050 and 10.9 billion by 2100. In the high scenario, population increases to 10.9 billion by 2050 and 16.6 billion by 2100. All three scenarios are based on a fertility rate which is declining, relative to those witnessed in the second half of the twentieth century.

This chapter looks at historical and contemporary thinking about the population-environment nexus and charts the implications of this thinking for efforts to do conservation and development simultaneously. In so doing, it considers the relationships between:

- the origins of Western scientific thinking about population, grounded in the work of Thomas Malthus, and a variety of responses to his work throughout the twentieth and into the twenty-first century. The influence of Malthus persists, but his ideas are also contested, and not necessarily the most helpful way to understand how and when development processes do or do not result in environmental degradation.

- population and a variety of other factors linked to environmental change, foremost amongst them, consumption patterns characteristic of contemporary development trajectories. Such an analysis makes clear that sometimes, the relationship between population growth and environmental degradation is by no means straightforward. Sometimes, it is consumption levels, not population growth, that cause environmental problems. Therefore, if we focus on areas in which there is significant population growth in conservation efforts, we may not be targeting the underlying causes of environmental degradation in those areas.
- population, environment and conflict, and the extent to which this relationship bears upon contemporary efforts to bring about conservation and development.

Historical and contemporary debates around population and environment

Historical roots of population–environment debates

Environmental historians take the view that the relationship between population size and environment has been topic of discussion or concern probably for as long as there has been human society. As far as contemporary scholarship into this relationship goes, a very clear starting point is the work of Thomas Malthus who published the enduringly influential *Essay on the Principle of Population as It Affects the Future Improvement of Society* (Malthus, 1999 [1798]). For Malthus, the concern was that human population, which grows exponentially, would eventually outstrip the pace of food production, which grows linearly. This would trigger episodes, principally famine, that would bring human population numbers back into line with the natural limits of the amount of food that could be provided. This thinking led Malthus, who drew his arguments from his observations of Britain in the late eighteenth and early nineteenth centuries, to oppose redistributive efforts to feed the hungriest, such as the Poor Laws, on the grounds that they were ineffective and exacerbated, rather than relieved, the suffering caused by famine. They were, in other words, a futile effort to ignore environmental limits.

Few people working on population and environment in the twenty-first century would fully subscribe to this mono-causal thesis, or seek to generalise it across time and space without substantial caveats. Nevertheless, in the 1960s, at a time when the notion of a global ecological crisis was becoming ever more current, it found expression in the work of commentators such as Paul Erlich (1975), who sought to popularise the idea of the 'population bomb', and also in the work of the Club of Rome on the 'Limits to Growth' thesis (Meadows et al., 1972 – see Chapter 2 for more details). To these commentators, the idea that sheer human numbers – especially in the developing world – would simply overwhelm global environmental limits was all too real a prospect, for which the most important remedy was population control. Proponents of this kind of logic often find themselves labelled as neo-Malthusians.

Despite their enduring popularity, the ideas of Malthus have been increasingly challenged. In particular, the idea that there is a simple relationship in which population growth induces adverse environmental change, and eventually ecological catastrophe, has come under considerable scrutiny (de Sherbinin et al., 2007). In part, this is because the assumption that the environment consists of finite stocks which humans will use up before they change the way they use them is not borne out by historical experience. The food production system in the twentieth century adapted in ways which met the provision of a rapidly growing population, at least in terms of producing sufficient food to feed everyone in the world. This does not mean that no-one went hungry in the twentieth century; nor has, in the twenty-first century, hunger been eradicated. But the reason for continued hunger is not 'famine' as a natural disaster in the Malthusian sense, a collapse in the food production system. It is a question of the inequitable distribution of available food resources, of some people not having access to food, as the work of Drèze and Sen (1991) has clearly demonstrated. This is no less catastrophic than famine induced by ecological crisis, but it is a social, political and economic problem at heart.

One of the most influential counterarguments to Malthus, in terms of explaining why it is that human population growth does not inevitably outstrip food production, is associated with Esther Boserup. Sometimes known as the Boserupian hypothesis, the idea is that growth in human populations induces intensification of agricultural production, through greater labour and capital inputs (Boserup, 1965). The principal difference between Malthus and Boserup was not so much about whether the intensification of agricultural production was possible (Malthus recognised intensification as a common

but insufficient response, whilst Boserup did not assume intensification would always occur), but about whether it would be sufficient to
feed a growing population. Other commentators, drawing on the
work of Boserup and sometimes known as Cornucopians or population optimists, stand in starker opposition to Malthus, maintaining
instead that human ingenuity and technological innovation will
always overcome 'the force of crude numbers' (Robbins, 2012:
16). Cornucopians argue that with increased population comes a
greater supply of creative people who will think of new technological
advances to intensify production or ways in which better to marshal
existing technologies. This capacity, in combination with market
substitution – that is, finding a replacement for a resource because its
scarcity drives its price up – will stave off crisis (Simon and Kahn,
1984). Taking this line of argument to its logical conclusion, the
corollary is that seemingly finite resources become infinitely available through a combination of reduced demand and more efficient
use. Arguably, this is as problematic as the assumption that efforts to
intensify food production can never keep pace with population
growth. Nevertheless, such thinking demonstrates that 'resources are
constructed rather than given' (Robbins 2012: 17).

Figure 7.1 *A comparison of Malthusian and Boserupian thinking on the implications of population growth for food production*

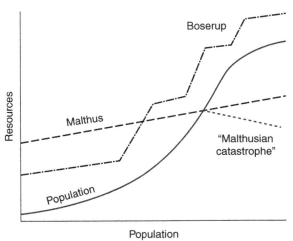

The evolution of thinking on the population–environment nexus

Thankfully, the brunt of work done on the population–environment nexus has moved considerably beyond the tendency to 'reduce environmental change to a function of population size or growth' (de Sherbinin et al., 2007: 2), or conversely to assume the relationship to be sufficiently elastic as not to prove a cause for concern. De Sherbinin et al. (2014, 2007) argue that monocausal explanations like these are simplistic, do not actually explain as much as they claim to and can sometimes be plain wrong. Demographers have become less ambitious in their claims and more nuanced in their questions, which broadly revolve around the following research agenda:

> *How do specific population changes (in density, composition, or numbers) relate to specific changes in the environment (such as deforestation, climate change or ambient concentrations of air and water pollutants)? How do environmental conditions and changes, in turn, affect population dynamics? How do intervening variables, such as institutions and markets, mediate the relationship and how do these relationships vary in time and space?*
> (de Sherbinin et al., 2007: 2)

The key idea here is the interaction between a range of variables – population is definitely in the mix, but it is only through understanding it in relation to other factors that its importance to processes of environmental change can be discerned.

Crucially, furthermore, it is now much more widely recognised and accepted that the consumption levels of a given population can be as or more important, in terms of adverse environmental impacts, as the size of that population. Perhaps just as importantly, the deeply political character of resource distribution and differentiated resource use between populations has received much more attention, raising fundamental questions about our assumptions concerning not just the effects of population growth on the environment but turning the spotlight more clearly on who is ultimately responsible for these effects. The issue of consumption is considered in the following section; subsequently, we turn to environmental impacts resulting from or associated with population growth that are most relevant to biodiversity conservation, in particular, soil erosion and deforestation. These impacts have often been seen as fundamental to conservation and development processes, because, on the one hand, of the implications for the degradation or destruction of ecosystem services/functions and biodiversity, and on the other because poverty is often considered to be a core driver. However, unless we understand this relationship properly, it will not be possible to target our efforts towards conservation effectively, with potentially adverse implications for poor populations.

Disentangling the effects of population growth and consumption

Population, consumption and environment in China

A focus on consumption is revealing precisely because it reminds us that not all populations are the same; nor do all populations, or subgroups within populations, behave in ways with similar implications for environmental change. In short, some people consume more material resources than other people – be it food, petrol, gas, clothing, etc. – whilst some people are more exposed to environmental problems resulting, to a greater or lesser extent, from consumption. For these reasons, it is important to look beyond average population trends if we want to understand both the causes and the consequences of forms of environmental change linked to consumption (and, of course, other variables). At the same time, if a greater portion of a given population consumes more material resources, the implications for environmental change can be correspondingly greater. Perhaps the most significant example of these dynamics in the twenty-first-century so far is China, which since the late 1970s has experienced historically unprecedented economic growth. This is in part due to the exponential increase in China's exports of consumable goods, and in part due to the expansion of its own domestic economy and markets. For this reason, this section focuses solely on China in order to illustrate the interplay between population, consumption and the distribution of environmental externalities.

China has long been a populous country. For much of the twentieth-century, it was also characterised by rapid population growth: according to the World Bank (World Bank, 2011), population levels rose from approximately 600 million people in 1949 to 1.3 billion people by the twenty-first century. By 2030, the population is projected to reach almost 1.5 billion people (ibid.). Under Chairman Mao, the

Communist Party positively encouraged population growth, not least in order to ensure the availability of troops for a large army (Shapiro, 2012). It was only with the reforms of the late 1970s that concerns about the difficulties of supporting such a large population were reflected in national policy. It was at this point that the famously controversial one-child policy, prohibiting families from having more than one child, was introduced; but by then, China was locked in to a population growth pathway that is still playing out today. However socially unjust – a preference for boys led to a rise in abortions of female foetuses and killings of newborn female babies, thus skewing the gender balance towards males – the one child policy has been very effective in terms of its own objectives. China's fertility rate is now low, compared with the global average, population is projected to stabilise by the 2030s and even to decrease after that (World Bank, 2011). This has led the government to relax restrictions on childbirth, for fear of a scenario in which an ageing population is overwhelmingly large for younger generations to support (Shapiro, 2012).

Predictably, some lines of analysis have directly linked environmental problems to China's rapid population growth. Perhaps the best-known and most influential work in this line is Qu and Li's *Population and the Environment in China* (Qu and Li, 1994). This book took the view that population growth was at the root of almost all of China's environmental woes, including land, water, minerals and energy scarcity, in addition to forest and grassland degradation. Certainly, population growth has been heavily implicated in recent concerns about China's ability to feed its people owing to environmental degradation. The almost fivefold increase in China's agricultural productivity throughout the 1950s and 1990s (You et al., 2011) brought about what has been called 'the greatest increase in economic wellbeing within a 15-year period in all of human history' (Sachs et al., 1994: 102). It led, effectively, to the use of 7% of the world's land mass to feed a little over a fifth of its human population (You et al., 2011). However, concerns have been raised about how sustainable this level of production is: for instance, between 1998 and 2003, China's grain production fell from 434 million tons to 363 million tons (Liu et al., 2010). Whilst it subsequently rose to 440 million tons in 2008, the decrease was attributed to environmental stress by some commentators (ibid.), in the form of soil degradation, water scarcity and environmental pollution. You et al. (2011) argue that, through use of panel data and geographical Information Systems (GIS), a correlation exists between grain-producing provinces in the north and east and higher levels of environmental degradation, characterised as severe in existing grain producing areas and projected to become a problem in the new grain producing areas. Extrapolating from current trends, Ye and Van Ranst (Ye and van Ranst, 2009) project a 9% loss in food crops productivity by 2030 if soil degradation were to continue at the current rate, under a business as usual scenario (i.e. without any intervention to address the issue). They also project an increase in productivity losses of 30% by 2050 under conditions of a doubling in the rate of soil degradation. They conclude that without policy and technical change now, China will experience food insecurity by 2030, with knock-on effects for the world's food systems, given the importance of China as an exporter of grains.

However, as Judith Shapiro succinctly puts it, 'sheer numbers of people may not be as important as their consumption and production capabilities' (2012: 37). Growing prosperity for an increasing proportion of the Chinese population has been accompanied by more consumptive, energy intensive lifestyles, characterised by the purchase of larger homes and vehicles, a switch from grain to meat centred diets. In 2010, the Chinese car market was the world's largest, with 17.2 million cars and commercial vehicles sold domestically; 11.6 million were sold in the USA (the second largest market), by comparison (J.D. Power & Associates, 2011). The market for luxury goods in China in the same year was worth US$10 billion and projected to account for 20% of global demand by 2015 (Atsmon et al., 2011). At least for the middle classes in China, changes in consumptive behaviour are driving concerns that China's ecological footprint is becoming too large in per capita terms. WWF calculated that in 2012, China's ecological footprint was 2.1gha (global hectares) per capita (Gaodi et al., 2012). This is 20% less than the global average, but still above 1.8 gha per capita, which WWF considers to be the global bio-capacity available to support human development.
The idea of global bio capacity is in many ways an update on the 'limits to growth' thesis. In this and other chapters (especially Chapter 2) we have seen that, whilst everyone can agree that at some point, it is possible to reach biophysical limits, the continued uncertainty around where these are and how elastic they may be means that it is prudent to treat these figures with caution. Nevertheless, the direction of travel is towards substantially more consumption in China, as indeed it is globally, and this may well make it likelier that such limits will be encountered at some point.

Chinese carbon emissions

Another indicator which would appear to suggest that China's vertiginous economic growth in recent decades has produced environmentally irresponsible behaviour is the fact that it has been since the early 2000s the largest emitter of greenhouse gases in the world (see Figure 7.2). Increasingly, however, it has been argued that it is far too simplistic to lay the blame for all of these emissions at China's door.

There are a number of issues to take into consideration here, the first of which being that even if China does have the highest gross emissions, this is because it has a population in excess of 1.3 billion people, and therefore to compare it (as Figure 7.2 does) with other countries – or even regions – which are much less populous is misleading. It is increasingly agreed that a fairer way of calculating carbon emissions is on a per capita basis, that is, per person in the country. This calculation shows that the consumption levels behind carbon emissions are much lower in China than they are in the United States (6.2 metric tons in China as opposed to 17.6 metric tons in the USA, in 2010) and remain lower than those of the

Figure 7.2 *China's overall CO_{2e} emissions 1990–2009, in comparison with other major economic powers*

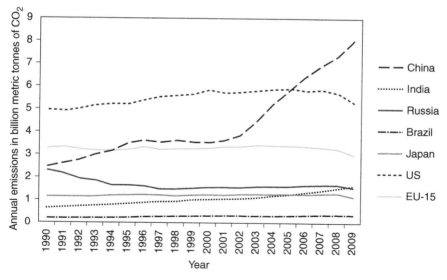

Source: www.alcs.ch/tag/china-carbon-emissions (accessed 1 August 2014)

European Union – see Figure 7.3 for a breakdown. Second, much of
what is bought and sold in the global economy is made in China, and
there is a distinct sense in which this permits the countries which buy
Chinese made goods to 'outsource' their carbon emissions to China as
the site of production, but frequently not consumption, of such goods
(Guan et al., 2008). Blaming China for these emissions is to leave
unexamined the role of countries and consumers around the world in
the generation of the carbon emissions linked to these goods. Taking it
into account helps us see that it is particular forms of consumption by
particular populations, rather than the size of those populations in
themselves, that underlie the biggest global environmental problems.
The implication for conservation efforts is that consumption, rather than
the size of poor populations, may be what we need to tackle in order to
reduce environmental degradation. From this perspective, a focus on
population growth in poor parts of the world as a driver of degradation
may be misdirected, or at least, missing a very important part of the
story. It raises the question of whether there is a mismatch underlying
drivers of environmental degradation and the focus of efforts to achieve

Figure 7.3 *Per capita CO$_{2e}$ emissions in metric tons 1960–2010, for China, the EU, least
developed countries composite, the UK and the USA*

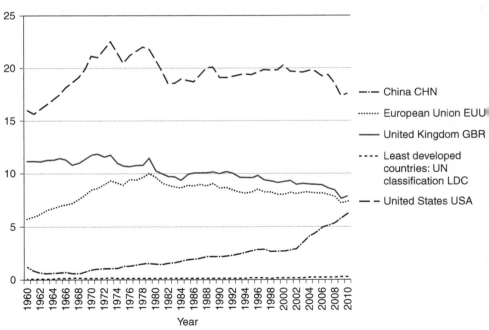

Source: Created by A. Newsham using World Bank Data

conservation and development simultaneously, around protected areas and community conservation (see chapters 9 and 10).

If responsibility for the carbon emissions is to be assigned, then it will be necessary to do so on the basis of the more nuanced and complex picture painted by the per capita breakdown. Even bearing these considerations in mind, though, it remains the case that carbon emissions are increasing in China on a per capita basis precisely because of higher levels of more energy intensive consumption behaviours. Guan et al. (2008) argue that even though there is a lot of scope for emissions reductions through efficiency improvements in China's economy (for instance less carbon intensive energy generation technologies), these would not be enough to offset increases in carbon emissions if Chinese consumption patterns converged to those of current US levels. If the majority of developing countries follow China's path and increase their own consumption patterns to a similar extent, it is hard to see how current development trajectories can be sustainable in terms of the impact that they have on the environment.

Yet even breaking down carbon emissions on a per capita basis obscures important differences in consumption and production dynamics within the overall Chinese population, with important implications for environmental justice. Feng et al. (2013) highlight the highly variable range in the levels of carbon emissions that exists between Chinese provinces. They estimate that 57% of China's emissions derive from goods consumed outside the province of origin. Their research suggests that as much as 80% of the emissions related to goods that richer, coastal provinces consume are imported from poorer provinces in central and western China, where there is a high concentration of low-value-added but carbon-intensive goods produced. Given that China is currently setting targets for the reduction of carbon intensity in its economy, richer provinces currently enjoy an advantage over poorer ones in the ease with which they can meet these targets, precisely through having 'outsourced' their industrial production. This phenomenon is reflective of a wider trend of relocating industrial production from China's rich eastern seaboard to its relatively poor central and western provinces. The result is that poor people in China bear the brunt of the huge range of environmental problems that have been caused by China's rapid industrial progress (see Shapiro, 2012; and Geall, 2013 for wide-ranging accounts of environmental problems related to industrial processes in China). In addition, their provinces may be expected to make the greatest reductions in carbon emissions. Outside of their work in carbon-intensive factories, however, such people may well have a very low 'carbon footprint'.

Taking into account of all of these complex dynamics demonstrates vividly the dangers of reducing processes of environmental change to a function of population growth. Instead, what we see is a situation in which economic growth comes at the expense of environmental conservation, and that the development pathway linked to this form of economic growth distributes both prosperity and exposure to environmental externalities very unevenly in the Chinese case. In the twenty-first century, this is both one of the most fundamental characteristics of, and greatest challenges for, efforts to achieve conservation and development simultaneously. It is in one sense, then, astonishing that people working, thinking and writing on the relationship between conservation and development do not currently appear to be looking at this relationship within this kind of context. Rather, the focus is on rural poverty and biodiversity conventionally valued by conservationists (Adams, 2013). This point is elaborated in some detail in the concluding chapter of the book. Yet these are the kinds of issues, processes and dynamics that are urgently in need of greater attention if we are to have a full understanding of the prospects for reconciling conservation and development objectives.

Population, poverty and environmental degradation

Population, environment and soil degradation/erosion

A highly contested arena of debate within population and environment scholarship is the posited link between increasing population in areas in which subsistence agriculture is practised widely and soil degradation/erosion (see Box 7.1 for definitions). One mode of exploring this link has been through work related to the vicious circle model (VCM) (i.e. Marcoux, 1999). The model posits that poverty results in high fertility because of requirements for farm labour, practices such as 'insurance births' to compensate for high mortality rates in infants in poor families, as well as an absence of contraceptive rights for poor women

Higher fertility drives population growth, which in turn stimulates demand for food, placing greater pressure on a limited resource base. The consequence is a loss of soil fertility, leading to declining yields and poor environmental sanitation, all of which reinforce the persistence of poverty and lead to long-term degradation in the absence of intervention to reverse the cycle. In development policy and practice, this logic has been profoundly influential, often running far beyond the available evidence base (Keeley and Scoones, 2003). Indeed, this influence is all

Figure 7.4 *An extreme example of soil erosion within a river catchment*

Source: Dr Andrew Brooks

the more remarkable given that there is only qualified support for the VCM hypothesis in the small number of existing qualitative studies that have been conducted to test it (de Sherbinin et al., 2007, 2014).

Soil fertility has also been extensively researched by political ecologists who, whilst also claiming a link between poverty and environmental degradation, seek to move beyond blaming poor farmers directly for issues such as soil erosion. Instead, they implicate state development intervention and/or increasing connections to regional and global markets in processes of degradation. Cash crops in West Africa, in particular cotton, are often promoted by states and donor agencies as a means for achieving poverty reduction. At the same time, cotton, in part because of its high nutrient requirements, leads growers to employ inorganic fertilisers and pesticides and in some cases to set aside the production of subsistence crops which are better adapted to local conditions and which can be grown without such intensive recourse to environmentally harmful agricultural inputs (Blaikie, 1985; Moseley, 2005). This can leave farmers with a double bind. On the one hand, they are exposed to considerable volatility in international market prices for commodities such as

---

Box 7.1

Soil degradation and soil erosion – definitions and differences

It is quite tempting to use soil erosion and soil degradation interchangeably, because they sound so similar. However, there are subtle but important differences between them. Soil erosion is often taken to refer to soil degradation more widely, but technically, it 'refers only to absolute soil losses in terms of topsoil and nutrients'. Soil degradation, on the other hand, can be defined as 'a change in the soil health status resulting in a diminished capacity of the ecosystem' (FAO, n.d.).

Given the focus of this definition on beneficiaries, an important caveat is that 'for a hunter or herder, the replacement of forest by Savannah with a greater capacity to carry ruminants would not be considered degradation. Nor would forest replacement by agricultural land be seen as degradation by a colonising farmer' (Blaikie and Brookfield, 1987: 4). In other words, degradation is in the eye of the beholder, and often involves the exchange of one capacity for another, depending on the perspective of the person perceiving/commenting on the change.

---

cotton, whilst on the other, they are growing a crop which is more vulnerable to climate impacts (Eakin, 2005; Leichenko and O'Brien, 2008). Worst of all, disguised in the value flowing from the export of the cotton is the cost of the soil nutrients which are 'spirited away from local producers', whilst the significant profits are made by the cotton traders and shirt manufacturers, little of which is passed on to the farmer (Robbins, 2012: 161). In this way, the workings of global capital accumulation are held to induce both poverty and environmental degradation

Interestingly, research in Mali attempting to trace this process empirically has drawn inconclusive results. The work of William Moseley (2005), charting the sharp fall in the productivity of cotton yields between 1990 and the early 2000s appears to point to erosion resulting from ever more intensive cultivation and shorter periods of resting the land. Moseley's soil analysis found that land worked by poorer producers – often the ones most heavily blamed for soil erosion – had similar soil quality measures to that of wealthier producers. Nevertheless, another analysis of Mali and cotton zone soils across the country concluded that fallowed or rested land was not in better condition than heavily and intensely cultivated cotton soils (Benjaminsen et al. 2010). From this they inferred (although no analysis of this inference has been done) that the cotton yields decline was a consequence of extended periods are cultivating marginal lands, in combination with limited farm labour availability. Other research from Kenya and West Africa posits that at least in some cases, population growth can also lead to the improvement of the land, even in poor countries, although these findings are contested (see Box 7.1).

Box 7.2

Population growth and land improvement in Kenya and West Africa?

An iconic study which sought to test Boserup's hypothesis that population growth would result in agricultural intensification is Tiffen et al.'s *More People, Less Erosion* (1994), which looked at farming practice between 1930 and 1990 in the Machakos district of Kenya. During that period population increased from 240,000 to 1.4 million people. In the 1930s, this mountainous and semi-arid region was characterised by significant soil erosion, isolated from national markets and facing restrictions on access to land and crops. During the 1950s and 60s farmers adopted terracing techniques, moved away from livestock and towards more intensive farming of high-value crops, through a mix of local innovation and state assistance. At the same time, feeder roads increased access to market towns, with the result that by 1990, in per capita terms, the value of agricultural production had doubled. Although food aid was also available in times of drought, the changes in agricultural practice towards intensification were held to have allowed people to escape from a Malthusian scenario of inevitable famine. At the same time, soil erosion decreased and levels of soil fertility rose. One of the study's authors, Michael Mortimore (2005), went on to document other 'success stories' from Nigeria, Senegal and Niger, which he defined in terms of improved ecosystem management, land investments, productivity and personal incomes (albeit that in none of these cases were these indicators positive across the board).

However, these findings – and especially those from Kenya – have been contested. Rocheleau (1995) argued that beneath the average figures presented by Tiffen et al. (1994), the benefits of the changes were distributed quite unevenly across Machakos residents, some of whom were more vulnerable to land alienation and state imposed restrictions on mobility. In effect, some of the reduction in soil erosion could have been explained not just by technical innovation, but also because some farmers were finding themselves with no choice but to give up farming and look for livelihood options elsewhere (ibid.). Moreover, in a study encompassing Burkina Faso, Ghana, Nigeria, Senegal, Tanzania and Uganda, Boyd and Slaymaker (2000) found little evidence of the success story of the Machakos district, raising the question as to how widespread a phenomenon it was, and to what extent it was dependent upon very local and specific conditions. Nevertheless, one of the pivotal achievements of the Tiffen et al. (1994) research was to bring empirical coverage of local agency and capacity, and a much finer level of nuance to the debate around the extent to which poor people are likely to degrade the environment.

Population, environment and conflict

Population, 'environmental scarcity' and conflict

Just as questions of poverty and population growth have been linked to environmental degradation, so too they have been linked to conflict. Since the 1990s, a body of rather disparate scholars, clustered loosely under the banner of environmental security, have developed differing, often competing, accounts of the relationship between poverty, population, environment and conflict. The research which kick-started this agenda tends to be associated principally with a research group based at the University of Toronto and led by Thomas Homer-Dixon, but also with researchers based in Switzerland clustered around the work of Günther Baechler (1999). Both groups were interested, essentially, in the extent to which environmental and demographic change contribute to violent conflict. Both started, broadly speaking, from the premise that human pressure on the environment could adversely affect societies in ways which heightened the risk of conflict (Deligiannis, 2013).

This starting point lead Homer-Dixon and colleagues to hypothesise that in particular, what he termed "environmental scarcity" (Homer-Dixon, 1999) could be amongst the causes of conflict. Using a qualitative case study approach, in which conflict featured as the dependent variable, environmental scarcity was taken as the independent variable whose effect was to be measured. Environmental scarcity was deemed to be present in contexts in which there was evidence of change in:

1   the level of degradation or depletion of an environmental resource;
2   the level of resource demand owing either to population growth, consumption or both;
3   levels of access to the resource, often in favour of one group to the exclusion of others (Homer-Dixon 1999).

Research from the Toronto Group found no evidence that environmental and demographic scarcity was linked to inter-state conflict. Instead, they suggested that environmental scarcity acted upon a range of intermediate factors which more directly influenced the likelihood of violent conflict, such as tensions and divisions between different groups, institutional fragility and state incapacity to deliver public goods. Homer-Dixon maintained that environmental scarcity was never the lone cause of conflict, but its effects could be seen through its interaction with a broad variety of contextual factors at different levels and scales. Specifically, environmental scarcity could be seen as a cause of two key processes which could in themselves engender violent conflict:

1   resource capture – in which environmental degradation, interacting with population growth, leads powerful groups to consolidate their own control over and access to an increasingly scarce resource (Scwartz et al., 2001);
2   ecological marginalisation – when unequal access to a resource, in conjunction with population growth, leads people to migrate to regions of ecological fragility, i.e. steep upland slopes, public urban lands exposed to flooding, etc. (ibid.).

However, this initially prolific line of research largely evaporated at the start of the twenty-first century, at least partly in the face of critiques which, arguably, misrepresented as much as problematised the concept of environmental scarcity and its causal propensities in relation to conflict. A common but perhaps unfair critique made of the Toronto school, and in particular Thomas Homer-Dixon, is that it is too simplistic to make a causal attribution between environmental scarcity and conflict. In offering this mono-causal explanation, these scholars lapse into precisely the kind of specious neo-Malthusianism that environment-population studies have been trying to move away from for decades. Whilst there is something distinctly neo-Malthusian about the focus on scarcity, and therefore some legitimacy in the use of this term in relation to the work of Homer-Dixon, it does not follow logically that he or colleagues working on the same agenda are positing a mono-causal model of the kind often associated with neo-Malthusianism. Even a cursory perusal of Homer-Dixon's core text in this area, *Environment, Scarcity, and Violence* (1999) – or more simply still, a visit to his website[2] is sufficient to show that they do not posit that environmental scarcity alone can cause conflict.

More nuanced and better-targeted critiques have been made, however. Some of these focus on the conceptual and methodological implications of bringing together resource degradation, population growth and access to resources in one composite variable. For instance, Gleditsch and Urdal (2002) suggest that unequal distribution of resources, and the inequalities that surround this, are much more important, causally, than degradation or population growth. If distributional dynamics are excluded from the independent variable, then there is little evidence to link supply and demand scarcities, a core focus in Homer-Dixon's work, to conflict. A broader variant of this critique is offered by political ecologists. Their argument is that Homer-Dixon and colleagues focus too heavily on the composite variable of environmental scarcity, and thus draw attention away from a cluster of underlying factors which are much more important in explaining the variegated causes of conflict. Peluso and Watts (Peluso and Watts, 2001) point to culture, political economy and power – the structural determinants of inequality – as processes with much greater causal significance. In other words, Homer-Dixon is too fixated with scarcity produced and the immediate, proximate factors relating to this. As a result, he undertheorises the "mechanisms of appropriation and exclusion from access" which create scarcity in the first place. Whilst these dynamics are covered, arguably, within Homer-Dixon's theoretical framework under the heading of 'ideational factors', they are much less richly conceptualised or treated empirically than they are in the field of political ecology.

Taking a different tack, other researchers have argued that resource scarcity is in fact a red herring as a way to explain conflict; it is, rather, the abundance of resources – particularly non-renewable ones, such as oil, diamonds and coal with high export values – which is much more heavily implicated in generating conflict (le Billon, 2005; Collier, 2007; de Soysa, 2013). Sometimes referred to as the resource curse thesis, conflict arises between different groups each wanting to keep for themselves the benefits (principally financial) of the proceed from exports of these abundant resources, either to prop up or over-throw an existing regime (Sachs and Warner, 2000). Another way in which this kind of conflict is conceived is as driven by greed, that is, when actors perceive that the benefits of resource capture outweigh the costs of conflict. This is in contrast with, though not unrelated to, conflict motivated by grievance, that is, when people are moved to engage in conflict because of a real or perceived state of deprivation (Collier, 2007). Diamonds in Angola and Sierra Leone (le Billon, 2005; Mamdani, 2002; Samset, 2002), are some of the more well-studied examples of the resource curse. A more recent example might be taken from Syria and Iraq, where the attempts by an armed group known as the Islamic State of Iraq and the Levant (ISIS) to establish new Islamic caliphate through taking lands by force from both of these countries, has been funded in part by gaining control of oil resources in both states. Indeed, by some estimates, by June 2014, (ISIS) was considered to be 'worth' as much as US$2 billion[3].

Nevertheless, Tom Deligiannis (2013) has argued that some of these differences are overstated and at risk of creating false dichotomies. The political ecology approach, he contends, is not without its own shortcomings. First, he argues that political ecology frequently fails to account sufficiently for the impact of environmental processes in conflict and other situations. Second, the focus on the causal primacy of political-economic factors runs the risk of downplaying or under-analysing the importance of environmental factors or of population, as if these could be explained solely with reference to the underlying political economy, rather than to explicit analysis of their own characteristics and interactions in relation to the outbreak of conflict. Deligiannis takes Peluso and Watts to task for their assertion, 'to say that environmental scarcity can contribute to civil violence is to state the obvious' (Peluso and Watts, 2001: 27). For Deligiannis, the specific ways in which environmental scarcity can contribute to violence are neither obvious nor well understood, and the tendency of political ecologists to use very broad definitions of violence may make it even harder to understand these processes better. In respect of the scarcity/abundance question, he argues that both can coexist within specific contexts and frequently interact with each other in ways which generate or exacerbate conflict. This assessment of disputes in the field lead him to conclude that there is a great deal more room for consensus than is generally acknowledged and that, by implication, whilst factors such as population growth should not be overstated as causes of conflict – and especially not in their own right – they need to be taken into account as part of a more holistic analysis rooted in an understanding of coupled human–environmental systems.

In summary, it is probably fair to say that our understanding of the role of population in generating environmental scarcity, to the extent where this, in combination with other factors, triggers conflict, remains contested, even elusive. The implications of these debates for efforts to realise conservation and development remain, therefore, difficult to state conclusively. Nevertheless, we might suggest, relatively uncontroversially, that the conservation of non-renewable and renewable natural resources could contribute to the avoidance of conflict if the benefits deriving from their use and continued existence were distributed according to a more equitable development model. A larger (human) population has implications for the conservation of the natural resource base and the development prospects linked to it, but will not automatically lead to degradation or conflict – this is a proposition which, it turns out, ironically enough, all commentators in this some-what fractured debate can actually subscribe to.

This much can be surmised, then, by exploring the broader literature on environment, security and conflict. Little work to date has explicitly considered the relationship between conflict, conservation and development, let alone the potential role (or lack therein) of population within this relationship. However, it is starting to receive attention, and in ways which already demonstrate how useful it is for conservation and development practitioners and scholars to apply a broader conceptualisation of conflict to their own work. Indeed, the generation of conflict in relation to the international trade in illegal wildlife is becoming such a central theme in conservation that we return to it in the concluding chapter, where we survey important trends and issues for the future of conservation and development.

Summary

- Population–environment scholarship has moved on beyond simplistic, mono-causal analysis of the effects of population growth on environmental change. Demographers are now less ambitious in their claims about the role of population growth, and much better at linking it to other variables which intersect and also play important roles in environmental change. Not least among these is consumption, a consideration of which shifts the focus away from the role of poor people in environmental degradation towards the role of the affluent populations of the global North and the growing middle classes across the global South. Focusing just on population growth in poor contexts does not help us to understand the extent to which consumption is a fundamental driver of global environmental degradation. Our thinking about the relationship between conservation and development needs to reflect this if we are to achieve conservation and development objectives simultaneously and to target activities to address the causes of environmental degradation.

- Development trajectories based on unequal distribution of wealth are implicated both in the major challenges for environmental conservation of the twenty-first century and in the higher likelihood of poor people, in parts of the world such as China, being exposed to adverse effects of environmental degradation and contamination.

Conservation and development debates, however, with their rural focus on areas rich in the biodiversity conventionally valued by conservationists, simply fail to recognise the need for engagement in this kind of context.

- Poverty can, of course, drive environmental degradation, but this characterisation often gets in the way of paying sufficient attention to practices by poor people which actually conserve environments and valued biodiversity.

- The relationship between population, environment and conflict is highly contested and the research on it somewhat inconclusive. There is consensus around the proposition that whilst population growth and environmental degradation can contribute to conflict, they do so no more than, and frequently less than, a number of other variables and processes related to greed and grievance.

Discussion questions

1. What is the relationship between population growth and environmental degradation?

2. If the consumption patterns of specific populations are heavily implicated in global environmental degradation, to what extent do our efforts to reconcile conservation and development address this phenomenon?

3. How useful would it be to focus conservation and development efforts towards conflicts which are linked to control over environmental resources?

Notes

1. The (total) fertility rate refers to the number of children a woman would give birth to if she did not die before the end of child-bearing years and bore the number of children typical for the age-specific fertility rates of her area or country (UNDESA, 2013).

2. On the 'Research' tab of his personal webpage, Homer-Dixon describes his book, *Environment, Scarcity, and Violence*, in the following terms: "This book explained that environmental stress by itself does not cause violence. It must combine with other factors,

such as the failure of economic institutions and government".
http://www.homerdixon.com/research/

3. http://www.theguardian.com/world/2014/jun/16/terrifying-rise-of-isis-iraq-executions

Further reading

de Sherbinin et al. (2007) Population and environment. *Annual Review of Environmental Resources* 32: 345–373.

Douglas, L.R. and Alie, K. (2014) High-value natural resources: linking wildlife conservation to international conflict, insecurity, and development concerns. *Biological Conservation* 171: 270–277.

Duffy, R. (2014) Waging a war to save biodiversity: the rise of militarized conservation. *International Affairs* 90: 819–34. doi:10.1111/1468-2346.12142

Floyd, R., and Matthew, R.A. eds. (2013) *Environmental Security: approaches and issues*. Abingdon: Routledge.

Shapiro, Judith (2012) *China's Environmental Challenges*. Cambridge: Polity.

Robins, Paul (2012) *Political Ecology*. Malden, MA; Oxford: Wiley-Blackwell

Tiffen, M., Mortimore, M., and Gichuki, F. (1994) *More People, Less Erosion: environmental recovery in Kenya*. Chichester: J. Wiley.

Online resources

United Nations Department of economic and social affairs: population division
http://www.un.org/en/development/desa/population/

World Bank world development indicators (carbon emissions per capita): http://data.worldbank.org/indicator/EN.ATM.CO2E.PC

Centre for International Earth science information network: http://www.ciesin.columbia.edu/

Thomas Homer-Dixon personal website (good for sources on the Toronto school work on environmental scarcity): http://www.homerdixon.com/research/

just conservation (for blogs on conservation, justice and issues of community rights): http://www.justconservation.org/

8 Management of natural resources in a globalised world

- Globalisation of land and sea
- Localisation of resource management

Introduction

The quest for food, fuel and shelter throughout human history has led to the extraction and management of nature in the form of hunting, gathering, agriculture, pastoralism, fisheries and forestry. Innovation in the use of natural resources was a key factor in the success of human societies. When human societies were smaller, they practised these activities at a subsistence level, exploiting the abundant natural resources. But when the human population increased, these natural resources came under increasing pressure. One recent and major change that has affected the scale of exploitation of these natural resources is globalisation. As we have seen in Chapter 7, population growth alone does not lead to the unsustainable exploitation of natural resources; but when population growth is combined with increasingly global consumption, it can lead to a sharp depletion of natural resources. This chapter explores the ramifications of this powerful combination. With globalisation, the exploitation of these resources has become possible far beyond the areas where they are geographically. As a result, the entire earth system has started to feel the pressure from resource exploitation. This chapter traces the effects of globalisation in the context of agriculture, fisheries, wildlife and forestry, and considers the consequences for the ecosystems that feed development but require conservation.

Globalisation of land and sea

Globalisation promotes an economic system in which raw materials and manufactured goods flow freely across international borders

Figure 8.1 *Mechanised farming in Kerala, India*

Source: Photographer Jayanunni; used under Creative Commons licence

under the supervision only of an international trade authority
(Ehrenfield, 2003). Globalisation has consequences for natural
resources because many of the raw materials and goods that are transported across international borders are derived from natural resources.
On the one hand, globalisation of these resources has generated
wealth, providing new livelihoods and opportunities for new enterprises that would not have been possible without globalisation. But
on the other hand, globalisation has also caused degradation of
resources, often due to their overexploitation or unsustainable use by
the global consumer. Expansion of industrial agriculture in many
cases has led to the loss of natural forests while expansion of indus-
trial-scale fisheries has led to degradation of ocean ecosystems. Similarly, the expanding agricultural frontier has pushed back against
tropical forests, which have also come under pressure from logging
for timber. Similarly, expansion of fisheries has altered the ocean ecosystem including coastal and estuarine habitats and coral reefs around
the world. Commercial forestry has caused reduction in natural forests and wildlife trade has endangered several species. The speed of
changes in natural ecosystems and their negative effect on species
which inhabit those ecosystems has increased since the end of the
Second World War and the beginning of globalisation. This chapter

seeks to examine the use of land and sea and the influence of globalisation on it. We cover agriculture, fisheries, forestry and wildlife trade as key areas of natural resource exploitation and examine the benefits and drawbacks of globalisation.

Agriculture

With more than a third of the world's land area under some form of cultivation, agriculture plays an important role in providing livelihoods to a large proportion of the world's population. In some African countries, over 80% of the population is composed of smallholder farmers, many of them women (IAASTD, 2009). Growing world population also demands enhanced food production and the post-Second World War 'green revolution' has brought about dramatic increases in agricultural yields. Globalisation of agriculture has also brought with it technological advances in storage and distribution of commodities. While agrochemicals have improved yields, refrigeration has improved the ability to store perishable products and air transport has enabled their rapid distribution across the globe. These processes have led to specialisation in agriculture and facilitated economies of scale in the production, storage and distribution of agricultural commodities. This has also enabled the reduction in cost of many agricultural products and rapid growth in their trade.

In many less developed countries, globalisation of agriculture enhanced economic growth because export-oriented agriculture grew considerably faster than domestic consumption between 1960s and 1980s (Mellor, 1992). For example, the cut-flower industry in Kenya has contributed to Kenya's economic growth while in other countries industrialised agriculture has enhanced the food security of a large majority of the rural poor whose subsistence farming had failed because of impoverished soil or lack of irrigation. It has been suggested therefore that the globalisation of agriculture makes food production more efficient because countries that are deficient in the production of certain types of food will have their requirements fulfilled by other countries – meaning that countries can invest in agricultural products that they are best equipped to grow. Secondly, it has been suggested that the globalisation of agriculture will lead to increased food production, an enhanced standard of living and improved economic conditions for farmers throughout the world because of participation of the farming sector in global trade in commodities. Thirdly, supporters of the globalisation of agriculture suggest that with increased financial investment in agriculture, the low productivity of some of the small farmers will be eliminated and therefore even the poor farmers, who have been deprived thus far will gain from the globalisation.

The opponents of the globalisation of agriculture, however, raise a number of counter-arguments as to why small farmers stand to lose from the industrialisation of food production. The global trade of commodities is often driven and controlled by large corporations, which means that individual farmers have little power in deciding the nature of this trade. Large corporations are profit-oriented and when markets are not particularly buoyant, these corporations tend to squeeze profit margins from small farmers in order to reach their corporate profit targets. In globalised markets, large corporations hold enormous power and determine prices. For example, in 2006 PepsiCo, India contracted farmers in Punjab state to grow tomatoes. Although the market price for tomatoes at that time was Rs 2.00 (USD 0.03) per kg, the farmers received only Rs. 0.75 (USD 0.01) per kg. When the farmers protested, PepsiCo abandoned Punjab state and moved somewhere else meaning that the Punjab farmers lost all the revenue from their crop (Singh, 2002). Furthermore, investment in farming from large corporations has changed farming practices substantially. Previously, farmers had the power to decide which crops to plant and how to plant them, but with corporate interests in agriculture, the corporations prescribe farming practices including the seed varieties and agrochemicals. In many cases, the focus is on increasing yield with disregard to the natural fertility of soil. In the long term, therefore, farming becomes entirely reliant on agrochemicals, forcing farmers to put all their profits into purchasing expensive seeds or large quantities of agrochemicals, squeezing their profits even further. Despite the noble intention of the globalisation of agriculture, its quest to achieve efficiency in farming and to improve the quality of life of poor farmers, experience has suggested that with globalisation agriculture becomes monopolised by large corporations leading to the reduction in the quality of life of farmers as well as the productivity of the soil.

Fisheries

The globalisation of fisheries has also had similar, often detrimental effects on ecosystems and communities who are dependent on small-scale fisheries for their livelihoods. With the emergence of global markets and the commercialisation of fisheries, marine resources are now extracted from hitherto inaccessible areas of the world's oceans. Simultaneously, the global demand for seafood has put pressure on coastal areas, particularly mangrove swamps and coral reefs. These demands have resulted in widespread ecological degradation of oceans (Roberts, 2009). Berkes et al. (2006) argue that the key problem lies in global markets, which fail to generate self-interest that arises from attachment to place. Commercial fisheries have therefore no interest or incentive to conserve. Also, the speed with which new markets develop in the globalised fisheries also means that resource exploitation overwhelms the ability of local institutions to respond. The ecological consequences of this overexploitation include the simplification of marine food webs and loss of biodiversity, which has a further impact on the resilience of marine ecosystems to climate change (Hughes et al., 2005). Campling and Havice (2014) argue from a Marxist perspective that the privatisation of fisheries is closely linked to the social struggle among some of the poorest people in the world dependent on small-scale fisheries for their livelihoods. In tropical countries, where more than 200 million people depend on small-scale fisheries for their livelihoods, exploitation of coastal resources is also on the rise due to the increasing demand for seafood. In these developing countries, export-oriented fisheries generate wealth and employment for some sections of the society, but also expose small-scale fisheries to fluctuating markets. Furthermore, the distribution of wealth generated by these markets might exclude small-scale fishers, with the bulk of the profit made by fish processors and other agents in the global supply chain. As the poor fishers get squeezed further by the globalisation of fisheries, conflicts between fleets also increase. These between-fleet conflicts normally occur where the fleets share the same fishing ground, which has limited fish stocks.

The problems of worsening economic situation of the fishers are compounded by poor functioning of formal and informal institutions that would otherwise be able to resolve such conflicts. Berkes et al. (2006) argue that to combat the harmful effects of globalisation of fisheries on marine and coastal ecosystems, institutions with broad authority and a global perspective are needed to create a system with incentives for conservation – a global solution to a global problem. In other quarters, however, there are calls for the 'localisation' of fisheries. For example, in May 2012, Scotland's Fisheries Secretary Richard Lochhead argued that fisheries management should be decentralised away from the European Union's (EU) Common Fisheries Policy (The Fraserburgh Herald, 2012). The rationale for localisation stems from the disappointment for many EU member states from undue interference from the EU Headquarters in Brussels, as a result of which the local fisheries in Scotland are known to have suffered disastrous consequences. Therefore, in the same fashion as agriculture, fisheries management is also contested in a globalisation vs localisation debate. The argument for globalisation would include the 'connected' nature of oceans meaning that the effects of overfishing in one part are felt in another, requiring solutions to problems at global scale. The argument for localisation would include the disenfranchisement of local fisheries when 'one-size-fits-all' policies are introduced at a higher level of governance.

Figure 8.2 *Essaouira fish market. The intensification of commercial fisheries to feed the growing demand for fish in the global marketplace has impacted on fish stocks*

Source: Photographer Donar Reiskoffer; used under Creative Commons licence

Forestry

During the first decade of the twenty-first century, FAO (2012) reported the loss of over 130 million hectares of forest globally, out of which 40 million hectares were primary forests (FAO 2012). This forest loss and degradation are estimated to cost the global economy between USD 2–4.5 trillion a year (TEEB 2010). The forest sector, which includes forest management, timber harvesting and processing represents a relatively small component of most national economies. At the global level, forestry contributes about 1% of GDP (FAO, 2012). Although these numbers are small in comparison with other sectors of economy, forestry operations have disproportionate and often dramatic impact on the natural environment. Furthermore, estimates suggest that at least ten million people across the world are employed in forest management and conservation (FAO, 2012) and an estimated 1 billion people depend on forests for subsistence, as an economic safety net or as a direct source of income (Scherr et al., 2004). Forests therefore play an important role in provision of ecosystem services as well as providing livelihoods to some of the poorest communities in the world.

Traditionally, forestry operations have relied on extractive methods in natural forests. This has led to large-scale deforestation and degradation of forests in addition to the substantial environmental cost of transporting timber from its origin to destination. This has also meant that forest processing locations are also where there are natural forests, causing an increase in human population around these hubs of activity and leading to the degradation of many forest ecosystems which are in serious need of conservation. The globalisation of forest industry over the last two decades has had negative as well as positive impacts. The negative impact pertains to increasing demand from expanding urban centres hungry for construction timber. This, combined with long-distance transportation of timber, has had detrimental effects on many tropical forests, particularly in Africa and South-east Asia; although some of the assertions about levels of deforestation in Africa have been shown to be problematic; a point to which we shall return. On the other hand, the globalisation of the timber industry has had beneficial effects on natural forests in that timber is now routinely grown in forestry plantations that are owned and managed by the private sector (Paquette and Messier, 2010). This has meant that natural forests are spared if timber can be grown in forestry plantations and processed and distributed with transport networks (Bael and Sedjo, 2006). The rising demand for timber and slow-growing nature of many hardwood timber trees, however, is likely to cause gaps in supply vs demand.

Furthermore, the timber industry has also suffered from monopolies in ways similar to agriculture and fisheries. Therefore, more local solutions are sought for. These local solutions would include production, distribution and use of timber at much smaller scales. In theory, such localisation can benefit communities, who might be able to participate more actively in forestry operations – plantation management, harvest and distribution. In practice, however, governance regimes in many tropical countries are corrupt meaning that when responsibility for forest management is transferred to local authorities, they see new opportunities to mine the forest and generate new revenues. As such, the localisation of the timber industry in these countries often has detrimental effects on the status of forests. Among the solutions proposed for addressing the local vs global management of forests, hybrid solutions are often seen to be successful. For instance, partnerships for co-management of forest resources and institutional structures that support such co-management, including international donors, government agencies, national and international NGOs and the private sector, are considered necessary for more effective forest management in developing countries (Ros-Tonen, 2003). These multi-scale and multi-stakeholder partnerships in forest management might link global conservation objectives with local needs whilst conserving forest resources.

Deforestation and biodiversity loss feature heavily also in the population–environment literature, but not necessarily in the way one might expect. This literature tends to complicate, rather than support, the simple notion that population growth leads to higher rates of deforestation. For instance, in much of the global South, whilst fertility rates are rapidly decreasing, especially in urban areas, it does not necessarily lead to a corresponding decrease in deforestation (de Sherbinin et al., 2007, 2014). One of the reasons for this unstraightforward relationship is a prominent concern for conservationists, because often, the highest fertility rates are to be found in populations living in remotely settled lands, in which the agricultural frontier continues to be extended. These areas are frequently biodiverse and/or feature sensitive ecologies, and at risk of being lost in the process of land conversion (Geist and Lambin, 2001, Rudel and

Horowitz, 2013). An even more important driver of deforestation, however, is migration into forest frontier zones, which often has a knock-on effect as younger members of established farm households migrate to new frontiers. Deforestation rates can also rise in rural areas where fertility rates are falling, such as the Ecuadorian Amazon region (Rudel and Horowitz, 2013). However, this is not all down to small agricultural land holders expanding the available area for agriculture; it is just as much, and possibly more, linked to demand in international markets for forest and agricultural products, which depend upon large-scale logging and agricultural activities (Richards et al., 2012).

At the same time, there are also instances in which population grows and deforestation rates fall (Carr, 2002), which demonstrates that great care needs to be taken in assessing the relationship between population and afforestation. Political ecologists have taken issue with the blame often heaped on poor people for deforestation, which in some cases has hidden a trend towards reforestation. In this regard, the work of Melissa Leach and James Fairhead (see Box 8.1), along with the broader anthropological, ecological and political ecology literatures that they draw upon, has been as seminal for forestry as the research of Tiffen et al. (reviewed in Chapter 7) has been for soil erosion. More broadly, political ecologists have sought to move beyond simple population explanations of deforestation, and to uncover the underlying political and economic processes which are at play.

Box 8.1

More people, more trees? Reframing deforestation in West Africa

West Africa has in recent decades been identified as a region characterised by extreme levels of deforestation with some accounts estimating a 90% decline in forest cover in the twentieth century (i.e. Bryant et al., 1997). This trend has been attributed largely to population growth, especially since the 1950s. At first sight, therefore, West Africa would not appear to be a place to go looking for examples in which farming practices lead to increases in forest cover. However, the work of James Fairhead and Melissa Leach has done precisely that, questioning in the process assumptions underlying some of the scientific studies on which claims of such high deforestation rates are made.

In their most widely cited works on the topic (Fairhead and Leach, 1998), these scholars make two fundamental claims about forest cover, population and settlement. First, they contend that some farming practices contribute to afforestation by modifying soils and related ecological processes. In forest–savannah transition zones across Ghana, Ivory Coast, Guinea and other West African countries, mounding, mulching, gardening and techniques related to fire protection have led savannah land to become populated with woody vegetation. These practices have also contributed to full-blown forest generation in previously cultivated savannah, once human population numbers have declined, including those contained in many forest reserves.

Second, the existence of 'forest islands' is actually a hallmark of settlement in forest and forest–savannah transition zones across West Africa. Previously, these patches have been interpreted as relics of undisturbed primary forest, but Fairhead and Leach maintain that they result from the presence, not the absence of humans. Therefore, in places where such practices are maintained, population growth could lead to afforestation, not deforestation. Elsewhere, Leach and Fairhead (2000) interrogate prevailing framings of deforestation in West Africa. They argue that these are over-simplistic in attributing deforestation rates principally to population growth, in particular from the 1950s onwards. Additionally, they take issue with the science on which accounts of deforestation have been based. For instance, studies done by organisations such as the Food and Agriculture Organization (FAO) and the World Conservation Monitoring Centre (WCMC) assume that at the start of the twentieth century, the forest zones of West African countries were fully forested and not inhabited. Using the accounts of travellers in the region, as well as estimates of forest cover made by colonial administrations early in the twentieth century, Leach and Fairhead provide a plausible counter evidence base for suggesting that the size of 'original' forest zones have been significantly overestimated. If they have, deforestation rates must also have been overestimated, and by up to two thirds, on the baseline that they put together. Whilst they do not dispute that significant amounts of deforestation have taken place, they argue that framing the debate in such distinctly neo-Malthusian terms 'has deterred consideration of other possible processes and relationships, including those by which people may be enhancing tree and forest cover' (Leach and Fairhead, 2000: 29).

Wildlife

The globalisation and urbanisation of the world along with increased levels of wealth have promoted increased levels of wildlife trade. For example, Nijman (2010) reports, based on data on international trade in CITES-listed animals in South-east Asia between 1998 and 2007, that a total of 35 million animals (0.3 million butterflies; 16.0 million seahorses; 0.1 million other fish; 17.4 million reptiles; 0.4 million mammals; 1.0 million birds) were exported during this period, out of which 30 million (c. 300 species) were wild-caught. In addition

18 million pieces and 2 million kg of live corals were exported. These data also suggest that Malaysia, Vietnam, Indonesia and China are the major exporters of wild-caught animals and the European Union and Japan are the most significant importers. Although a large majority of countries in the world, particularly developing countries are signatory to the Convention on International Trade in Endangered Species of wild fauna and flora (CITES), illegal wildlife trade is rampant. There-fore, the true scale of the international wildlife trade might be even greater than what data suggest. Similarly, concerns have been raised about the scale of bush meat hunting. Wilkie and Carpenter (1999), for example, estimated that over 1 million metric tons of bush meat is consumed annually in the Congo basin.

In the same region, estimates suggest that nearly 1000kg of bush meat is harvested per sq. km per year. As Fa et al.'s (2006) subsequent

Figure 8.3 *UK Border Force wildlife crime seizures. Illegal wildlife trade has been fuelled due to new demand for animal body parts in the emerging economies in the world along-side the developed Western nations*

Source: UK Home Office, www.flickr.com/photos/49956354@N04/8756269537 (accessed 8 June 2015); used under Creative Commons licence

estimates suggest that this figure might be nearer 5 million tons annually. Over 93% of bush meat harvested comes from ungulates, 63% from primates and carnivores and 11% from rodents, indicating unsustainable levels of harvest (Fa et al., 2006). It has been suggested that such a 'bush meat crisis' stems from land conversion to agriculture, logging concessions, population growth surrounding protected areas and increasing urbanisation (Poulsen et al., 2009). These factors often lead to creation of local markets for bush meat, which eventually begin to supply bushmeat to growing urban centres. Globalisation and urbanisation, therefore, can be considered as most prominent drivers of the loss of endangered species. Despite the expansion of global protected area network, the scale of wildlife trade and bush meat hunting might jeopardise some of the most threatened species in the world.

Localisation of resource management?

Globalisation and the spread of technology for development

Globalisation of resources poses important questions for conservation and development. For many developing countries, their natural resources are an important trump card in their development because these can be exchanged for investment in their countries from more developed nations. Emerging economies such as China, for example, have invested heavily in the commercial exploitation of natural resources in African countries. Such investments have brought to Africa more efficient technologies and provided employment to many Africans. Such investment, on the face of it, therefore, leads to economic development to the benefit of the rural poor. Some have argued that advanced technologies accompanied by technological innovation would also enable developing countries to bypass the development trajectory that many developed countries have taken. This phenomenon, described as 'leapfrogging', might allow more sustainable exploitation and use of resources (Munasinghe, 1999). For example, advanced technologies in the renewable energy sector might allow developing countries to grow their economies at a fraction of the energy use that would have been required to achieve economic growth at the same level in the past. It can be argued therefore that advanced technologies might allow more efficient use of resources and in effect their conservation. With respect to natural resources, this would mean more efficient use of land for agriculture or reduction in bycatch for the more efficient harvest of marine resources. This would also mean more sustainable exploitation of timber and wildlife as well as better technologies for tracking and monitoring. The key question is: do technological advances directly benefit the rural poor?

Although humans have bred plants and animals throughout the history of agriculture and pastoralism, genetically modified organisms go beyond this selection of traits for breeding and offer a more powerful technological solution. GM crops have been developed over the last two decades to enhance food production and this technology has been adopted by many developing countries. In developed countries, GM crops often cause an outcry due to a strong anti-GM lobby, who argues that there might be unintended health and environmental consequences of GM crops and therefore the precautionary principle should be observed. In the developing world, however, there is a genuine need for GM crops, provided they do indeed enhance food production.

While GM crops do help to increase the quantity of food produced, there are a number of other issues that need to be considered in order to assess whether or not these are suitable for achieving the development goals. GM technologies, and as a result GM seeds, are often patented and these patents are held by large multinational corporations. This means that farmers cannot save seeds and they have to be purchased for every planting. In developing countries, this amounts to a substantial cost to the extent that many farmers face chronic debt. Owing to commercial interests in GM crops, only certain plant traits are modified. These often include fast-growing and high-yielding traits which increase yields in experimental conditions. In the real world, however, these modified traits often fail to deliver because of harsh climatic conditions, unsuitable soil type or the lack of irrigation. There is therefore disjunction in plant traits that should be improved for the benefit of poor farmers as opposed to those that are improved for commercial gains. This has caused a widespread feeling that the GM technology has not lived up to its initial expectations of enhancing the livelihoods of poor farmers. Climate change is likely to pose additional challenges for poor farmers in the future, meaning that if the GM technology is to be useful for these farmers, plant traits that enhance the ability of crops to produce high yields in a changing climate would be necessary. Another technology that has been put forward for development of the rural poor is biofuels. Biofuels in the form of cereal grains or oil palm are often grown on prime agricultural land, meaning that food production has to be displaced to sub-prime land or 'outsourced' all together. Such displacement or outsourcing of food production has serious consequences for the ability of local communities to be self-sufficient in terms of food production and ultimately for their resilience in the face of widespread changes in environmental, political and socio-economic spheres. There are, however, biofuel crops such as Jatropha that are grown to 'wasteland'. However, what governments classify as wasteland often has use value for the local communities. New and emerging technologies in agriculture can have the potential to benefit the rural poor, but experience has suggested that these technological advances take a long time to trickle down to the poor and are more likely to benefit large businesses or corporations, who have greater control over technologies and disproportionate influence on the modus operandi of resource use and exploitation.

Localisation of farming systems as an alternative to globalisation?

In view of this, what are the alternatives for the rural poor? Can globalisation be made to work in their favour or can alternative approaches that veer towards 'localisation' be more effective? One example that has attempted to chart the way forward for such localisation is that of rural social movements in agroecology. These movements have started a discourse on 'food sovereignty' as opposed to the global discourse on 'food security'. During the first two decades of the twenty-first century, neoliberal policies which consist of deregulation, free trade, open markets and privatisation have provided opportunities for multinational and transnational corporations to invest in enterprises across the world (see Chapter 5). The financial crisis in the Western world during the first decade of the twenty-first century has given rise to a desperate search for new investment opportunities and developed countries are increasingly looking at the developing countries to make these investments. In many cases, such investments are made in rural natural resources, including biofuels and export-oriented agricultural commodities. These investments in rural resources are causing conflict between rural farming practices and transnational agribusinesses over rural spaces.

Rosset and Martínez-Torres (2012) argue that the rural social movements are also becoming more multinational and transnational because of their opponents, global agribusinesses, also operate across nations. La Vía Campesina is one such movement, which has turned into a global alliance of at least 200 million farming families worldwide, including indigenous people, landless peasants and farm workers, and rural women and youth. Based on the successful model of this rural social movement and many other similar examples from around the world, Rosset and Martínez-Torres (2012) argue that there is growing interest in agroecology – diversified farming systems based on the integrated management of functional biodiversity on farm – among rural social movements. Growing corporate investment in land and rising input costs have prompted farmers to look for alternatives and experience has suggested that in many cases agro-ecology has helped farmers build autonomy from unfavourable markets and restore degraded soils as well as social processes. This approach to the 'localisation' of farming systems provides a robust alternative to the 'globalisation' of agriculture.

While the global discourse on agriculture tends to focus on the issue of 'food security', localisation movements emphasise 'food sovereignty', a greater local control and management of the food system. This is of particular relevance to smallholder farmers because for them farming also addresses issues about health and nutrition as much as the quantity of food. As such, a diversified range of crops also addresses the need for a diversity of food groups in the diet. In many smallholder systems such as homegardens, farmers tend to grow a much wider variety of food crops including fruits, vegetables and medicinal and aromatic plants than they would be able to in conventional farming systems. In addition to addressing the health and nutrition of poor farmers, these homegardens are also known to protect a wide variety of species by providing complex ecological habitat (e.g. Bhagwat et al., 2008).

Furthermore, a congregation of such smallholder systems in the landscape provides an ecological network of habitats that is complex and heterogeneous. This is important for conservation of biodiversity outside protected areas and for creating a wildlife-friendly habitat that can support a wide variety of species in otherwise human-dominated landscapes. Such systems might also provide a solution to reducing deforestation by altering the way in which farming is practised, namely by keeping more trees on farms. It is likely that such agroforestry systems will enable reconciling biodiversity conservation with development by addressing concerns about health, nutrition and human wellbeing. There are, however, a number of preconditions to bring about such a change in farming systems. First, the global disparities in the distribution and trade of food need to be addressed alongside a transformation of farming systems. Second, many more farmers need to remain or enter in food production and need strong incentives to do so. Third, the food system needs to have a much greater democratic control and an acknowledgement that access to food is an issue of basic human right. With these changes, it might be possible to square conservation with development.

Case studies suggest that such localisation of resource use has also benefited people and biodiversity in fisheries, forestry and wildlife management sectors. The Philippines experienced in the early 1990s levelling-off of its fish catch despite expansion of commercial fisheries. In addition, changes in the species composition of fish catch were also observed with anchovies partially replacing sardines, scads and mackerels, leading to a gradual stock collapse (Green et al., 2003). As a result of catastrophic declines in the fish stock, the Philippines government established the Local Government Code in 1991, which devolved authority over the management of municipal waters to Local Government Units within the parameters set by national fisheries legislation and policies (FAO, 2015). These are areas where many small-scale fishers depend on fish catch to earn their livelihoods. The collapse of fish stocks in the 1990s also resulted in severe depletion of fish stocks in municipal waters, affecting the livelihoods of many fishers. The 'localisation' of fisheries was accompanied by the Philippine Community-Based Coastal Resource Management programme, which has been very successful and the country is now considered a leader in the devolution of authority for coastal resource management with outcomes beneficial to biodiversity conservation and poverty alleviation.

In many developing countries, the devolution of power in the forestry sector started in the 1980s and 1990s. In many countries such a change was brought about by the limitations of the government forestry departments to manage resources at the local level, resulting in the large-scale degradation of forests. Joint forest management initiatives in India, in particular, were at the helm of collaborative management of forest resources between the government forestry departments and local communities. In 1988, India introduced a new forest policy, which for the first time recognised the importance of the involvement of local communities in forest management. This policy was strengthened in the 1990s with the transfer of rights for local forest management to local communities in many Indian states (Kumar, 2002). Although empirical evidence on the success or failure of joint forest management is patchy, Singh et al. (2011) identify factors that determine the success of such joint arrangements. They identify two main factors that are responsible for the success of joint forest management:

1 higher levels of local monitoring and enforcement of locally made rules;
2 strong autonomy of rule-making at the local level.

As such, they emphasize the need for translating knowledge to action at village level and argue that joint forest management committees that have local rule-making, local monitoring and local enforcement are more likely to succeed in their efforts to improve the quality of the forest resource and to improve people's livelihoods simultaneously.

Similar to joint forest management, community-based wildlife management has also been examined as an alternative to formal wildlife management by the state. The important aspect of such a management approach is that stewardship of local wildlife resides at the local rather than the state level. It has also been suggested that this approach makes it possible to improve rural livelihoods, conserve the environment and promote economic growth at the same time (Roe et al., 2009). Africa has a history of such community-based initiatives and they involve some degree of co-management of resources between central authorities, local government, and local communities which share rights and responsibilities through diverse institutional arrangements. Such joint management initiatives have diversified approaches to natural resource governance in sub-Saharan Africa. Such initiatives have led to successful wildlife conservation in various countries: in Namibia key wildlife resources have recovered and illegal use of wildlife has fallen and similar success has been reported from Zimbabwe, Tanzania, Kenya, Cameroon and Ghana (Roe et al., 2009). As Chapter 10 explores in greater depth, community conservation, whether in relation to wildlife or other elements of nature, is far from straightforwardly simple, to the extent that it has fallen out of favour relative to other approaches. Nevertheless, experiences like these suggest that localised perspectives inherent in approaches like community-based wildlife management can yield conservation and development benefits, with perseverance.

Summary

- Globalisation has had substantial impacts on the consumption of natural resources. As the boundaries across countries and continents have dissolved with increased transport links, there are more opportunities to use natural resources in one part of the world to fulfil the demands of people in another part of the world.

- The globalisation of natural resource consumption has led to the transformation of agriculture in many less developed countries; precipitated decline in fisheries; promoted deforestation in some of the remotest parts of the world; and put endangered species under further threat.

- A process common to all these environmental problems is that these resources originate in one geographic location, but are exploited to fulfil the demands of people in another geographic location. In other words, the environmental externalities of consumption in one part of the world often manifest on the other side of it.

- There are examples where the 'localisation' of resource use has shown promise through a greater participation of local communities. Such participation has empowered local people to make decisions at a local level of governance.

- The greatest challenge for local governance is perhaps in continuing to persist in a world that continues its trajectory of globalisation.

Discussion questions

1. Is globalisation linked to a sharp depletion of natural resources?

2. How are agriculture, fisheries, forestry and wildlife trade impacted by the globalisation of natural resource use?

3. Can a greater participation of local communities in natural resource management benefit conservation of resources as well as the development of these communities?

Key reading

Berkes, F., T. P. Hughes, R. S. Steneck, J. A. Wilson, D. R. Bellwood, B. Cron, C. Folke, L. H. Gunderson, H. M. Leslie, J. Norberg, M. Nyström, P. Olsson, H. Österblom, M. Scheffer, and B. Worm (2006) Globalisation, roving bandits, and marine resources. *Science* 311: 1557–1558.

Ehrenfeld, D. (2003) Globalisation: effects on biodiversity, environment and society. *Conservation and Society* 1:99–111.

Hughes, T.P., D.R. Bellwood, C. Folke, et al. (2005) New paradigms for supporting the resilience of marine ecosystems. *Trends in Ecology & Evolution* 20:380–386.

Nijman, V. (2010) An overview of international wildlife trade from Southeast Asia. *Biodiversity and Conservation* 19:1101–1114.

Roe D., Nelson, F., and Sandbrook, C. (eds.) 2009. Community management of natural resources in Africa: Impacts, experiences and future directions, Natural Resource Issues No. 18, International Institute for Environment and Development, London, UK.

Ros-Tonen, M. A. F. (2003) Globalisation, localisation and tropical forest management: introducing the challenge of new markets and partnerships. *ETFRN News* 39–40 (Autumn/Winter 2003):7–9.

Online resources

The Fraserburgh Herald (2012) Localisation: The way forward for fishing? http://www.fraserburghherald.co.uk/news/local-news/localisation-the-way-forward-for-fishing-1-2275380 [accessed 10 February 2015]

Rosset, P. M., and M. E. Martínez-Torres. (2012) Rural social movements and agroecology: context, theory, and process. *Ecology and Society* 17(3): 17. http://dx.doi.org/10.5751/ES-05000-170317

FAO (2015) Search Geographical Information: Fishery and Aquaculture Country Profiles. http://www.fao.org/fishery/countryprofiles/search/en [accessed 10 February 2015]

PART 3
Conservation and development in practice

9 Protected areas, conservation and development

- Protected areas as instruments
- Beyond protected areas
- Landscape-scale approaches to conservation and development

Introduction

Over the last two centuries, protected areas have increasingly become a key instrument for the conservation of biological diversity. There are various types of protected areas with various objectives for protection – from strict protection for scientific research to community management to support human needs. However, protected areas have not always been able to join up conservation and development as some of them have compromised people's livelihoods and in the process antagonised people. The attention has therefore focused on what can be done beyond protected areas that will help integrate the conservation of nature and development of people. To this end, newer categories of protected areas have emerged where people are actively engaged in their protection. Simultaneously, joint management of resources by people and the state has shown promise. This chapter gives an overview of protected area types and extent and then focuses on conservation outside protected areas, where challenges as well as opportunities lie in joining up conservation and development.

Protected areas as instruments

Growth, types and extent of protected areas

A protected area is defined by the IUCN (2007) as 'an area of land and/or sea especially dedicated to the protection and maintenance of biological diversity, and of natural and associated cultural resources, and managed through legal or other effective means'. Yellowstone National Park in the USA, created in 1872, was the first-ever protected area. There are nearly 160,000 protected areas in the world now, covering an area of nearly 25 million km^2 or 13% of the Earth's surface (Protected Planet, 2015). Protected areas are considered an important instrument of biodiversity conservation and the lofty aims of these areas are summarised in the IUCN's (2007) statement that 'the world's protected areas are the greatest legacy we can leave to future generations'. There are also high expectations attached to protected areas and their role in the conservation of biological diversity as Dudley (2008) suggests: 'Today they are often the only hope we have of stopping many threatened or endemic species from becoming extinct.' But what does the designation of protected areas mean to communities who are dependent on natural resources for their livelihoods? What are the conflicts between the global agenda of preventing extinction of species and the local needs to make a living from the land? Can the conflicts be squared up to protect biodiversity and to provide livelihoods to people simultaneously?

Categories of protected areas and their role in conservation and development

Early protected areas were set up to preserve 'pristine wilderness' but, as Chapter 3 explores, now there is an increasing body of evidence which suggested that wilderness as a landscape free from human habitation or influence is actually a myth (Adams and McShane, 1992). For instance, Denevan (1992) argues that, at the time of the arrival of Columbus to what are now called the Americas, the ostensibly 'pristine' forests of the Amazon played host to a human population of eight million people. Wood (1993) proposes that a feature of each of the world's tropical forest regions was human settlement. Some have concluded that the concept of wilderness as the untouched or untamed land is mostly an urban perception, the view of people who are far removed from the natural environment they depend on for raw resources (Gomezpompa and Kaus, 1992). The critique of the notion of wilderness as a non-human landscape infused (some) early proponents of community conservation with an acute sense of the social (in)justice implications of protected area conservation. It was indeed the awareness of the human costs of preservationism which prompted many conservationists – notably in the work of Jonathan Adams and Thomas McShane – to reject indignantly the idea that environments had to be defended 'even against the people who have lived there for thousands of years' (Adams and McShane, 1992: xviii). As Chapter 10 demonstrates, this reappraisal of the separateness of humans from nature has also led to greater appreciation, within conservation circles, of the ability of people with no grounding in conservation biology to act as wise environmental stewards.

Protected areas today have come to be seen as instruments of recreation, carbon capture or provision of 'ecosystem services'. These multiple functions are often achieved under the broad umbrella of integrated conservation and development around protected areas. The purpose and meaning of protected areas has therefore shifted over the last 150 years. IUCN now recognises six categories of protected areas (Table 9.1) and these categories have become 'gold standard' in nature conservation.

The approach of protected areas to separate people from nature has led to tensions between conservation and development. Although some forms of protected areas promised benefits to communities living on the margins of these areas through, for example, tourism or other opportunities for livelihood provision, more often than not communities have been left resentful. Furthermore, whilst striving to protect nature in most natural condition (left-hand side of Figure 9.1), the protected area model of nature conservation has often overlooked conservation needs in least natural conditions (right-hand side of Figure 9.1).

Integrating conservation and development

Adams et al. (2004) analysed a gradient of policy responses with regard to the conservation-development nexus. These responses include:

1 treatment of conservation and development as separate policy realms;
2 identification of poverty as a critical constraint on conservation;
3 recognition that conservation should not compromises poverty reduction;
4 the assertion that effective conservation depends on poverty reduction.

Table 9.1 *IUCN Protected Area categories (adapted from Dudley et al., 2008)*

IUCN Protected Area categories	Description
Category Ia: Strict nature reserve	These are strictly protected areas set aside to protect biodiversity and also possibly geological/geomorphological features, where human visitation, use and impacts are strictly controlled and limited to ensure protection of the conservation values. Such protected areas can serve as indispensable reference areas for scientific research and monitoring.
Category Ib: Wilderness area	These are usually large unmodified or slightly modified areas, retaining their natural character and influence, without permanent or significant human habitation, which are protected and managed so as to preserve their natural condition.
Category II: National park	These are large natural or near natural areas set aside to protect large-scale ecological processes, along with the complement of species and ecosystems characteristic of the area, which also provide a foundation for environmentally and culturally compatible spiritual, scientific, educational, recreational and visitor opportunities.
Category III: Natural monument or feature	These are set aside to protect a specific natural monument, which can be a landform, sea mount, submarine cavern, a geological feature such as a cave or even a living feature such as an ancient grove. They are generally quite small protected areas and often have high visitor value.
Category IV: Habitat/species management area	These aim to protect particular species or habitats and management reflects this priority. Many category IV protected areas will need regular, active interventions to address the requirements of particular species or to maintain habitats, but this is not a requirement of the category.
Category V: Protected landscape/seascape	A protected area where the interaction of people and nature over time has produced an area of distinct character with significant ecological, biological, cultural and scenic value: and where safeguarding the integrity of this interaction is vital to protecting and sustaining the area and its associated nature conservation and other values.
Category VI: Protected area with sustainable use of natural resources	These areas conserve ecosystems and habitats, together with associated cultural values and traditional natural resource management systems. They are generally large, with most of the area in a natural condition, where a proportion is under sustainable natural resource management and where low-level non-industrial use of natural resources compatible with nature conservation is seen as one of the main aims of the area.

There is a gradient of wilderness or naturalness in these categories, with categories Ia and Ib being at one spectrum of naturalness and category VI at the other end (Figure 9.1). All categories of protected areas, however, strive to protect landscapes and seascapes that are less modified by humans and are more natural. In most cases, therefore, the goal is to keep humans out of protected areas.

The protected area model of conservation, with its lofty goals of preserving nature without interference from humans, has traditionally treated conservation and development as separate policy realms. This is mainly because the founding principles of protected areas had at their core the ideal of wilderness preservation. The IUCN Category Ia (Strict nature reserve) and Category Ib (Wilderness area) protected

Figure 9.1 *Protected areas and naturalness*

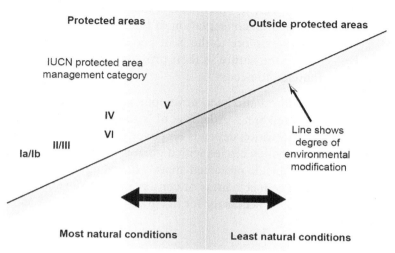

Source: Dudley, 2008

areas and the policies that preserve them are influenced by the wilderness preservation ideal. IUCN Category II (National park), Category III (Natural monument or feature) and Category IV (Habitat/species management area) protected areas, on the other hand facilitate public engagement with nature or manage nature in a way that enhances its public appeal, whether through landscape design or species management. These categories of protected areas promote tourism, which attempts to create new forms of livelihoods for the local poor. At the heart of promoting tourism is the hope that uplifting people out of poverty can make conservation outcomes better. IUCN Category V (Protected landscape/seascape) and Category VI (Protected area with sustainable use of natural resources) protected areas recognise that conservation should not compromise poverty reduction. As such there is a recognition that people are part of nature and the interaction between people and nature over generations has shaped the character of habitats and landscapes that we see today.

The protected area categories, however, miss out on nature that is found in everyday landscapes outside protected areas. Increasingly, moreover, the focus on 'undisturbed' biodiversity central to the rationale for protected areas may hinder, rather than help us, in recognising the value of 'novel ecosystems' that are emerging in 'unprotected areas' across the world, and the contribution they can make to development (see Chapter 14). Many rural poor depend on this nature for their livelihoods and therefore the alleviation of poverty depends on conservation of this nature. It is the nature 'in here' that matters more to the poor people than the nature 'out there' because the nature in their proximity takes care of many of their immediate needs.

Integrating conservation and development makes it necessary to look outside protected areas. Adams et al.'s fourth policy position, that effective conservation depends on poverty reduction, is increasingly recognised in policy circles with a greater effort made towards conservation outside protected areas. This recognition of the interdependence and connectedness of conservation inside and outside protected areas is necessary for integrating conservation and development effectively.

Beyond protected areas

Landscapes within and outside protected areas

The aim of increasing the global protected area coverage to 17% by 2020 under the Convention of Biological Diversity's Aichi targets is ambitious. Even if this target is met, 83% or more land will always remain outside of formally protected areas. This land includes agriculture and farmland, urban and peri-urban land, and 'wasteland' or land without any other obvious use. Management of nature on these diverse types of land requires approaches very different to those enshrined in protected area conservation. As Rosenzweig (2003) describes, the land outside protected areas is where people 'live, work and play'. In these types of land, particularly in developing countries, there is no other option but to integrate conservation and development. This is because the economic wellbeing of the rural poor in these countries is closely linked to the livelihoods they derive from the presence of nature in their vicinity. For example, tropical agroforestry systems, where trees are planted on farms, provide a variety of tree-based products that are necessary to supplement outputs from agricultural farms (Bhagwat et al., 2008). The need for the integration of conservation and development also demands a greater engagement with and participation of local people in making decision about conservation instruments.

Joint management of resources has been promoted in such situations. This approach to management, where state forest departments and local communities manage forests together, has been successful in India (e.g. Kumar, 2002). This form of management is often practised on land that is outside protected areas, but retains some form of vegetation cover, usually consisting of trees. In joint forest management, villagers from local communities assist in safeguarding forest resources by protecting them from fire, grazing or encroachment. In return, the communities receive rights to non-timber forest products from their community forest and where these products are sold, a share of the revenue from such sale. This 'benefit sharing' arrangement has been successful in a number of places to provide livelihoods to the rural poor.

'New conservation' vs 'old conservation'

Integrating conservation and development has been dubbed as the 'new conservation', which Soule (2013) describes as an approach that 'promotes economic development, poverty alleviation, and corporate partnerships as surrogates or substitutes for endangered species listings, protected areas, and other mainstream conservation tools' (p. 895). Soule goes on to contend that because the goal of this approach is to 'supplant the biological diversity-based model of traditional conservation with something entirely different, namely an economic growth-based or humanitarian movement, it does not deserve to be labelled conservation' (p. 896). In defence of the new conservation approach, Marvier (2013) argues that '[The new conservationists] advocate building a solid foundation from the bottom up and providing alternative livelihoods to the poor so that they are not forced to illegally harvest resources or otherwise work against protected areas' (p. 1). This debate is symptomatic of changes to the backdrop of the conservation movement since its inception. While the old conservationists were motivated by wilderness preservation ideals, the new conservationists are starting to accept the argument (as presented above and in Chapter 3) that humans have had disproportionate influence on the planet's natural ecosystems. As a result, some commentators argue that the preservation-focused conservation movement is giving way to a movement that accepts people as part of natural landscapes whist also acknowledging that humans have a certain self-interest in protecting nature – something which conservation needs to recognise to find new solutions to the loss of biodiversity. Miller et al. (2011) argue that some of the novel models of conservation planning and new policy instruments that attempt to integrate biodiversity protection with human wellbeing might provide fresh solutions to the new conservation movement. For example, conservation efforts are increasingly moving away from species- or habitat-focused conservation to landscape-level conservation. Such widening of scale captures landscapes beyond protected areas to include a full mosaic of land-uses. Furthermore, such a landscape-scale approach also helps in recognising social and cultural values of these land uses including their key role in supporting human livelihoods (e.g. Wiens, 2009).

Against this optimisim, however, signs of a resurgence of 'old conservation' have been spotted. In the mid 1990s even as community conservation initiatives (see Chapter 10) were claiming policy terrain previously held by 'fences and fines' or 'fortress' conservation approaches, advocates of strict enforcement of protected areas were developing a strident counter-critique of their own. A number of publications set out the case for reaffirming strong commitment to strictly protected areas as both a moral and ecological imperative (Brandon and Wells, 1992; Oates, 1999; Terborgh, 2004). This reaction against setting up community conservation as a means of achieving sustainable development was in large measure driven by a deep scepticism at the feasibility of this objective. But for these critics, the concern was that the mandate of conservation had been stretched too far. If conservationists were given too many social objectives to achieve on top of their existing responsibilities, then the result would be merely a 'recipe for ecological and social failure' (Redford et al. 1998: 546). Conservation, in other words, should 'stop trying to be all things to all people and simply focus on the central goal of nature protection' (Wilshusen et al., 2002: 21). Only rational, scientific principles should serve as the basis for decision making, and only those people who genuinely worked in the interests of biodiversity conservation should be involved in making decisions about its management (Terborgh, 2004). This approach to conservation can be characterised as 'old conservation'. Table 9.2 surveys the broad features of the wave of thinking characterised by (Hutton et al., 2005) as the 'back to the barriers narrative'. It also captures the critical response from commentators who were alarmed at this return to the kinds of conservation whose legitimacy had been so thoroughly contested in the final decades of the twentieth century.

Table 9.2 *Arguments in favour and against strictly protected areas (adapted from Wilshusen et al., 2002; Hutton et al., 2005)*

Strictly Protected

Arguments in favour	Arguments against
Protected areas require strict protection: PAs are the only remaining barrier against global environmental degradation, they too are disintegrating	Strict protection is characterised by colonial (and post-colonial) efforts to assert control over territory at the expense and/or expulsion of its occupants; strict protection thinking ignores rather than defuses the politics of conservation
Biodiversity protection is a moral imperative: the scale of human-caused biodiversity loss is unprecedented in the planet's history	Protecting biodiversity carries huge human costs, especially for those who live in or close to PAS, and can de-legitimise their moral claims; masks elite interests served by appeals to the 'common good'
Integrated conservation and development has failed: ICDPs have not engendered wise resource use as a springboard for sustainable development	ICDPs are very imperfect but critics over-generalise against them. In rejecting them outright there is no space for learning lessons or addressing the root causes (the decision-making, organisational and governance dynamics) which drive unwise resource use. The common-pool resource literature offers many examples both of sustainable resource use and the difficulties of maintaining such use.
Assumptions about the efficacy of 'traditional' 'community' resources practices are romanticised and misleading: 'indigenous' and 'traditional' institutions and practices often work poorly and cannot stem damaging phenomena such as the extension of the agricultural frontier	Again, the critique over-generalises in implying that no 'traditional' management works, in oversimplifying local resource use dynamics, an in overlooking local capacity to adapt practice.
ICDPS divert resources away from more effective instruments (i.e. PAs): the funding of community conservation is not justified by its minimal biodiversity protection outcomes; and even worse, this squeezes more funds available for PAs	ICDP donor funding came from development budgets, representing new spend for conservation activities. ICDPs took conservation into the agricultural frontier and built alliances that it would in fact be wasteful to lose.
Emergencies are a 'special case' and require extreme responses: biodiversity protection is in the common good; measures to protect it need to be proportionate to the scale of the challenge faced (i.e. involving the military).	Strict protection which is not socially and morally just is illegitimate and, in practice, often ineffective. The militarisation of conservation has undermined its legitimacy.

Focus on ecosystem services

The new conservation is also characterised by a focus on ecosystem services, the benefits humans derive from nature. In Chapter 5 we discuss in detail the arguments for and against ecosystem services, against a background of neoliberal globalisation. Here we revisit those in the context of conservation outside protected areas. Ecosystem service benefits are particularly relevant outside of protected areas where people share landscapes with other species. The proponents of ecosystem services argue that in order to make a case for conservation of those species and habitat in which they live, it is necessary to value the benefits that nature provides to people. The argument goes that by valuing nature and by translating that value into a 'currency', it should be possible to get people to pay for ecosystem services that are all too often taken for granted. The critiques of the ecosystem services approach indicate that conservation has had a history of valorising magic solutions to the problem of biodiversity loss and ecosystem services is one such solution (Redford and Adams, 2009). Critiques go on to argue that the provision of ecosystem services and conservation of biodiversity are not always congruent. For example, the ecosystem service of carbon sequestration does not necessarily require a biodiversity stand of forest; a monoculture plantation can provide that service as effectively as near-natural forest. Critiques also argue that the valorisation of ecosystem services disregard 'ecosystem disservices', the inconveniences posed by living in the vicinity of nature, e.g. conflict between people and wild animals. While conservation of large and charismatic wild animals has aesthetic appeal to urban tourists, these same animals pose threats to the livelihoods of people who come in frequent contact with these animals. Payments for ecosystem services do not always account for the disservices that are closely tied to each other from the perspective of people who live close to nature. As such, the recipients or beneficiaries of ecosystem services and those who have to put up with inconvenience in order to provide those benefits are often very different. The new conservation, despite its lofty aims of treating people as part of nature, does not recognise the disparity between those who benefit and those who suffer as a result of ecosystem services. We return to and expand upon these concerns in Chapter 13, which surveys the evidence for the contribution of PES to conservation and development outcomes.

Landscape-scale approaches to conservation and development

The tensions between conservation and development have promoted a move towards integrating conservation and development by considering these issues at a landscape scale. Such landscapes are composed of protected, non-protected or semi-protected land parcels and can include a mosaic of productive and conservation areas. The need for integrating conservation and development in one landscape has also given rise to new forms of protected areas or new ways to manage conservation areas. This section provides an overview of some of those means of management.

Newer categories of protected areas

One of the recommendations that came out of the fifth IUCN World Parks Congress held in Durban in 2003 was about Community Conserved Areas, which subsequently became known as 'Indigenous and Community Conserved Areas' (ICCAs). These areas are defined as 'natural and modified ecosystems including significant biodiversity, ecological services and cultural values voluntarily conserved by indigenous and local communities through customary laws or other effective means' (IUCN, 2003). The definition of ICCAs recognises the 'multi-functional' nature of these landscapes, which not only consist of multiple types of land uses, but also possess a variety of cultural values and provide a wide range of ecological services. What distinguishes these sites from protected areas is that they are predominantly controlled and managed by communities; and that there is often a commitment on part of the stakeholder communities to conserve biodiversity at these sites. This commitment takes the form of various informal means such as cultural taboos or community protocols regarding access to or harvesting of resources in these sites. Although this form of informal conservation had been recognised through other international instruments such as The Convention on Biological Diversity (which emphasised biodiversity-relevant knowledge, skills, innovations, and community practices) or the UN Declaration of the Rights of Indigenous Peoples (which acknowledges the right of indigenous peoples to control and manage their territories). The recognition of these informal practices as an instrument for conservation of biological diversity was proposed for the first time at the 2003 World Parks Congress.

ICCAs, however, are still struggling to be recognised by their state authorities because they are largely governed by customary tenure systems, and norms and institutions that are not formally or legally recognised in many countries. As a result, ICCAs are facing a number of threats:

- the land ownership of these lands is unclear and insecure;
- the state authorities plan community development projects on these lands that often fail;
- the state authorities can easily de-legitimise of customary rights associated with these lands;
- the state authorities do not consult communities when making decisions about the future of these areas;
- the stakeholders of these areas are often seriously disadvantaged on social, economic and political terms;
- the lack of long-term security to these areas is leading to the loss of knowledge and cultural change;
- the commercialisation of resources is challenging some of the traditional modes of resource extraction practised at these sites.

These continuing threats have been attributed to differences in the interests of indigenous peoples and those of state authorities. These differences are reflected in the approach to management of conservation areas by the indigenous peoples and by the state authorities. The lack of legal recognition of many of these areas also means that they may not always be recognised and respected beyond the stakeholder community. Despite calls for their recognition in international conservation policies, therefore, there are still serious hurdles to their effective and appropriate recognition in national policies and practices.

There is, however, a growing recognition that a large portion of the Earth's surface might be covered by ICCAs and these sites are privileged over others in providing relatively secure spaces for the Earth's biological diversity, often acting as 'refugia' in otherwise highly human-dominated landscapes (Bhagwat and Rutte, 2006). Furthermore, despite the small size of these sites, they are widely distributed; and as unmanaged habitats within highly managed landscapes they provide the so-called 'ecosystem services' – benefits that people derive from natural ecosystems. For example, a distributed network of sacred groves might provide the ecosystem service of pollination, beneficial for many agricultural crops. As unmanaged habitats, they might also provide benefits in the form of groundwater storage. In addition, they are also repositories of non-wood forest products, particularly medicinal herbs, which are very important in local health care systems. Therefore, as 'informal' protected areas, ICCAs offer a large number of benefits to local people. Whilst complementing the formal protected area networks, they might also provide refugia for species, habitats and ecosystems that are not sufficiently protected within the formal reserve networks.

Joint management between communities and the state

Despite the fact that some practices of indigenous communities (such as maintenance of ICCAs) are an important instrument for the conservation of biological diversity, these communities have often been affected by the global instruments of conservation, i.e. formally protected areas. Such conflicts between protected areas and people often lead to resentment on part of many communities. In response to such resentment, 'joint management' has often been suggested as a way forward. Borrini-Feyerabend et al. (2007: 69) define joint management as 'a partnership by which two or more relevant social actors collectively negotiate, agree upon, guarantee and implement a fair share of management functions, benefits and responsibilities for a particular territory, area or set of natural resources'. Such joint management is also often referred to as 'co-management', 'collaborative management' or 'participatory management' and literature has indicated that this form of management can be effective in reconciling cultural practices and biodiversity conservation in landscapes where protected areas occupy prominent position (Carlson and Berkes, 2005, Plummer and Arai, 2005, Berkes, 2009). In theory, such arrangement can work to the benefit of both the communities and the state. However, the motivations for such joint management are often quite different. From the state authority's perspective, joint management of a means to achieve conservation goals, while for indigenous communities, the aim of joint management is to have a control over their traditional land. Their aims are motivated by the need for a stronger cultural identity and self-determination while the state authorities are interested in safeguarding natural habitats with minimal human intervention (e.g. Lawrence, 2000). Even in joint management context, therefore, motivations for co-management are quite different.

Given these apparent conflicts, what are the factors that make joint management successful? Stacey et al. (2013) compared the performance of four jointly managed protected areas in northern Australia. Working with indigenous communities, they identified a number of indicators that assessed performance of joint management in governance and decision making; application and interpretation of cultural heritage and traditional ecological knowledge; expansion of social capital, and human and financial resources; and visitors. Overall, they found that indigenous communities placed greatest emphasis on development of social and human capital, indicating that recognition of this kind of capital is necessary before developing conventional measures of protected area performance. Also, they found that the indigenous communities put significant emphasis on the process, or doing things in the right way, suggesting that building strong relationships with local communities is essential for the success of joint management. These findings from the four Australian protected areas compare well with those from other parts of the world. For example, Timko and Satterfield's (2008) study from national parks in Canada and South Africa indicates that improving conditions for the conservation of biological diversity are highly dependent on good relationships, communication, and levels of decision making among the partners. These studies suggest that in joint-management context, more needs to be done to improve the social and economic benefits for indigenous stakeholders of protected areas. Where the local communities are affected by protected areas, consideration needs to be given to alternative livelihoods.

Provision of alternative livelihoods

A number of studies over the last decade have suggested that protected areas are increasingly surrounded by human settlement (e.g. Wittemyer et al. 2008). This means that these nature reserves are turning into isolated islands surrounded by the sea of human settlement. The human settlement also brings with it resource extraction from the fringes of protected areas and often from inside their boundaries. Research has suggested that such resource extraction has an effect on the ecological functions of protected areas (Hansen and DeFries, 2007). Hansen and DeFries (2007) identify four mechanisms that alter the ecosystem functioning of protected areas due to human impact:

1   changes in effective size of protected areas, with direct implications for biological diversity;
2   altered flows of materials and disturbances into and out of reserve
3   loss of crucial habitats for seasonal migrations and population source areas;
4   exposure to human activity through hunting, poaching, exotic species, and disease.

Such alteration of ecosystem function has implications for the livelihoods of local people who live in the vicinity of protected areas. Maintaining ecosystem functions and thereby ensuring the provision of ecosystem services is therefore essential for the provision of alternative livelihoods to these people.

A 'livelihoods approach' to nature conservation, however, encompasses more than maintaining ecosystem functions. It is a people-centred approach focusing on the different types of assets that people need to have access to in order to make a living. Such an approach therefore includes the policies, institutions and processes that influence people's livelihoods; it recognises that there are a range of activities that individuals may engage in to make their overall 'livelihood strategy', and that these strategies affect both financial security of the household and environmental quality of the surrounding area (The Nature Conservancy 2014). The Nature Conservancy's (TNC) work in marine conservation has indicated that the provision of successful alternative livelihoods is an important factor in the successful implementation of marine protected areas. This work has also highlighted that the projects on alternative livelihoods that create access to additional forms of income generation and increase expertise within the community, also contribute to increased social resilience. Some examples of livelihoods projects that TNC has successfully implemented include: training fishermen to become sport-fishing guides in Belize; support for seaweed farming in East Africa to address coral reef degradation in the Indian Ocean; creation of a large-scale mariculture enterprise for grouper and snapper in Komodo National Park in Indonesia; and bee-keeping activities to produce organic honey in Sian Ka'an Biosphere Reserve in Mexico as part of the United Nations COMPACT program (The Nature Conservancy, 2014). Despite a wide range of alternative livelihoods that could potentially be made available for any local community, experiences from TNC's projects around the world have suggested that it is not the type of alternative activity that is the most important factor influencing success, but rather researching and understanding existing community livelihood systems, and their influences and constraints. This is because in addition to asset creation, alternative livelihoods projects also need to be well aware of the local cultural context. Otherwise, there is a risk that such projects may be seen by the local community as a 'top-down' intervention from outside, and indeed may consolidate existing power relations not conducive to local empowerment objectives. An example from Costa Rica, showing both the potential benefits and problems with the provision of alternative livelihoods, is explored in Box 9.1.

Box 9.1

Lynn Horton's (2009) richly nuanced account of the Osa Peninsula, Costa Rica, provides an apposite example of the consequences of introducing alternative livelihoods, namely ecotourism, to mitigate the adverse human impacts of the establishment of a protected area. In the late 1970s and early 1980s squatters started to settle in an estate of mostly forested land owned by a private company, Osa Forest Products, which had done very little in the way of extractive activities in that area. The arrival of the squatters caused conflict with the company, which persisted until the Costa Rican president designated most of the estate as the Corcovado National Park. The squatters resisted this move, but were unable to stop the environmentalists who were pushing for the declaration of the park, and who had the ear of people at the highest levels of government. In many ways, the actions of the Costa Rican government looked to be another instance of top-down environmental policy that did nothing to address the root causes of poverty, social injustice, inequality, etc., which had led to the incidence of squatting in Osa in the first place. Yet the severity of continued clashes and encroachment into the National Park, Horton argues, brought about a level of conflict visibility that changed the country's conservation policy environment. Measures which entailed such harsh impacts for people living in or adjacent to protected areas became politically unviable. It effectively paved the way for approaches which attempted to deliver development as well as conservation benefits through the provision of alternative livelihoods, chief amongst these being ecotourism.

The environmental benefits from this change are straightforward: the creation of Corcovado National Park is generally accepted to have halted deforestation. Whilst problems such as illegal hunting and agrochemical run-off persist, it is thought that ecotourism opportunities have indirectly helped to draw people away from rice cultivation and cattle-ranching. From a development perspective, Horton maintains that this change has fostered potentially empowering exchanges between local residents and ecotourists. Moreover, ecotourism has expanded important livelihood opportunities for women as well as men, albeit within the context of an extension of existing gender roles.

However, the change in conservation policy has not altered Costa Rica's historical pattern of concentrated landholdings in the hands of the few. Wealthy International investors purchased land in Osa on which to build eco-lodges, and had much greater wherewithal to develop an ecotourism project. Local people, who could not afford to purchase land, took a lot longer to 'get in on the act' and, in purely financial terms, benefitted less. Nevertheless, Horton maintains, ecotourism does generate important economic benefits locally. (Some) local people can get jobs which are often, if not always, less exploitative and better paid than equivalents in the mass tourism sector. Nevertheless, the 'market exclusion' of poorer Costa Ricans from the acquisition of land means that their participation in key ecotourism activities has been limited. Whilst the government has talked of supporting small-scale ecotourism activities, with a view to addressing this inequity, state intervention has done very little to change it. The provision of alternative livelihoods came, on one view, on the back of a neoliberal economic and political model which unevenly distributes the fruits of ecotourism and entrenches, rather than challenges, modes of economic activity implicated in the production of global environmental problems.

Summary

- Protected areas have been around for a long time, but whether or not they help join up conservation and development has been (and continues to be) debated endlessly.

- If the primary purpose of protected areas is conservation of nature, then should they be expected also to deliver the development of communities who live adjacent to their boundaries? Or, if the protected areas focus entirely on nature conservation and leave aside concerns such as poverty, is that likely to backfire on the protection status of these areas? If conservation and development are to go hand-in-hand, how should protected areas operate?

- It is now increasingly acknowledged that protected areas are not islands of biological richness surrounded by extensive stretches of highly human-dominated landscapes, but in fact they are connected with what is around them. As such, the 'new conservation' is increasingly looking beyond reserves. This is where nature's services are most relevant because this is where nature and people must coexist.

- The new conservation can no longer focus exclusively on the conservation of biological richness, but it has to pay attention to providing livelihoods to people who are ultimately in the close vicinity of places upheld for their biological richness.

- In response to the pressure to reconcile conservation and development, newer categories of protected areas have emerged, which are managed jointly by local communities and the state. These new categories show promise in conserving biodiversity whilst also providing alternative livelihoods to people who depend on natural resources.

Discussion questions

1. Has the exponential growth in protected areas over the past 150 years been effective in saving nature?

2. What are the limitations of protected areas in integrating conservation and development in places where both are important?

3. How can new categories of protected areas help in joining up conservation and development?

Further reading

Berkes, F. (2009) Evolution of co-management: role of knowledge generation, bridging organisations and social learning. *Journal of Environmental Management* 90:1692–1702.

Borrini-Feyerabend, G., M. Pimbert, M. Taghi Farvar, A. Kothari, and Y. Renard. (2007) *Sharing Power: A Global Guide to Collaborative Management of Natural Resources*. Earthscan, London, UK.

Hansen, A. J., and R. DeFries. (2007) Ecological mechanisms linking protected areas to surrounding lands. *Ecological Applications* 17:974–988.

Marvier, Michelle (2013) New conservation is true conservation. *Conservation Biology*. 28(1): 1–3.

Rosenzweig M. (2003) Reconciliation ecology and the future of species diversity. *Oryx* 37: 194–205.

Stacey, N., A. Izurieta, and S. T. Garnett. (2013) Collaborative measurement of performance of jointly managed protected areas in northern Australia. *Ecology and Society* 18(1): 19.

Online resources

IUCN (2003) International Union for Conservation of Nature. World Parks Congress recommendation V.26 URL: http://cmsdata.iucn.org/downloads/| recommendationen.pdf [accessed 11 February 2015]

Protected Planet (2015) Discover and learn about protected areas. URL: http://www.protectedplanet.net [accessed 11 February 2015]

The Nature Conservancy (2014) Livelihoods Approaches as a Conservation Tool URL: http://www.reefresilience.org/pdf/LivelihoodsApproachShortVersion.pdf [accessed 11 February 2015]

10 Integrated conservation and development through community conservation

- Antecedents and elements of integrated conservation and development
- The rise of integrated conservation and development (the wedding of the decade)
- A marriage on the rocks: critiques of integrated conservation and development
- Acrimonious or amicable divorce? Conservation and development in the twenty-first century
- Space and perspective for community conservation?

Introduction

As Chapter 2 demonstrates, conservation has made ever greater efforts to engage with the process of development, in the second half of the twentieth century, largely on the back of the discourse around sustainable development. Conservation became more important – or at least somewhat more recognised – within the development arena because it seemed to offer tools for sustainability. Conversely, conservationists were switching their focus towards what was happening along the borders of, and indeed beyond, the protected area. What emerged from this mutual regard was a policy narrative and a corresponding set of interventions around the concept of integrated conservation and development. And as much from the conservation as from the development point of view, a preoccupation with the local, with the "community", was at the core of this approach. Because of this meeting of minds, what has been termed 'community conservation' (Adams et al., 2001) became the poster child for the ideological project of integrated conservation and development.

This chapter explores the life cycle of the integrated conservation and development 'paradigm'. It therefore charts the early enthusiasm which accompanied its emergence, the scepticism and slide into disillusion which followed the initial excitement, and its legacy, once the paradigm of integrated conservation and development had lost its influence.

Antecedents and elements of integrated conservation and development

The economics of conservation: 'use it or lose it'

For much of the twentieth century, the economics of natural resource use had been viewed by conservationists (and many others) as the root cause of environmental degradation. Yet the rise of the sustainable development discourse shifted the focus beyond the protected area and rehabilitated the 'wise use' school of conservationist thought (see Chapter 2), reaffirming the idea that the economics mattered. If conservationists wanted people to use nature differently (and to consume less of it), then it was going to be necessary to give them good reasons to change their behaviour (Adams and McShane, 1992).

In the idiom of sustainable development, it was critical to get the incentives right; and again, economics provided answers. It was precisely the economic value that could be gained from biodiversity that provided a prime incentive for their conservation. The bottom line was, if the contribution biodiversity could make to economic development was not clear, then there would not be sufficient incentive for its conservation. This held as much for governments faced with tough choices about allocating limited resources, as for people at the local level with an urgent need to make a living from their environments (Emerton et al., 2001). This thinking permeated a raft of studies that aimed to quantify the economic contribution of biodiversity and show how the benefits generated would ensure its conservation, for instance in relation to wildlife-related tourism in Southern Africa (Muir et al., 1996; Norton-Griffiths and Southey, 1995; Wells, 1996).

The message, therefore, was to get the economics right, generate the benefits, and conservation will follow. This logic was the basis of the 'win-win' scenario that integrating conservation and development

Figure 10.1 *Springbok near Uis, Erongo Region, Northwestern Namibia*

Source: A. Newsham

would, it was held with increasing enthusiasm, bring about. Community conservation would become a charismatic poster child for this approach (Campbell and Vainio-Mattila, 2003). As time went on, commentators would increasingly ask a) whether other incentives, be they cultural or spiritual, were more important but sidelined by the focus on economics (Bond et al., 2001; Emerton et al., 2001b); and b) what the implications were for conservation and development of the influence of market mechanisms (Sullivan, 2006), a point to which we will return later on in the chapter. But whatever the misgivings, it is impossible to understand integrated conservation and development in isolation from the consideration of the importance of economic incentives.

Common pool resource management

Of course, getting the economics of integrated conservation and development right was not an automatic process. It was critical to understand *how* natural resources could be managed in ways which struck the right balance between conservation and development incentives. What were the rules that should be applied to ensure sustainable use of natural resources? Through which institutions? These were the questions which economists, anthropologists and natural resource managers alike sought to understand through the study of common-pool resources; that is, a set of resources that it is difficult to stop people from using, and which diminish in quantity or quality through use (Berkes, 1989; Ostrom, 1990). The important point to grasp in relation to community conservation is that this work undermined the idea that only the state or the market could resolve the collective action problems at the heart of sustainable common-pool resource use. Scholars such as Elinor Ostrom, Fikret Berkes, M. Taghi Farvar, James Acheson and David Feeny amongst others documented examples at the local level under which common-pool resources had been used sustainably. This body of literature brought much greater recognition of and respect for the capabilities of local people, whose resource management systems had been presumed inferior to systems derived from scientific knowledge (Berkes, 2008; Holling et al., 1998; Pimbert and Pretty, 1995a). In many ways common-pool resource management thinking provided the intellectual cover and confidence necessary for community conservation to be taken seriously.

Locally-held knowledge (of 'ecological' and 'indigenous' persuasions)

In effect, common-pool resource theorists helped to draw much greater attention to the pre-existence of vast bodies of what here we term locally-held knowledge, and the ways in which this knowledge had been used to construct common pool resource regimes which had proved themselves enduringly sustainable. There is a lot of potential for confusion given the large number of terms that are used to refer to such knowledge. So it is worth straightening out the relationship between all of them, and why we use 'locally-held' knowledge. For us, locally-held knowledge is knowledge strongly connected to a place and culture, contested, changed and transmitted inter-generationally through learning and practice, and through (sometimes imposed) exposure to other ways of knowing – most commonly science. There are many other terms – local, indigenous, traditional (ecological), hybrid, endogenous. 'Locally-held' is synonymous with 'hybrid' or 'endogenous' but we propose it as a more simple and accessible framing of the same ideas, and one capable of acting as an umbrella term for two which refer to more specific domains of knowledge and/or holders of it: traditional ecological and indigenous forms of knowledge.

There is no way to define 'traditional ecological knowledge' that will please everyone. Some have sought refuge in the term 'indigenous' knowledge on the grounds that 'traditional' connotes 'the nineteenth-century attitudes of simple, savage and static' (Slikkerveer et al., 1995: 13). Others prefer 'local' knowledge (Ruddle, 1994) as a way around the objection that indigenous knowledge is not always indigenous, or distinct as a knowledge category (Ellen and Harris, 2000; Pottier et al., 2003). Berkes (2008) argues that it is worth maintaining 'traditional ecological knowledge' for two reasons. First, it is more specifically focussed on the environment than are terms such as 'local' or 'indigenous' knowledge, and it is this particular type of knowledge which is of most relevance to and has found most resonance in conservation circles. Second, a robust, if not a unanimous definition, is available: 'a cumulative body of knowledge, practice and belief, evolving by adaptive processes and handed down through generations by cultural transmission, about the relationship of living beings (including humans) with one another and with their environment' (ibid.: 7). This definition is very similar to the one we suggest for 'locally-held' knowledge, but more specific to the environment. In both definitions, tradition is as much about change as it is about continuity, and is as much about the process of knowing (how we come to know something) as it is about what is known of any particular object or thing (Berkes, 2009, 2008).

Anthropologists, of the biological and social varieties, were important in the inception of traditional ecological knowledge, particularly those working on ethno-ecology in the 1970s; that is, the study of how different cultures classify objects, practices and events (Berkes, 2008; Hardesty, 1977). They sought to bring into sharper focus the variety of human-ecological interactions across cultures (Toledo, 1992). As such their work was complemented by scholars working on different aspects of indigenous environmental knowledge, such as the agricultural knowledge of different cultures (Brokensha et al., 1980) and water engineering (Groenfelt, 1991). Some drew on examples from arrangements that emerged in (pre-industrial) Europe and North America. Spain's *huerta* irrigation scheme (Maass and Anderson, 1986) and the Swiss Alpine commons both featured at length in the late Elinor Ostrom's landmark text, *Governing the Commons* (1990).

Others, however, highlighted examples from around the world, in a bid to find alternative common institutions, practices and traditions to those most common in the globalised North; to expand, as it were, the available repertoire (Berkes et al., 2000a; Colding and Folke, 1997; Turner, 2007).

Whilst conservationists find more in traditional ecological knowledge of direct relevance to their subject area, within the development arena 'indigenous' or 'local' knowledge often in a broader, looser sense has become of increasing interest as part of the reappraisal of the role of poor people within the development process. Since the 1980s, there have been energetic efforts from scholars of development to 'rescue' or highlight local or indigenous forms of knowledge, in part stemming from a concern for the adverse consequences of transferring 'Western' knowledge without long-standing analysis of the context in which it is being applied (Warren, 1998). The anxiety that development remained an ethnocentric enterprise gave rise in some instances to efforts which attempted to draw upon local skills, traditions and practices. The most famous instance of this kind of intervention focussed on rural agriculture, using what came to be known as 'indigenous technical knowledge' (ITK), which was best accessed through the 'farmer first' approach. This kind of intervention was very much at odds with conventional agricultural assistance, which took the form of providing farmers with techniques and technologies derived from agricultural science. 'Indigenous technical knowledge' was often deemed complementary or even 'better', applied in the context of its origin, than the knowledge carried by scientific experts sent to work as extensionists with farmers in developing countries (Richards, 1985). 'Farmer first' approaches have been used by Robert Chambers to argue for the need to renegotiate the roles between development professionals and the poor people with whom they work and as an antidote to what he saw as the arrogance and ignorance of 'top-down' approaches to development (Chambers, 1997, 1983; Chambers et al., 1989).

The resurgence of interest in locally-held knowledge, whether ecological, agricultural, hydrological etc., was one manifestation of the re-evaluation of people beyond the richest, most industrialised parts of the world, which recast them not as oppressed, colonised "primitives", but as newly-independent, capable people with rights to a stake in determining their future. Within this broader enthusiasm for local people and places, there are three key contributions of the focus on locally-held knowledge to the genesis of community conservation in particular:

1 Empirical evidence against the separation of humans from nature as the basis for biodiversity conservation (see Chapter 3).
2 Examples of classification of biodiversity and conservation practice.
3 As a repository of wisdom, in respect of humans' relationship with nature, that had been lost amid centuries of industrialisation and globalisation, a 'holistic' theory of knowledge which chimed with the critique of a reductionist approach to science which could see the trees but not the forest, so to speak (Berkes et al., 2008). Part of this knowledge base resides in people's religious beliefs; but this is a topic so large and significant, not least in relation to the potential contribution to conservation and development, that all of Chapter 12 is devoted to exploring it.

What emerged from this reconsideration of the value of locally-held knowledge was an understanding of how humans had contributed to biodiversity which had become a conservation priority (Pimbert and Pretty, 1995). Indeed, some human actions, of the kind conservationists have been known to rail against, were found in some instances to be necessary to the continued existence of particular forms of valued biodiversity. In other words, the ostensible enemy of conservation was at least in some instances its friend.

An increasing number of examples of traditional ecological knowledge started to be documented over time (see Table 10.1). That protected area conservation had left little space for understanding the contributions of local environmental knowledge came itself to be seen as a flaw in the conservation paradigm.

Scientific knowledge, credibility gaps and holistic approaches

Another reason for the rise to prominence of locally-held knowledge was that the trust and credibility invested in scientific knowledge has been increasingly brought into question, on a variety of fronts. There is not the space to explore this important sphere of enquiry (landmark texts are: Barnes, 1983; Barnes et al., 1996; Bloor, 1991; Collins et al., 1992; Jasanoff, 2004; Knorr-Cetina, 1985; Latour, 1993, 1987; Wynne et al., 1994). But Phillips captures well the reaction against

Table 10.1 *Examples of traditional ecological knowledge used for conservation purposes*

Region	Examples (find variety of sources for tropical forests, grasslands, mountains, tropical fisheries and irrigation water, or other categories i.e. drylands)
Polar Arctic	1. Conservation of fish population resilience through the use of nets of mixed mesh size in Cree Indian subsistence fishery in James Bay, Canada (Berkes, 2008, 1981; Berkes et al., 2000b) 2. Inuit customs, practices and restrictions governing the hunting of Caribou (Ferguson et al., 2007).
Africa	1. Enrichment of agricultural plots and forest fallows with perennial plants (Dubois, 1990) 2. Creation of 'anthropogenic dark earths' which contribute to soil fertility and greater forest cover in West Africa (Fairhead and Leach, 2009) 3. Ecological recovery in Kenya despite higher population density (Tiffen et al., 1994) 4. Grazing rotation and alternation practices amongst herders in the Sahel, who follow the migration patterns of wild ungulates by moving their herds to pastures on the fringes of the Sahara during the rainy season (Niamir-Fuller, 1990). 5. Local soil conservation techniques in South Africa (Beinart, 1984, 1984; Beinart and McGregor, 2003)
Asia	1. The conservation of relict forest in Upland Banten, Indonesia (Ellen, 2007) 2. Uses of the Baduy Swidden system to respond to environmental stress in Java, Indonesia (Iskandar, 2007, p. 20) 3. The role of shifting cultivation in preserving wild resources across Asia (Alcorn, 1993; Pimbert and Pretty, 1995a) 4. The use of traditional ecological knowledge in India as a means of monitoring the use of non-timber forest products in Southern India (Rist et al., 2010)
Latin America	1. The human cultivation (through soil enrichment) of the babassu palm across 200,000km^2 of the Amazonia basin (Anderson et al., 1991) 2. The contribution of agricultural practice to the production of Amazonian 'dark earths' (nutrient rich soils) (Lehmann et al., 2003) 3. Shifting cultivation as a sustainable form of agro-forestry in Andean Northwest Argentina (Ramadori et al., 1995) 4. The contribution of farmers' traditional ecological knowledge to the biodiversity of an oasis in the Sonora desert, Mexico (Nabhan et al., 1991) 5. Cuyabena Reserve in Ecuador, managed by the Cofan and other indigenous groups, using indigenous guards, strict rules to limit resource utilisation and on-going wildlife inventories and evaluation programs (Ormaza and Bajana, 2008)

claims to scientific 'truth' that are one of the hallmarks of debates within science and technology studies:

> *No longer can it be claimed there are any absolutely authoritative foundations upon which scientific knowledge is based ... The fact is that many of our beliefs are warranted by rather weighty bodies of evidence and argument, and so we are justified in holding them; but they are not absolutely unchallengeable (Phillips, 1990: 130).*

In contrast, traditional ecological knowledge has in fact been held up as a source of inspiration for the approach to conservation and resource management envisaged by advocates of a more holistic, interdisciplinary scientific paradigm. Berkes has argued that the 'some indigenous groups have resource-use practices that suggest a sophisticated understanding of ecological relationships and dynamics. In particular, many of these examples seem to show an understanding of the key relationships on the land as a whole, that is, a holistic as opposed to a reductionistic view' (2008: 182). He argues that the knowledge held by the Inuit people of (modern day) northern Canada is an instance of a holistic understanding of human-environment interactions (2008). The significance of this argument is that it is not just that TEK is something we might invest with the same level of credibility as we would for resource use practice based upon the scientific method. It is, instead, one source of input and inspiration for a new scientific paradigm. The increasing prevalence of this kind of thinking in the 1980s and 1990s is one factor which explains the enthusiasm amongst proponents of community conservation for extolling the virtues of other sources of wisdom beyond the science which had been the bedrock of conservation interventions from the eighteenth century onwards.

The rise of integrated conservation and development (the wedding of the decade)

The confluence of these interlinked, overlapping strands of thought and activity profoundly influenced people working in both conservation and development spheres, especially in the 1980s and 1990s. The discourse on sustainable development, that had been gaining critical mass from the 1970s onwards, provided the background against which the contours of integrated conservation and development would be drawn. By the 1980s, integrated conservation and development projects (ICDPs) were starting to be posited as an important way to operationalise sustainable development, a form of 'showcase' to demonstrate how, tangibly, development aspiration could be reconciled with caring for the earth. For instance, they were highlighted in *Our Common Future* (Brundtland, 1987).

ICDPs had their roots in protected area conservation. They emerged from the increasing conviction that protected area models could only be viable with the acquiescence of the people who surrounded them, if they were not to become isolated islands of biodiversity that could not survive the pressure put on the areas which surrounded their fenced boundaries (Wells et al., 1992). As such, their origins were in the notion of a surrounding 'buffer zone', the sustainable use of which would ensure the viability of the protected area whilst permitting people to practise livelihood activities, albeit often in modified form.

Community conservation: the poster child of integrated conservation and development

A plethora of terms have emerged to describe the varied approaches that have been clustered under the banner of 'community conservation'. Unsurprisingly, some of the terms overlap: all are in one way or another prefigured by the concept of linking conservation and development processes through the notion of sustainability. Barrow and Murphree (2001) offer a good starting point in this respect, with their typology of different initiatives, as seen in Table 10.2 below. There is one significant omission from this categorisation, namely 'indigenous and community-conserved areas' (ICCAs) – protected areas which have been established and/or enforced locally. These are dealt with in Chapter 9.

Protected area outreach is often an attempt to make some reparation for the problems that the establishment of national parks and other types of protected areas can cause to people who live within or

Table 10.2 *Varieties of community conservation (adapted from Barrow and Murphree, 2001: 3*

	Protected area outreach	Collaborative management	Community-based conservation
Objectives	Conservation of ecosystems, biodiversity and species	Conservation with some rural livelihood benefit	Sustainable rural livelihoods
Ownership/tenure status	State-owned land and resources, i.e. national parks	State-owned land with mechanisms for collaborative management with community. Complex tenure and ownership arrangements	Local resource users own lan and resources either *de jure* de facto. State may have 'las resort' control
Management characteristics	State determines all decisions about natural resource management	Agreement between state and user groups about managing some state owned resources. Management arrangements critical	Conservation as element of land use. Emphasis on rural economic development
*Prevalence in Southern Africa**	Low	Middling	High

*(in comparison with East Africa)

adjacent to them. Although the state retains ownership of the area, affected communities, at least in theory, are to be permitted some minimal level of rights to resource use or some form of compensation if usage rights are prohibited or curtailed. They seek to recognise the problems facing people living near protected areas, generate some benefit for them from the protected area, and solve conflicts between people and management/protection authorities (ibid.). Protected area outreach initiatives are also often known as integrated conservation and development projects (ICDPs).

Collaborative management, entails, ostensibly, the meaningful involvement of all interested parties in management functions and activities related to conservation. In the Southern African context this translates into something resembling the joint management of resources between state and communities, state and the private sector or communities and the private sector. This category describes rather well the kind of arrangements that are often found – and incorporated into national policy – in, for instance South Africa (see Mahony and Van Zyl, 2001; Poultney and Spenceley, 2001; Spenceley and Seif, 2003 for examples of private–public partnerships and collaborations).

Community-based conservation tends to be defined in terms of four main assumptions. First (and perhaps foremost), local people are by and large to be seen as capable and knowledgeable common-pool resource users who are often as well-placed, if not better-placed, as government or private sector actors to manage the resources on which they depend. The second emphasises the need to offer an incentive, be that economic, social or whatever, to people who live with the costs of conservation. Asking someone to continue to bear existing costs or even increased ones is unlikely to meet with success if they have no incentive to do so (Martin, 1994; Murphree, 1997a). The third is that devolution of ownership, responsibilities for and owner-ship over wildlife to local people is the best way to provide this incentive (Murphree, 1997). The fourth is that communal institutions and management structures need either to be supported, resurrected or created in order to allow communities to benefit from wildlife conservation (Barrow and Murphree, 2001).

Probably the most famous example of community-based conservation, and highly influential in the 1990s, was Zimbabwe's Campfire – the Com-munal Areas Management Programme for Indigenous Resources, with an acronym so handy that it quickly became naturalised (see Box 10.1).

Box 10.1

Zimbabwe's Campfire programme

In essence, Zimbabwe's Communal Areas Management Programme for Indigenous Resources (Campfire) became in the 1990s the 'flag bearer' for community-based conservation' (Adams and Hulme, 2001). Campfire was based on two core premises. First, giving people an incentive (principally economic) to live with wildlife could contribute to conservation and development objectives simultaneously. Second, people in communal areas of Zimbabwe, far from being irresponsible and indiscriminate hunters, were well capable of looking after wildlife and in fact had traditions of environmental custodianship that had been interrupted by European colonisation (Murombedzi, 1992).

Both of these ideas chimed with the ZANU-PF governments of the 1980s who were attracted by the 'black empowerment' objectives implicit in the approach (Newsham, 2007, 2002). Campfire sought, then, to reconcile conservation with development through devolving rights to manage wildlife and, crucially, to earn income from management at the local level. In order that wildlife might 'pay its way', Campfire was premised on the notion of sustainable utilisation, which sanctioned a variety of activities – and chief amongst them trophy hunting – which proved controversial in the international conservation arena. Proceeds from trophy hunting were substantial, and in some cases they could pay for school buildings – such as in the oft-cited case of Masoka Ward, with revenues left over for payouts to households which comprised a 56% increase in gross income from the commonly grown cotton crop (Murphree, 1997).

Even though these characteristics made Campfire the 'flag-bearer' for community based conservation – and for ICDPs more broadly – it was also seen as a flawed approach by many of those most involved with it. Principally, some commentators argued, devolution of rights, which were left at the local government level – did not go far enough, because it meant that the people who bore the costs of living with wildlife only received 50% of the revenues, the rest of which stayed with local government – which also received all of the revenues from activities such as trophy hunting and safari tourism (Murombedzi et al., 2001; Murphree, 1997). Ownership did not rest at the local level as a result of what came to be known as 'Murphree's law', namely that every level in an organisation tries to gain more power from the levels above it and tries to resist passing that power down to levels below it (Murphree, 1997). Rather than centralisation, devolution in Zimbabwe's communal areas resulted in what James Murombedzi (1992) christened 'recentralisation'. As instability from farm invasions and land resettlement decimated the tourism industry throughout the 2000s, Campfire lost key revenue sources. Even before interruptions to the tourism industry, however, some were arguing that the economic incentives were, by and large, insufficient to guarantee that people would choose to conserve wildlife in their area (Bond et al., 2001; Emerton et al., 2001).

Despite these difficulties, Campfire continues to exist and to change, not least with the introduction of 'direct payments' of revenues from activities such as tourism being paid directly to local people, in wards such as Masoka (Campfire Association n.d.).

A marriage on the rocks: critiques of integrated conservation and development

The enthusiasm around integrated conservation and development projects was not to last, dwindling throughout and especially towards the end of the 1990s. It seemed the 'happy couple', perhaps blinded by the eagerness of the first encounter, had less in common than they had at first thought. Certainly, people working both in conservation and development increasingly questioned how realistic the assumptions that underpinned the ICDP 'logic' were, and particularly in relation to the feasibility of 'win-win' outcomes. This was chiefly because the accumulated efforts over the course of the 1980s and 1990s had yielded few examples of initiatives which achieved both their conservation and development objectives (Newmark and Hough, 2000; Roe, 2008; Wells et al., 1999, 1992). The problems that the conservation and development communities identified with ICDPs reflected their own priorities; and it is these differences in starting point and objectives that lie at the heart of the difficulties of making ICDPs work. In practice, a trade-off between conservation and development objectives – and one which tended to satisfy neither conservation nor development practitioners – was the norm, whilst the win-win scenario remained an isolated occurrence (Brown, 2004; McShane et al., 2011; McShane and Wells, 2004; Minteer and Miller, 2011; Salafsky, 2011). To be sure, the conclusion was *not* that 'win-win' scenarios were impossible *per se;* it was simply that they were improbable and correspondingly infrequent.

Conservationist critiques

For their part, conservationists did not find ICDPs to be sufficiently robust vehicles for conservation. Biodiversity was often not conserved because the development incentives offered for its conservation were insufficient (Emerton et al., 2001b) or were available regardless of whether conservation objectives were met or not (Newmark and Hough 2000). Often, the livelihoods-based activities through which development objectives were pursued were in fact seen as harmful to conservation (Newmark and Hough 2000). Sometimes it was difficult even to tell whether conservation objectives were being met because ICDPs frequently lacked mechanisms for monitoring biodiversity levels before and after intervention had occurred (Brandon & Wells

1992; Wells et al., 1992). Something of a backlash occurred in relation to the capacity of 'indigenous' conservation and to local common-pool resource management more broadly (Minteer and Miller, 2011; Wells et al., 1992). Was it wise to devolve control over conservation decisions to people who might not have sufficient capacity to make them or who might have priorities which were not readily compatible with conservation objectives?

To add insult to injury, it was becoming increasingly difficult to access conservation funding for a project which was not able to demonstrate development benefits (Wells et al., 1999; Roe, 2008). Some conservationists started to voice the concern that they were now burdened with doing poverty reduction whether they wanted to or not, taking them far beyond their core competencies and raising fundamental questions about the purpose of conservation (Redford et al., 2008). Worse still, biodiversity had fallen out of the development agenda owing to the rise of poverty reduction as the overarching priority for development intervention. Development projects were not automatically expected to deliver conservation; nor had they tended to become, in the round, much more concerned with environmental protection as a result of their contact with the world of conservation (Roe, 2008; Minteer and Miller, 2011).

Development disillusion

People starting from a development perspective shared the concern that the benefits generated by ICDPs were frequently insufficient to provide incentives for conservation or to become financially sustainable (Brandon and Wells, 1992, Newmark and Hough, 2000, Wells et al., 1999). Yet their focus lay more squarely on the human costs entailed by this lack of incentives. Across Africa and Asia examples abounded of ICDPs which required people living in buffer zones to curb or forgo staple livelihood activities, such as shifting cultivation or livestock herding, in order to meet conservation objectives. The alternatives offered, such as revenue sharing or training in making arts and crafts, often did not adequately compensate this loss of livelihood (Brockington et al., 2006; Emerton et al., 2001; Homewood et al., 2013; Neves and Igoe, 2012).

Even where the revenues generated through conservation activities were significant, there was a well-documented tendency for them to be captured either by people who did not have to live with the costs of conservation or local elites. For instance, in Kenya, Maasai herders have captured very little of the revenue generated by the protected areas which restrict their pastoralist livelihoods: 95% of tourism revenues, for instance went to tour operators, service industry workers and the state (Norton-Griffiths, 2007; Norton-Griffiths and Said, 2009). Similar concerns were raised about who received the economic benefits from ICDPs more broadly across Africa and Asia (Roe and Elliott, 2004). From very early on in the ICDP experience, it was clear that local elites were also easily able to capture benefits (Brandon and Wells, 1992), to the detriment of the broader 'communities' they represented. Indeed, the very notion of communities as stable, homogenous entities was criticised for oversimplifying both the level of diversity between different groups and for underplaying the extent to which these differences determined who gained from ICDPs and who did not (Brockington et al., 1996; Chapin, 2004; Wolmer, 2007). For these and other reasons, the ability of ICDPs to contribute to development was increasingly called into question (Newmark and Hough, 2000; Roe, 2008).

Other commentators offered a more bristling, political reading of the proliferation of integrated conservation and development projects. In a review of ICDPs in Africa, Roderick Neumann argued that the rhetoric of participation and benefits sharing masked an effort to extend the range of 'coercive forms of conservation practice' and to consolidate the state's presence in remote areas (Neumann, 1997: 559). For Neumann, ICDPs in the African context were intimately bound up with a) the politics of land, tenure and access; and b) Western notions of 'the other', which unhelpfully categorised people living adjacent to protected areas either as 'good natives' (nature conserving) and 'bad natives' (modern, destructive) (ibid.). In Latin America, Mac Chapin (2004) infamously attacked the work of three large international conservation NGOs for their attitudes towards and adverse impacts upon the lives of indigenous peoples in Latin America and elsewhere; though conservation commentators derided Chapin for what they held to be a poor grasp on the facts (Misc., 2005a, 2005b).

Exceptions to the ICDP 'failure narrative'? Namibia's conservancy programme

There are some important exceptions which, arguably (and it remains a contested field), do generate conservation and development benefits simultaneously, through a genuine engagement with people at the local level, and in ways which contribute substantially to local livelihoods. One of the most widely acclaimed examples of this is Namibia's conservancy programme.

In their review of ICDPs, Newmark and Hough (2000) argued that whilst many were deeply flawed, there was another 'generation' of project which was offered better prospects for achieving conservation and development simultaneously. They named Namibia's conservancy programme as a prime example of this better approach. The conservancy programme's own public reports appear to offer rich evidence of the conservation and development benefits that it offers (NACSO, 2013, 2011). One of its most influential proponents has stated, 'I'd challenge anyone to find a development programme anywhere in the world that delivers the same conservation and economic value over the same time period as the conservancy programme has.'[1] Nevertheless, not all commentators share this bullish outlook.

The conservancy programme is an example of community based natural resource management (CBNRM). It is essentially is a nationwide initiative which permits Namibians living on state-owned communal land to request the devolution from central government of limited rights to manage wildlife and benefit from tourism operations. Private landowners already possessed these rights, but the legislation which made this possible was modified in 1996, in order to extend them to communal land inhabitants,[2] with a view to ensuring an 'economically based system of sustainable land management and utilisation of game' (GRN, 1996: 14). As such, the conservancy programme was from the outset in the mould of 'wise use' conservation outside of the protected area, an attempt to give the people who live with wildlife (and the costs of so doing) sufficient reason to view it as an asset, not a hindrance. In the round, the incentives derive from income generated largely through wildlife tourism and trophy hunting, but other activities, such as sustainable plant harvesting and craft-making, are also common. Another key conservancy function is to monitor wildlife numbers and illegal hunting through setting up a 'community game guard' network.

Forming a conservancy is voluntary. If a group of communal land residents complies with the requirements of the legislation, the government will declare it a legal entity. It will then receive assistance from NACSO (the Namibian Association of CBNRM Support

Organisations), to: help form the conservancy committee; explore and attract revenue sources; set up community game guard structures; and devise natural resource use plans based on land zoning according to activity (farming, trophy hunting, etc.). This has proved a very popular formula, with 82 conservancies having been registered by March 2014 (mainly in the Northwest and Northeast), covering a surface area of 160,092km^2 – roughly 19% of the total land area of Namibia (NACSO 2011) – see Figure 10.2. By 2011, it was estimated that total revenues flowing into conservancies totalled

Figure 10.2 *MET Map of gazetted (legally established) communal area conservancies in Namibia 2013*

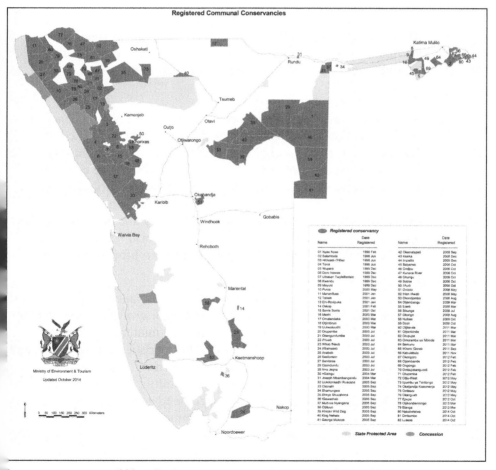

N$49,859,433 (Namibian dollars), or US$4,995,935.17 (by 2013 exchange rates).

On this basis, the conservancy programme can lay claim to many achievements. Yet it has also come in for criticism on a number of fronts. Perhaps chief amongst these is whether the development benefits are as strong as the conservation ones, and the extent to which a focus on the former constrain the scope for generating the latter. Findings from WILD (Wildlife Integration for Livelihoods Diversification) a multi-year research initiative, concluded that

a) whilst the conservation impact was impressive, the revenues gener-
    ated by the conservancy programme did not trickle down in great
    quantities to conservancy residents, or convert into significant live-
    lihood benefits;
b) there were important issues with elite capture of benefits;
c) 'addressing issues of development and livelihood support
    will require some reorganisation of support priorities'
    (Long, 2004: 4).

More recent research appears not to contradict the WILD conclusions on the livelihood/development benefits. In a study conducted in Kwandu conservancy, Suich (2013) 'found no positive impacts on the multiple dimensions of poverty'. Jones et al. (2013) are considerably less dismissive of the conservancy programme's contribution on this front. They dispute the charge of elite capture of benefits, but con-clude, nevertheless, that it is limited to poverty alleviation – i.e. short term relief from the symptoms of poverty (World Bank, 2001) but does not reduce poverty, i.e. is not sufficient to lift people out of poverty in the longer term.

Other critics have questioned whom the conservancy programme best serves. Sian Sullivan argues that 'CBNRM *in practice* maintains the interests of conservationists, tour operators, hunters and tourists; i.e. those conventionally associated with 'touristic' enjoyment of, and financial benefits from, wildlife and 'wilderness' (Sullivan et al., 2003: 165). The predominant focus on wildlife, for Sullivan, means also that other resource practices, specific to gender and ethnicity, do not make it onto the conservancy programme's agenda, such as the resource-gathering practised by Damara women across North-west Namibia. Key conservancy programme actors have vociferously contested this argument.

Finally, in research conducted in 2004, one of the authors of this book argues that participation in the conservancy programme cannot accurately be characterised as 'grass-roots', in the way that it frequently has been (Newsham, 2007). This is chiefly because the knowledge required for planning, establishing and running a conservancy is not held by conservancy residents, would-be or actual. It is held, instead, by the actors who know not just about the practicalities of implementation, but have been able to use international discourses – principally related to sustainability – to make the integrated conservation and development paradigm so influential in Namibian policy circles and to attract the funds necessary to operationalise the programme. This dynamic renders these actors indispensable and, simultaneously, to a significant extent defines what there is to participate in, even before a conservancy has been established. This is not automatically to delegitimise such actors or the conservancy programme, and does not mean that local preferences are not taken into account as a matter of course. Yet understanding how best to respond to the asymmetry, in terms of who has the necessary knowledge,

Figure 10.3 Twyfelfontein Rock art in Uibasen Twyfelfontein conservancy, Kunene Region, Namibia

Source: A. Newsham

between implementers and conservancy residents is central to the efficacy and legitimacy of the programme as a whole.

Acrimonious or amicable divorce? Conservation and development in the twenty-first century

By the mid-to-late 2000s, integrated conservation and development projects had become less common. By the early 2000s, even funders, such as USAID and the UK DFID, who had acquired a reputation for expertise and commitment to initiatives which used wildlife to reduce poverty moved away from offering financial support, perhaps vindicating those conservationists convinced that development actors had given up on biodiversity (Hutton et al., 2005). But whilst the debate has moved on, many of the core issues remain. Once the momentum around integrated conservation and development fragmented, the links between community and conservation were less easy to spot, but nevertheless re-surfaced in a number of different discourses and interventions. By the dawn of the twenty-first century, the ICDP concept had lost much its appeal for development funders and conservation practitioners, and both communities of practice effectively moved on, into different approaches, such as payment for ecosystem services, ecosystems-based adaptation and the tendency towards a 'back to the barriers' style of conservation. These are explored variously in different chapters of this book, but to a greater or lesser extent such approaches contribute to our understanding of, or are variously affected by, the relationship between conservation and poverty, which in the twenty-first century has received increasing attention, which we proceed to consider in this chapter.

Poverty and conservation: what is the link?

In large measure, the interest in the relationship between poverty and conservation was spurred by the emergence of and totemic status accorded to the millennium development goals. It was adopted, by and large, by actors who accepted that the efforts to do conservation and development simultaneously were likelier to be characterised by trade-offs than by win-win outcomes. It was driven in many ways by an effort to gain a more precise understanding of the relationship(s) between poverty and biodiversity. That there was a relationship, and a fairly straightforward one at that, had been taken as a given in earlier debates around integrated conservation and development, but it had been poorly specified, by and large. Perhaps the best-known and most widely-accepted description of this relationship is that offered by the Millennium Ecosystems Assessment (Hassan et al., 2005) – see Figure 10.4. Yet this relationship had been discussed in over-general terms which overemphasised some elements of it whilst underplaying others (Rangarajan and Shahabuddin, 2006).

Figure 10.4 *Relationship between ecosystems and poverty*

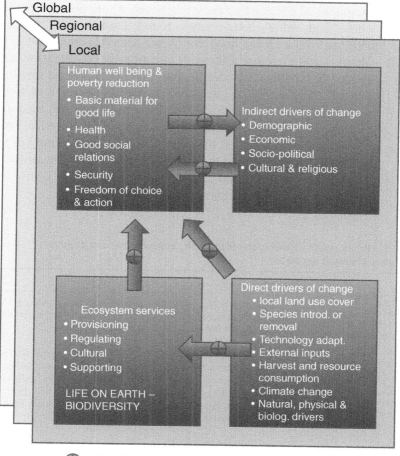

Source: Hassan et al., 2005

Roe and Elliott (2010) argue that a key reason for this over-generalising tendency stems from the variety of actors and starting points – all underpinned by differing objectives, values and ethics – of those contributing to the debate. Adams et al. (2004) have identified four broad positions within the conservation-poverty discourse, described in Table 10.3.

Aside, though, from questions of starting points and related values, more (though probably not enough) work has been done in the first decades of the twenty-first century to inject greater precision into our understanding of the links between poverty and conservation.

One area of research on these links has been around the extent of geographical overlap that occurs between where poor people live and where 'high value' biodiversity is found. Turner et al. (2013) present findings from a global study which suggests that areas rich in biodiversity – and therefore a focus of attention for conservation, are also frequently inhabited by poor people. The implication is that it is difficult and possibly counterproductive to separate efforts to manage biodiversity from those aimed at poverty reduction, because they will necessarily take place, in broad terms, in the same physical space. Turner et al. also find that the land containing biodiversity of the highest value, in terms of the ecosystem services it underpins, provides 56% of the essential ecosystem services that are of use to the world's poorest people (ibid.).

Table 10.3 *Broad positions within the biodiversity conservation and poverty debate (after Adams et al., 2004)*

Position	Values	Instruments
1. It is better to pursue conservation and poverty objectives separately	Conservation does require strong protective measures but need only contribute indirectly to poverty objectives	Strictly enforcing existing PAs; bringing more land under protected status
2. Poverty constrains conservation	Conservation will not work if it does not address poverty	PA outreach; revenue sharing arrangements to compensate for restricting livelihood activities
3. Conservation should not constrain poverty reduction	Conservation is only legitimate when it does not contribute to poverty reduction	Full compensation of costs of conservation; revenue from ecotourism; limited harvesting of resources
4. Conservation is a tool for poverty reduction	Conservation can help lift biodiversity-dependent people out of poverty	Conservation outside PAs; focus on conservation of biodiversity that contributes to livelihoods; support for common-pool resource management

Another important part of the link is the greater dependence the poorest are frequently held to have on biodiversity in their livelihoods. Wittmer et al. (2013) argue that, because of this dependency, standard economic measures of GDP do not adequately capture the ways in which environmental degradation affects the poorest more than it does other groups. Yet, as Vira and Kontoleon (2013) point out, caution should be exercised in statements about the dependence of 'the poor' on biodiversity. They highlight the need to separate out the widely-varying relationships that the different people contained within the lumpen category of 'the poor' have with different forms (and states) of biodiversity to which they may have highly differentiated access. Agrawal and Kent draw attention to an even more fundamental difficulty. In their review of interventions that seek to achieve biodiversity conservation and poverty reduction objectives simultaneously, they conclude that 'it may be inappropriate even to pose a question such as "what is the relationship between conservation and poverty?"' (2006: 32). This is partly because they find there is no shared understanding across interventions as to what poverty and biodiversity actually are; and partly because at best, poverty alleviation and biodiversity conservation efforts can only address small elements of this overarching relationship. Action to improve a particular dimension of poverty, or a particular species, may have adverse implications for other dimensions or species. We still do not have the evidence base, they maintain, to understand how the 'bigger picture' is affected by these interventions which address only smaller pieces of it, each intervention in its own idiosyncratic way.

Space and perspective for community conservation?

By now, it should be clear how it is that integrated conservation and development in the twenty-first century has been superseded by other approaches and responses, such as the 'back to the barriers' tendency (see Chapter 9), payments for ecosystem services (see Chapter 13) and, of course, the focus on the relationship between poverty and conservation. This is not surprising, given that it was seen so widely to have 'failed' when measured up against both conservation and development yardsticks, leading many to conclude that it was perhaps simply too difficult to achieve. Yet whilst it has been 'abandoned' as an approach, the issues that gave rise to it remain. There continues to be a need to integrate conservation and development, and there continue to be trade-offs between conservation and development objectives which, in practice, frequently remain unresolved. Perhaps Marshall Murphree, ever the engaging wordsmith, comes closest to capturing this dilemma and our response to it to date. Community conservation, he suggests, 'has not been tried and found wanting: it has been found difficult and rarely tried' (Murphree, 2000, cited in Hutton et al., 2005: 363).

Summary

- Against a background of a rapidly ascending discourse around sustainability, people working on conservation and development started to think they might be able to cooperate and integrate; and to a much greater extent than either group had previously been able to envisage.

- The key concept through which to integrate conservation and development, at least when it came to implementing projects, was the 'community'.

- The rehabilitation of 'community' was prompted by a re-evaluation, both amongst conservation and development thinkers and practitioners, around the abilities of local people and institutions to conserve, as well as the development benefits that could be derived from community conservation.

- The enthusiasm for integrated conservation and development translated into a substantial number of interventions, across the globe.

- However, the majority of integrated conservation and development projects were not deemed to deliver enough of either. The idea of the 'win-win' outcome had by the turn of the century been displaced by a more sober series of reflections on the trade-offs entailed by trying to do conservation and development simultaneously. The backlash against the earlier optimism also led some conservationists 'back to the barriers', arguing that the disappointing outcomes of ICDPs vindicate a return to (and even stricter enforcement of) conventional protected area instruments.

- In the 'post-ICDP' landscape, there is a fragmentation of its basic premises across a number of different approaches – including payments for ecosystem services (see chapter 13), ecosystems-based adaptation and a reframed discourse around conservation and poverty. Whilst the ICDP approach has faded, there remains widespread acceptance of the idea that there needs to be a reconciliation between conservation and development, and that the 'communities' bearing the costs of conservation must be better treated. Whether the approaches that aim to continue this conceptual project have learned the lessons of ICDP experiences remains to be seen.

Discussion questions

1. Why did trying to reconcile conservation and development through community conservation come to seem like such a good idea?
2. How has the legacy of integrated conservation and development projects shaped the thinking of conservationists and practitioners in the twenty-first century?
3. Was Marshall Murphree right to argue that community conservation 'has not been tried and found wanting: it has been found difficult and rarely tried'?
4. Considering what happened to ICDPs, would it be better to try to keep conservation and development activities separate, after all?

Notes

1. Author (Newsham) interview with Chris Brown, director of the Namibia Nature Foundation, 3.12.2003, Windhoek.

2. As a result of land allocation policy under colonial rule in the twentieth century, Namibia – or erstwhile South–West Africa – had been divided into private land, given to or purchased largely by white settlers, and state owned land, in which the pre-colonial populations lived or were re-settled into, often against their will. State land was renamed communal land after independence in 1990.

Further reading

Adams, J.S., and McShane, T.O. (1992) *The Myth of Wild Africa: conservation without illusion.* University of California Press, Berkeley.

Adams, W.M., Aveling, R., Brockington, D., Dickson, B., Elliott, J., Hutton, J., Roe, D., Vira, B., and Wolmer, W. (2004) Biodiversity conservation and the eradication of poverty. *Science* 306: 1146–1149.

Brandon, K., and Wells, M. (1992) Planning for people and parks – design dilemmas *World Development* 20: 557–570.

Brokensha, D., Warren, D.M., and Werner, O. (1980) *Indigenous Knowledge System. and Development,* University of America: Maryland.

Chapin, M. (2004) A challenge to conservationists. *World Watch* November/ December.

Newmark, W.D., and Hough, J.L. (2000) Conserving wildlife in Africa: integrated conservation and development projects and beyond. *BioScience* 50.

Neumann, R.P. (1997) Primitive ideas: protected area buffer zones and the politics o land in Africa. *Development and Change* 28: 559–582.

Roe, D., Elliott, J., Sandbrook, C., and Walpole, M. (eds) (2013) *Biodiversity Conservation and Poverty Alleviation: Exploring the Evidence for a Link,* John Wiley & Sons.

Online resources

Millennium Ecosystem Assessment http://www.unep.org/maweb/en/index.aspx (accessed 25.03.2014)

Poverty and Conservation Learning Group - http://povertyandconservation.info/ (accessed 25.03.2014)

Ecosystems-based adaptation:

http://www.ebaflagship.org/ (accessed 25.03.2014)

http://www.unep.org/climatechange/adaptation/EcosystemBasedAdaptation/ tabid/29583/Default.aspx (accessed 25.03.2014)

The Conservancy Programme, Namibia

Namibia Ministry of Environment & Tourism http://www.met.gov.na/Documents/ Conservanciea%20simple%20guide.pdf (accessed 25.03.2014)

WILD Project – http://www.eldis.org/go/topics&id=16810&type=Document#. Ul6o8JSDTx4 (accessed 25.03.2014)

Indigenous/Traditional ecological knowledge

Arctic Circle http://arcticcircle.uconn.edu/HistoryCulture/TEK/ (accessed 25.03.2014)

IUCN – http://www.iucn.org/about/union/commissions/sustainable_use_and_livelihoods_specialist_group/sulinews/issue_2/sn2_tek/ (accessed 25.03.2014)

UNESCO – http://www.unesco.org/new/en/natural-sciences/priority-areas/links/ (accessed 25.03.2014)

ELDIS Indigenous knowledge and climate change – http://www.eldis.org/go/topics/resource-guides/climate-change/key-issues/indigenous-knowledge-and-climate-change#.Ul6rt5SDTx4 (accessed 25.03.2014)

Indigenous Peoples' and Community Conserved Territories and Areas (ICCAs) http://www.iccaforum.org/ (accessed 25.03.2014)

Integrated conservation and development

World Bank – http://go.worldbank.org/29R7O0RNI0 (accessed 25.03.2014)

WWF – http://wwf.panda.org/who_we_are/wwf_offices/bhutan/projects/index.cfm?uProjectID=BT0874 (accessed 25.03.2014)

⬤11 Ecotourism, conservation and development

- ⬤ Ecotourism: origins and definitions
- ⬤ Ecotourism in practice: conservation pay-offs and trade-offs
- ⬤ Ecotourism in practice: development pay-offs and trade-offs
- ⬤ Ecotourism, politics and contested possibilities

Introduction

Ecotourism, for some advocates, has the potential to reconcile conservation and development, for two broad reasons First, it can be used to bolster the competitive advantage of economically poor countries that are nonetheless rich in biodiversity and beautiful landscapes. Second, the revenues that can be generated from the conservation of such landscapes serve, in theory, as a powerful incentive for conservation.

This chapter explores the pay-offs and trade-offs which have made ecotourism – and relevant varieties such as responsible, pro-poor and community-based tourism – so variously compelling and frightening to people and organisations within and beyond conservation and development. It explores the origins of ecotourism in the bigger tourism sector with some people working in conservation and/or development camps became increasingly disillusioned, and surveys the greater conceptual sophistication and nuance which came to characterise definitions of and thinking about ecotourism as the term caught on and then stuck. It presents evidence on the effectiveness of ecotourism as a tool for both conservation and development objectives, before proceeding to explore the politics of who benefits from ecotourism and how.

Ecotourism: origins and definitions

Ecotourism and the rise of mass tourism

The roots of ecotourism can be found in the 1970s. They are bound up with changes that were occurring within conservation and development thinking, but also emerged in response to the staggering growth trajectory and increasingly behemoth-like status of mainstream tourism. The tourism industry had, by the 1970s, come a long way from its nineteenth-century origins. From the second half of the twentieth century onwards, the diminishing costs of air travel made it available to a much wider cross-section of the populations of industrialised countries. The exponential growth experienced by an increasingly global tourism industry lead it, by the 1990s, to vie with oil as the largest global industry. By 2006, receipts from tourism (including transport) reached US$8.8 trillion (UNWTO, 2006). By 2012, it accounted for 9% of GDP, one in eleven of all jobs globally (UNWTO, 2013). Tourist arrivals numbers have skyrocketed in recent decades, from 25 million in 1950 to 1,035 million in 2012 (UNWTO, 2013). Nor does this phenomenal growth pattern show signs of stopping any time soon: projections are for annual growth in tourism arrivals of 3.3% worldwide average up to 2030, which would mean 5 to 6 billion domestic tourists and 1.8 billion international tourists numbers by that time (UNWTO, 2013).

Because it had become such a powerful engine of economic growth, tourism started to be seen as an avenue for the development of poor countries, some of which were increasingly important destinations for tourists from wealthier nations. Whilst many embraced mass tourism as part of their development strategies, its downsides were starting to be more apparent. Mass tourism brought overdevelopment in some places, uneven development in others, with only small amounts of the revenues generated remaining in country. Frequently the only jobs that local people could get within the travel industry were low paid positions such as hotel cleaners (Honey, 2008). The idea that mass tourism was a clean industry was challenged by mounting evidence of the magnitude of its adverse environmental impacts. Such concerns became increasingly vocal and 1970s. They were not enough to arrest the increasing expansion of tourism in developing countries, which continue to vie for and attract an ever-greater number of international tourist arrivals, as evidenced by a 30% growth in their share of the market between 1980 and 2012 (UNWTO, 2013). However, space was prised open for a niche form of tourism, one which was more concerned with mitigating both environmental and cultural impacts. Moreover, actors and institutions working both in conservation and development arenas would become increasingly interested with what these days we term ecotourism.

Ecotourism and conservation origins

Predominantly, ecotourism activities take place within areas of natural beauty afforded some type of conservation protection status by governments NGOs, or private landowners; and most of all, national parks (Honey, 2008). Chapter 9 discusses the merits and shortcomings of protected area approaches to conservation and development in more detail, but tourism, be it adventure, nature-based or more recently ecotourism, has become of fundamental importance to protected area networks. Ecotourism, which is in theory as much about benefiting local people and demonstrating cultural respect as it is about conservation, has been seen as a vehicle for this project in many developing countries. It has become an important part of the re-evaluation of how to do protected area conservation in ways which have fewer social costs and are more effective.

Dovetailing with these changes, Latin America played host, in the 1970s, to a rethink of the relation between conservation and tourism. In response to the threat to tropical forests posed by logging, mining, forest clearance for settlement, etc., Latin American thinkers such as Gerardo Bukowski (1976) and Héctor Ceballos-Lascuráin (1988) viewed ecotourism as a means by which to set the development prospects offered by tourism into a more symbiotic relationship with conservation objectives. By raising revenue for conservation initiatives and enlisting tourists into the 'struggle' for environmental protection, they hoped to use tourism as part of an effort to head off excessive depletion of tropical forests (Honey, 2008). Broadening out this conservation focused vision, early pioneers of ecotourism ventures such as Michael Kaye argued that it had to be as much about cultural exchange, learning and respect for the local as it was about the quest to save biodiversity (ibid.). This chimed with the thinking of Kenton Miller (1982), encapsulated by the term 'ecodevelopment', which emphasised the contribution that protected areas could make to national development through more environmentally responsible forms of tourism. As a result of this emergence of thinking around ecotourism in the parallel contexts of Africa and Latin America, the term started to find its way into the wider discourse on sustainable development.

By the time the IUCN organised its Fourth World Congress on National Parks and Protected Areas in Caracas, Venezuela, in 1992, the relationship postulated between tourism and conservation has been widened significantly beyond the point of tourism serving to fund strict, protection-oriented objectives, as this policy recommendation demonstrates: 'In developing greater cooperation between the tourism industry and protected areas the primary consideration must be the conservation of the natural environment and the quality of life of local communities.'[1] At the same event, the IUCN set up a small consultancy unit, headed by Héctor Ceballos-Lascuráin, to offer advice and technical support on planning ecotourism activities into parks management, expanded in 1996 and renamed as the Task Force on Tourism and Protected Areas.

Ecotourism and development origins

Conservationists were not alone in turning first to tourism and then ecotourism as a means through which to achieve their objectives. As mentioned above, the tourism boom of the second half of the twentieth century came to be seen as integral to the development prospects for poorer countries. In this regard, the multilateral development institutions, and in particular the World Bank, led the way, founding in 1969 its Tourism Projects Department. The idea was that tourism would allow countries with fewer alternative economic opportunities to attract foreign investment and earn foreign exchange (Hawkins and Mann, 2007). Between the 1960s and 1990s, support from actors like the World Bank for tourism would wax and wane (Honey, 2008; Trương, 1990), in part owing to questions over whether tourism in fact was an effective way to reduce poverty and scepticism over its capacity to transform the economic prospects of poorer countries, but increasingly as a result of criticism of the environmental impacts of other large-scale World Bank projects (Honey, 2008).

From the viewpoint of developing countries, the idea that some forms of tourism could serve their objectives better – and might be able to serve conservation and development goals simultaneously – gained currency in view of often disappointing experiences with mass tourism between the 1960s and 1980s, as much in economic as in environmental terms. The attraction of ecotourism lay in its potential to generate revenues for developing states in ways which were less destructive than alternative land uses, such as mining, logging, agriculture, etc. An increasing body of literature pointed to the possibility that attempting to use nature in non-consumptive ways could, at least in some cases, be more lucrative than any other option. For instance, in South Africa, one study has suggested that net income from wildlife was more than ten times that generated by cattle ranching (Grossmann and Koch, 1995). Because nature-based tourism became the largest, or one of the largest, owners of foreign exchange for a number of countries, others became more interested. Early adopters, such as Costa Rica, where the entire country is often marketed as an ecotourism destination, were followed by states such as Cuba, Russia, Rwanda, South Africa, Uganda. These countries tended to offer ecotourism attractions in specific places, as opposed to billing themselves as an ecotourism destination (Honey, 2008).

However, whilst states have been important actors in the spread of ecotourism, they have largely been superseded by the private sector. This is because in many ways, ecotourism is a manifestation of the influence of processes of neoliberal globalisation. State-run tourism enterprises have in large measure been privatised, and the vast majority of medical tourism enterprises are privately owned (Honey, 2008). From the very outset, ecotourism has put economic incentives and market mechanisms at the centre of efforts to realise conservation and development objectives simultaneously.

Defining ecotourism

Perhaps the most accepted, one-sentence definition of ecotourism is that offered by the International Ecotourism Society: 'Responsible travel to natural areas that conserves the environment and improves the well-being of local people.'[2] Much like the term 'sustainable development', this definition comprises a broad church, housing a congregation whose members can differ quite wildly in the emphasis given to environmental or development objectives, and to how demanding the standards must be, in order for an initiative to count as genuine ecotourism. In other words, people can use the term to refer to tourism developments which pay lip service to environmental or development objectives, or which are fundamentally organised around the provision of these. To the extent that the majority of uses of the term describe ventures which only pay lip service to environmental conservation or contributing to local well-being, then ecotourism may only be another form of 'green-washing'. A more comprehensive way of understanding the meaning of ecotourism is to approach it as a set of principles, against which to assess the legitimacy and accuracy of the claims of initiatives that refer to themselves as ecotourism. Martha Honey (2008: 29–30) has given a comprehensive list of what ecotourism is supposed to entail, captured by the following headings:

- Involves travel to 'natural' destinations
- Minimises impact
- Builds environmental awareness
- Provides direct financial benefits for conservation
- Provides financial benefits and empowerment for local people
- Respects local culture
- Supports human rights and democratic movements

The advantage of such a wide ranging definition is that it is very comprehensive, covering the many different associations ecotourism has. However, it sets the bar very high and even 'the committed of the committed' (Costas Christ, cited in Honey 2008: 28) have found it very difficult to claim that their ecotourism initiatives adequately meet all of these goals.

However elusive genuine ecotourism may be when measured against these standards, this kind of definition allows us to make clear distinctions between ecotourism proper and some of the other kinds of tourism which are conflated with it. For instance, it is quite clear that nature-based tourism is not quite the same thing as ecotourism if it does not contribute to development, but is solely about taking tourists to sites of natural beauty. Nor does adventure tourism (tourism in natural environments based on high-adrenaline activities like white water rafting), nor even pro-poor tourism make the cut by these criteria. Furthermore, some of these kinds of tourism are more clearly relevant to the conservation and development agenda, with pro poor tourism being a case in point. The examples given in this chapter are restricted to those which give us clearer insights on the extent to which ecotourism is capable of achieving conservation and development objectives simultaneously.

Ecotourism in practice: conservation pay-offs and trade-offs

Conservation through private sector led ecotourism

The main attraction of ecotourism from a conservation point of view is that it holds out the prospect of delivering the revenues necessary for intervention, whilst giving incentives and benefits to people who might be adversely affected by conservation activities, sufficient that their activities or behaviour will not compromise conservation objectives. Overall, it is estimated, ecotourism may be worth up to US$210bn per annum (Kirkby and Yu, 2010). In the context of tropical rainforest conservation in Peru, Kirkby et al. (2010) have argued that it outcompetes most other land uses, and in particular agriculture and livestock farming, over the short (1 year) and medium (25 year) term. In this case, they suggest, what makes most sense in economic as well as conservation terms is maintaining the rainforest, and indeed the lodges that have most benefitted from its presence have, understandably, been central to efforts to stave off encroachments

Figure 11.1 *Parismina eco-lodge, Costa Rica*

Source: www.eco-hotel-costa-rica.com (accessed 8 June 2015)

and threats it has faced in recent years, whilst income generated has boosted revenues available to the Tambopata National Reserve (Kirkby et al., 2010).

The Peru example is very much about a particular model of ecotourism, in which well-resourced, knowledgeable and capable private-sector actors run ventures which are very good at attracting affluent visitors. The group of lodges and guest houses that Kirkby et al looked at included a significant amount of Peruvian ownership, which may mean, somewhat atypically, that a substantial portion of the revenues stayed within the country, and perhaps within the local area. It looks, therefore, like a good advert for private investment as a way to secure both conservation and development objectives. However, resorting to the private sector carries a significant risk of 'greenwash'; that is, companies who adopt the ecotourism brand but which actually make only tokenistic contributions to conservation or, worse, operate in ways which are environmentally unfriendly under this cover.

Indeed, it could be – and has been, argued – that looking to the private sector to ensure that genuine ecotourism becomes the norm is simply to ignore the way in which the broader tourism industry functions, and the extent to which this has little to do with conservation or development objectives. In short, it has for quite some time been the case that control over the tourism market is increasingly concentrated within the hands of relatively few actors who are best placed to capture the revenues from tourism (Truong, 1990; Honey, 2008a). Such actors tend to be supportive of a global economic order built on free trade and economic (neo) liberalisation in the context of ever expanding economic growth (Honey, 2008a). The extent to which continued economic growth and sustainable development are in direct contradiction is, of course, at the very heart of the debate over whether conservation and development objectives truly can be reconciled. Some commentators are concerned that the free trade agenda poses the greatest risk to the prospect for ecotourism, because international treaties such as the General Agreement on Trade and Services (GATS) favour transnational interests with little commitment to conservation over the kinds of smaller, independent enterprises which are associated with a greater commitment to the principles of ecotourism (Thompson and Coyle, 2005). Martha Honey (2008a) notes that genuine ecotourism remains 'the road less-travelled', a form of tourism which, for all its impressive growth rates in recent decades, remains at the margins of the tourism industry. What is more likely: that ecotourism becomes the norm for the whole global tourism industry, or that it comes over time to resemble its larger, older sibling? This is surely the most troublesome question for ecotourism advocates, and perhaps especially those who want to see it achieve conservation objectives through a reliance on the private sector.

One way to make it easier to distinguish between ventures which comprise more genuine efforts towards ecotourism principles and those which are 'greenwash' is through certification. This may be defined as a 'formal process under which a nominally independent body certifies to other interested parties, such as tourists, marketing agencies and regulators, that a tourism provider complies with a specified standard' (Buckley, 2002: 197). Mirroring the growth rates of the ecotourism sector, green tourism certification schemes mushroomed in the early 2000s, such that by 2008, there were as many as 80 of them, some local, some regional and some global (Honey, 2008). Because of the potential for confusion resulting from the proliferation of different standards, the credibility of accreditation – that is, procedures for certifying the certification schemes themselves – has become increasingly important. Through a process which started in the late 1990s and which continues through international meetings such as the Mohonk workshop of 2000 and the World Ecotourism Summit in Quebec, 2002, efforts have been made to involve a wide variety of actors to agree on the principles for accreditation (Honey, 2002; Buckley, 2002). On the back of this, a global body, the Sustainable Tourism Stewardship Council was formed in 2009, but quickly merged with the Global Sustainable Tourism Council.[3] Certification schemes may follow one of two different methodologies: process-based, that is, evaluating the management procedures and practices of a particular initiative; and performance-based, that is, using checklists of environmental, social and economic criteria that have to be satisfied, benchmarks against which to measure performance (Honey, 2001). Despite these differences, certification schemes tend to have five characteristics in common:

1   enrolment is voluntary, not obligatory;
2   initiatives which meets the certification criteria can use the logo of the certifiers, as effectively a brand that can be identified by consumers;
3   their criteria meets or exceeds government regulations;
4   they require assessment and audit procedures to be conducted;
5   they bestow membership upon certified initiatives and charge fees for their services.

In many ways, certification looks to be a way of pinning down what ecotourism is, so that greenwashing can be avoided; or at least easily identified. However, it may cause harm if governments end up abdicating responsibility to untrustworthy organisations (Gunningham and Grabosky, 1998), or if they are used to delay or block stronger forms of legislation (Buckley, 2013). Critically, moreover, it is not clear that they modify the behaviour of actors who do not sign up to the scheme and as such, they may be less effective than government regulation (Buckley, 2002). Perhaps most crucial of all, to the extent that ecotourism remains a marginal activity within the broader tourism sector, the conservation effects of even genuine, certified ecotourism initiatives remain comparatively limited. These concerns are common within thinking on voluntary, self-regulating modes of environmental governance (see Chapter 6).

Community-based ecotourism and conservation

Partly because of scepticism towards the mainstream tourism industry and the role of the private sector in it, community-based ecotourism has received a lot of attention, and as much from conservation organisations – like WWF (see, for instance, their guidelines on community based ecotourism – Denman, 2001) or Conservation International – as from development actors such as USAID and the World Bank, both of whom have funded ecotourism initiatives handsomely (Kiss, 2004). The logic is appealing: use ecotourism as a means to reduce local level poverty, thereby incentivising communities to conserve the biodiversity. Additionally, community-based ecotourism holds out the prospect of financial sustainability as it can be made to work properly, in contrast with the finite resources of project aid and the difficulties which initiatives face when the funding for them runs out (Ferraro and Kiss, 2002). Claims made about successful ecotourism projects often coincided with the point at which when enthusiasm for community-based ecotourism as a conservation strategy was at its peak. For instance, Greenquist (1997) reported that in Garífuna, a small village within the Río Plátano Biosphere, Honduras, local people set up patrols on the beach to locate and hide loggerhead and leatherback turtle eggs from those who were illegally harvesting them, with a view to ensuring that the reserve would remain a tourist attraction. In Kenya, Waithaka (Waithaka, 2002) found that the Il Ngwesi wildlife sanctuary, set up by a group of Maasai, played host to higher numbers of trees, herbs and wildlife, when compared with the surrounding ranch land. Concerns over the validity of the claims made in these examples have been voiced (Kiss, 2004). A more robust example may be the conservancy program in Namibia, which runs annual game counts, which replicates the same methodology from one year to the next, thus providing a longitudinal dataset. On the strength of these claims, practitioners involved can more credibly claim rises in the number of wildlife spotted in conservancy areas over a 12-year period (NACSO, 2013). On the back of this, conservancy program supporters contend that Namibia has the largest free roaming (i.e. outside of national parks) black rhino population in the world (NACSO, 2013).

Nevertheless, it is surprisingly difficult to find unambiguous conservation success stories within community-based ecotourism. Less surprisingly, evaluations of community-based ecotourism projects often voice the disillusion which characterises evaluations of integrated conservation and development more broadly (see Chapter 10). A common problem is that ecotourism does not always provide people with sufficient incentive, in economic terms, to support biodiversity conservation, especially when it has adverse implication for livelihood strategies such as farming (Bookbinder et al., 1998; Coria and Calfucura, 2012). Frequently, this is because of a lack of capacity at the local level to establish and run professional ecotourism ventures (Aylward, 2003; Salafsky et al., 2001). Efforts to remedy this lack of capacity have come via joint ventures with private-sector actors well established in the tourism arena (Wunder, 2000). Whilst these can help, private-sector actors have quite frequently been deterred owing to lack of tenure security and control over resources, differences of priority relative to the communities they work with, and a reluctance to shoulder the social costs that accompany working in poor and marginal areas (Kiss, 2004; Magome et al., 2000). It has come to be commonly held that community-based ecotourism can only work under certain conditions and, crucially, in terms of delivering confirmation of objectives, rarely or never on its own (see Box 11.1).

Perhaps most crucial crucially of all, from a conservation perspective, community-based ecotourism is not necessarily a very good fit with modern conservation objectives. Because it is not necessarily suited to particular environments – for instance high mountains and deserts (Coria and Calfucura 2011) – ecotourism is not always able to contribute to the objective of protecting a wide range of ecosystems (Kiss, 2004). Moreover, ecotourism can cause environmental damage.

Box 11.1

Conditions under which community-based ecotourism (CBT) can contribute to conservation

In a review of 37 community-based tourism enterprises in Asia and the Pacific, Salafsky et al. (2001) identified the following conditions which need to be in place in order for conservation objectives to be achieved:

- Enterprise factors – an economically viable enterprise must be established. Out of the 37 enterprises that Salafsky et al. reviewed, four had no revenues, three had minimal revenues, 13 covered only the variable costs, 10 covered other variable and fixed costs and only seven made a profit. Overall, there was a weak association between enterprise and conservation success, but a very strong association between community involvement and conservation success.

- Enterprises must generate benefits to stakeholders in enterprises – surprisingly, Salafsky et al. find relatively little association between conservation achievements and the provision of cash-based incentives; non-cash benefits were sufficient to ensure conservation objectives were achieved.

- Stakeholders must have the capacity to recognise and counter threats to biodiversity – both leadership and some level of control over the land on which biodiversity is present, along with the ability to enforce usage rules, are crucial in this regard.

Kreg (2003) documents how management tools such as controlled burning, clearing of vegetation, and artificial water points are used in game ranches in KwaZulu Natal in order to enhance the tourist experience, but in ways which entail a loss of landscape resilience. Similarly, the astronomic expansion of ecotourism on the Galapagos Islands has brought a litany of problems to a fragile environment, in addition to numerous examples of greenwashing (Rozzi et al., 2010). Finally, community-based ecotourism is rarely suited to maintaining sufficiently large conservation areas, as they tend to be much more limited in size (Kiss, 2004; Rutten, 2002). Again, though, Namibia's conservancy programme may be an important exception, where an area of 152,686km^2 is covered by contiguous conservancies (NACSO, 2013).

Ecotourism's ecological footprint – a spanner in the works?

In recent years, even the environmentally friendly reputation of ecotourism has received a fundamental challenge. This is mainly because the way in which environmental harm is measured has changed considerably since the introduction of the ecological footprint measure. A variety of attempts to provide indicators for sustainable development have been made, which have been used to measure the impact of tourism on destination countries (see for instance Bossel, 1999). However, these indicators did not take into account the impact of travel between one country and another (Marzouki et al., 2012). In contrast, the ecological footprint tries to calculate the amount of biologically productive land and sea areas required to produce what people consume and to absorb waste and emissions, such as CO_2 and equivalent gases, that result from consumption (Wackernagel and Rees, 1998). Crucially, this measure does allow the impact of international travel to be incorporated. Marzouki et al. (2012) deploy the ecological footprint analysis to compare the impacts of mass tourism in Tunisia and ecotourism in the Seychelles.

They find that the ecological footprint for Tunisia is lower per capita than that for the Seychelles, chiefly because the international travel component vastly outweighs all other impacts from tourism, because of the amount of fossil fuels required to transport tourists to the destination. Given the paucity of data they admit to finding in Tunisia and the assumptions they need, therefore, to make in order to conduct their analysis, these results need to be treated with some caution. Moreover, the comparison is not of specific ecotourism initiatives and their local impacts with mass tourism ones, but rather with a country which does not claim to be an ecotourism destination and another which does. Therefore, the study may be an indication of levels of green-washing in the Seychelles, rather than an indictment of the environmental benefits of ecotourism ventures specifically. Nevertheless, other commentators make a similar argument, namely that without addressing the carbon emissions caused by flights – and in particular the long haul flights Northern eco-tourists travelling to remote, 'exotic' destinations in the South often take – and also lengthy overland trips in four-wheel-drive vehicles to reach remote spots, ecotourism cannot claim to be sustainable.[4] In purely environmental terms, the most obvious solution is simply for eco-tourists to reduce as much as possible the length of the journey that they make, perhaps chiefly by not travelling abroad on holiday. However, this drives a wedge between the conservation and development objectives central to most definitions of ecotourism. The southern countries seeking to use ecotourism to contribute to their development strategies would lose out if this were the direction the ecotourism industry were to take. On this reading, therefore, far from the panacea it is sometimes painted as, ecotourism may actually turn out to be part of the problem; at least for as long as it contributes to the increasing use of long-haul air travel.

The work of Marzouki takes its cue from the work of scholars such as Stefan Gössling and colleagues. Their seminal paper on tourism in the Seychelles concluded, 'Environmental conservation based on funds derived from long-distance tourism remains problematic and can at best be seen as a short-term solution to safeguard threatened ecosystems' (Gossling et al., 2002: 209).

Ecotourism in practice: development pay-offs and trade-offs

Development benefits from ecotourism

The case for ecotourism as a development strategy looks on the surface to be quite straightforward. The United Nations General Assembly unanimously adopted a resolution which recognises the role that ecotourism had to play in 'the fight against poverty and the protection of the environment' (UNWTO, 2013: 23). Put simply, the argument goes, ecotourism can, at least in some places with sufficiently lucrative natural attractions, outcompete other forms of land use both in terms of revenue generation and job creation. It is one of our best working examples of the economic incentives stacking up in favour of the conservation of nature, whilst holding the promise of a profound contribution to development objectives. In Kenya, revenues from wildlife viewing have been estimated to be 50 times greater than those generated by cattle ranching (Honey, 2008). A single lion is held to be worth US$575,000 (Honey, 2008). Furthermore, the ecotourism sector is growing, and at a rate which has been estimated to be three to four times faster than the tourism industry overall (UNWTO 2013). As such, it seems to have much to offer as a development option, and particularly for countries which are concerned to invest in a sustainable future. There are, indeed, instances of countries that have lined up to put ecotourism central to their development strategy. In the 1990s, Brazil committed to investing US$200 million into ecotourism in its nine Amazonian states.[5] In 2003, Malawi proposed using ecotourism to 'jump start' its economy.[6] Where some led, others appear to have followed: according to some sources, 83% of developing countries now depend on ecotourism as a major export (UNEP, 2013). Some caution may be required when assessing the veracity of such statements: this one probably incorporates nature-based and other kinds of tourism excluded from stricter ecotourism definitions. And as we have seen, claims that the tourism industry of entire countries, such as Costa Rica, is dominated by ecotourism are deeply problematic. Whilst it is probably fair to say that ecotourism has come to be seen as less of a panacea than it once was, there are still examples to be found which yield tangible development benefits.

The most important benefits at the local level tend to be in terms of income generated from alternative livelihood activities and funding for the provision of public goods. Income provided allows people to diversify their livelihood strategy and give them a source of cash to invest in human, physical or financial capital, and thereby in theory allowing them to deal better with or recover from shocks and stresses more easily (Chambers and Conway, 1992; Lepper and Goebel, 2010). An example of the provision of public goods comes via the Posadas Lodge in Ecuador, which provides revenue for schools, water supply and health services (Gordillo et al., 2008). However, even aside from the question of how significant the level of benefits is relative to the share taken by other actors (see below), the provision of benefits is rarely straightforward and indeed can often be contentious, as the case of the Makuleke people of South Africa amply demonstrates (see Box 11.2). Fundamental to the provision of benefits is the way in which the participation of local people in an ecotourism venture unfolds; yet despite the 'participatory turn' which swept both the conservation and development arenas in the 1990s, substantive participation of poor people is far from guaranteed (Cater et al., 1994).

Box 11.2

Contentious benefits amongst the Makuleke people in Kruger National Park, South Africa

The topsy-turvy story of the Makuleke people illustrates very well the difference that can be made in terms of development benefits from using ecotourism as a means of conservation instead of the approaches, based upon stricter enforcement, which characterised protected area conservation for the better part of the twentiethth century. Framing it as an unambiguous success story, however, would be misleading. When, in 1969, the land they were living on was declared part of Kruger National Park, the Makuleke people were forcibly evicted by armed soldiers and shunted into crowded, agriculturally marginal land (Robins and van der Waal, 2008). The precedent for this removal stretches back to the creation of the park itself, when an estimated 3000 Tsonga people were removed from a territory falling between the Sabie and Crocodile Rivers (Ramutsindela, 2002). In 1998, a landmark court case ordered that their lands be restored to the Makuleke people, who then proceeded to take full occupation of the area from which they had been rejected. However, this act of restitution was dependent on the condition that revenues could be generated only from ecotourism activities. This was a trade-off, because it ruled out any prospecting and potential mining activities in the area (ibid.). Moreover, whilst a hunting concession was permitted initially, residents were asked to forego the rights to hunt and to earn money from the concession when they agreed to the establishment of the luxury Pafuri eco-lodge. This was paid for with investment from Wilderness Safaris, a South African company with experience of tourism joint-ventures with local communities. The rationale was that, even though in the short term, it would lead to a decrease in revenues accruing to residents, the deal would establish a better revenue base and the opportunity to develop skills in the tourism industry over the longer term (Castley et al., 2012). This is not a perfect solution: it leaves the Makuleke as co-managers of land which is legally theirs. Moreover, it hinges much more upon land redistribution and restitution legal mechanisms created in post-apartheid South Africa than it does on the willingness of Kruger National Park authorities to ensure that it was managed in a way which generated most benefit for those who had borne the highest costs of its establishment. Yet it is clearly a great deal better than the settlement that preceded it. In this case, ecotourism appears to address, albeit partially, historical injustices brought about by the creation of Kruger National Park.

Furthermore, benefits can be rather volatile, leaving those who come to depend upon them at the mercy of the upturns and downturns of the world economy, on the back of which visitor numbers and tourism spending overall can rise and fall sharply (Coria and Calfucura, 2012). They can also be overtaken by events on the ground, not least in cases of conflict, which can send visitor numbers plummeting. Figure 11.2 demonstrates the effects of political turmoil upon visitor numbers to a site in Masoala National Park, Madagascar. Whilst some tour operators may be able to withstand dramatic drops in numbers, which then pick up the following year, as in this case, it can be more difficult for people who are attempting to use ecotourism to lift themselves out of poverty.

Another key factor which impinges upon the level of benefits that can be generated through ecotourism is the capacity, or lack therein, of governments to support its development. In Tanzania, for instance, a key reason why many of its national parks remain in a good environmental state is not because of careful management of visitor numbers or other such measures; it is simply that the infrastructure that would

Figure 11.2 *Effect of political uncertainty on visitor numbers to Nosy Mangabe, Masoala National Park, Madagascar.*

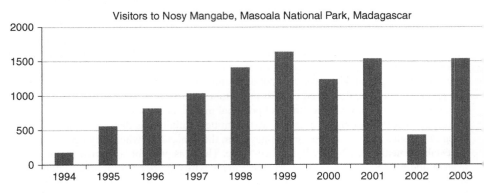

Visitors to Nosy Mangabe, Masoala National Park, Madagascar

Source: Reproduced from Ormsby and Mannle, 2006; with kind permission of Taylor & Francis

permit greater numbers of visitors to arrive is still being built (Honey 2008). Moreover, because Tanzania is increasingly locked into a more conventional beach tourism model which depends upon package tours, there is not always the space or the political will to engage with communities or to enforce environmental regulations (ibid.). Even in countries which are billed as ecotourism destinations, like Costa Rica, the capacity of local government to keep up with the pace of eco- and other forms of tourism has been outstripped (Koens et al., 2009).

When governments are so ill prepared to support the development of ecotourism, it should come as no surprise that local people often lack the skills and training required to run a successful ecotourism venture (Sakata and Prideaux, 2013; Schellhorn, 2010; Standish, 2009).

This is not to say that these kinds of support never yield benefits. For instance, in Ecuador, the Achuar community has gained management and marketing skills as well as financial capital to build a lodge, from Canodros S.A., a private company that specialises in nature-based and cultural tourism (Stronza, 2009). Similar arrangements are generating benefits in Botswana (Lepper and Goebel, 2010). Yet such examples are comparatively rare.

Who most benefits from ecotourism?

The biggest critique of ecotourism from a development perspective is that the benefits are very unevenly distributed and rarely reach the poorest. They tend, it is argued, to be captured by wealthier private sector actors, be they international investors, operators in the origin countries of ecotourists or those operating at the destination (Honey 2008). This process is commonly referred to as leakage. In mass tourism in developing countries, leakage rates have been estimated at anywhere between 40 and 90% of tourism revenue (Benavides Diaz, 2001). Ecotourism is supposed to run counter to this trend, but this argument is not always supported by the evidence. In a study of ecotourism in Chira Island, Costa Rica, Trejos and Chiang (2009) caution that positive economic impacts at the local level may be limited, polyvalent and sporadic in character. In the same country, Seales and Stein (2012) found that the extent to which tour operators educated and employed local people, sourced provisions locally, and used locally owned hotels and lodges did not correlate with commercial success. Whilst they did posit a relationship between the level of commercial success of a tourism enterprise and the provision of conservation benefits, they found little supporting evidence of a relationship between commercial success and community benefits. In a number of countries including the Bahamas and Nepal, Mowforth and Munt report that as little as 10% of the overall revenues from ecotourism stay within a community (Mowforth and Munt, 2009). Nevertheless, even with high levels of leakage, the benefits provided can be locally significant. Sandbrook (2010a) argues that leakage studies can employ flawed methods and overlook local benefits because they do not compare the revenues that are retained with other sources of income. In the context of Bwindi Impenetrable National Park, Uganda, he finds that whilst leakage was estimated at 75%, the remaining revenues exceeded all other sources of revenue in the area combined (ibid). Elsewhere, Sandbrook argues that high cost naturebased tourism may be no more effective than conventional tourism as a tool for generating local benefits from conservation (Sandbrook, 2010b). This was because whilst overall spending was greater than cheaper tourism packages, locally retained revenue was dependent on length of stay. This points to an important distinction to make in debates about leakage, namely that leakage is higher with some types of ecotourism than with others (Ceballos-Lascurain, 1996). For instance, in Bolivia, it was found that backpackers generated three times more revenue for local people than did those on a package tour (Meijer 1989, cited in Ceballos-Lascuráin 1996), because they stayed longer and used smaller, locally owned accommodation, restaurants, etc.

If it is important to distinguish between types of ecotourism, it is from a development perspective of even greater importance to differentiate between posited beneficiaries, as not all 'local' people are impacted equally by eco- or mass tourism. Some groups benefit more than others, as is made clear by the literature on indigenous people and ecotourism. Indigenous people frequently have a history of marginalisation and exclusion, and not least in relation to the establishment of protected areas (Zeppel, 2006). In the 1990s, international conservation organisations and bodies became much more receptive to engaging with and recognising the rights of indigenous people, especially to continue to live on and use the natural resources within lands to which they have historical claim – even when these lands overlap with protected areas. At the 1996 World Conservation Congress, the paramount body of the IUCN, adopted seven resolutions on indigenous peoples which formalised this recognition (IUCN, 1997). By the time of the World Summit on Sustainable Development in Johannesburg, 2003, the need for protected areas to be integrated 'equitably with the interests of all affected people' was enshrined in the Durban Accord (IUCN, 2003: 14).

On the back of this momentum the idea of the indigenous protected area rapidly gained currency (see Chapter 9 for more details). Likewise, ecotourism came to be seen as a means through which indigenous people could benefit from conservation. Despite these changes, it is frequently difficult for indigenous people to benefit from ecotourism. One of the principal reasons for this is the lack of legal status of the claim of many indigenous peoples to the land they inhabit. In Chile, for instance, Bonham et al. (2008) point out that fewer than 23% of the indigenous communities living within the Mapu Lahual Park are in possession of legally recognised land titles. The result is that they have been reluctant to exploit investment opportunities out of a fear of losing their lands. Despite efforts, especially in Latin America, to recognise the land claims of indigenous peoples, many still live on land whose ownership is legally disputed (Meza, 2009). The second significant reason relates to the capacity that indigenous people – along with many other groups – lack in relation to planning, business management, financial management, marketing, and other skills essential to the running of an economically viable ecotourism enterprise (Ashley and Jones, 2001; Zeppel, 2006). Because of this lack of skills and low levels of financial capital, it is very difficult for indigenous people to secure loans or other necessary investment for an ecotourism infrastructure (Fuller et al., 2005).

Aside from differences between groups, there are also important differences within groups when considering the distribution of ecotourism benefits. An insightful example of the importance of such differences is offered in Matthias Schellhorn's study (2010) of Lombok, Indonesia into the socio-economic outcomes of nature-based tourism within a heterogeneous community of migrant settlers and native residents. Schellhorn argues that although the majority of the area's tourism attractions derive from the indigenous heritage of the *wetu telu* group, few economic benefits reach them. Local women are particularly marginal to development opportunities from tourism, as benefits predominantly accrue to males, be they incoming migrant groups or *wetu telu*. Schellhorn identifies many barriers to greater integration of the *wetu telu* into the local tourism, not least amongst these that the local economy is only partially monetised and has a culture of hospitality that is at odds with defining encounters with visitors to the area in terms of payments for services rendered. For the *wetu telu* to engage more fully in tourism would require a fundamental change to their forms of social and economic organisation and would, therefore, be something of a cultural and economic imposition. This points to a more fundamental dilemma both within and beyond ecotourism and it is a common theme within debates around what development is, who it is for and on whose terms (Orlove and Brush, 1996; Weaver, 1999; West, 2008).

Looking in-depth at community-based ecotourism in Namibia

If one of the necessary attributes of genuine ecotourism is that it contributes to local-level empowerment by putting communities at the heart of the decision making process, the Daureb Mountain Guides Association (DMGA) of north-west Namibia amply fulfils this criterion. The DMGA was formed by a group of residents living in or near Uis, in the Erongo Region and offers guided tours of rock art – and in particular the White Lady – found at what is more commonly known as the Brandberg mountain, known in Damara-Nama (the local language) as Daures.

The roots of the Daureb Mountain Guides can be traced back to an environmental club at Petrus !Ganeb Secondary School in Uis. Students who would later become mountain guides first learned about the Brandberg Mountain at this club, and in particular about the tracks and trails which led to the White Lady and other famous instances of

Figure 11.3 *The Daureb (or Brandberg) Mountain in the rainy season*

Source: A. Newsham

rock art. It was at the environmental club that the idea of the potential
for deriving income from tourism was first mooted. In 1993 the six
founding members of the group started to offer tours (Newsham,
2004). Although rather loosely structured in its beginnings, the
DMGA slowly became a more professional group throughout the
1990s and into the 2000s. With support from the European Union,
Raleigh International, the Namibian Community-based Tourism
Association and private actors such as Wilderness Safaris, the DMGA
received tour guide training, acquired a uniform and started to be
marketed with national tour operators and travel agencies. A modest
site infrastructure was developed, including a reception building, car
park, environmentally friendly toilets, protection for the rock art and
better trails to guide tourists to it.

A study by Lapeyre (2010) demonstrates that the development benefits
generated by the guides are significant, locally, at four levels:

1   Average monthly income from guiding was estimated at
     N$1056 per month (including tips), which is over double the
     average monthly wage of N$410, and the average hourly rate of

N$3 for farm work, for an area in which the unemployment rate is estimated to be 46%.

2   Guiding offers a stability of income that few other local livelihood strategies can match, and which meant that the guides did not have to migrate to other regions in order to find work.

3   The revenue gave guides the opportunity to save and to invest in assets, both physical, such as donkeys and carts, and intangible, such as their children's education.

4   Revenues from guiding stayed in the area, with multiplier effects for the local economy.

The Daureb Mountain Guides Association can also claim quite forcefully to contribute to environmental objectives. On site, the guides have responsibilities to protect the rock art and the local environment from potentially damaging behaviour from tourism. Furthermore, the DMGA operates out of Tsiseb conservancy, Namibia's flagship, community-based, natural resource management initiative (see chapter 10 for more details on Namibia's conservancy programme) and contributes 10% of its revenues to the conservancy.

Figure 11.4   *Members of the Daureb Mountain Guides in 2004*

Source: A. Newsham

In spite of these unambiguous benefits, the DMGA is a vivid example of the dilemma faced by community-based tourism enterprises. Whilst it makes sense in local terms, it is considered much less impressive if judged by the standards of the Namibian tourism industry, in which, to the extent that it figures, it is a marginal player. Lapeyre's study of the Daureb Mountain Guides Association suggests that: 26% of visitors to the Brandberg Mountain are on a guided tour already; 38.4% report that a travel agent chose at least some of their destinations; and 50.2% used a travel agent or a tour operator to book travel services to or within Namibia. Like many other community-based tourism enterprises (CBTEs) in Southern Africa (Spenceley, 2010), the DMGA has struggled to gain broader market access and is advertised in only a very limited fashion. The guides are viewed as insufficiently professional by the major tour operators, owing to reports of guides not turning up to work on time and giving tours whilst drunk. Over time, revenues have stagnated, uniforms have become careworn and not been replaced, and greater conflict has characterised interactions within the Association itself. Arguably, the support that has been offered to the guides has not been good enough to help them transform into a commercially viable venture (Lapeyre, 2010). Perhaps it is also the case that the guides have not wanted to become the kind of organisation demanded by the Namibian tourism industry (Newsham, 2004). Partly owing to the perceived lack of professionalism, the National Heritage Council of Namibia has reasserted control over revenues accrued from the guiding activities, further reducing the amount available to the guides by taking a 50% cut of their revenue stream, albeit with a view to putting it towards site maintenance. The end result is that the kind of tourism that the DMGA stand for – which is genuinely appreciated by the community and well adjusted to local cultural norms – remains marginalised, especially in terms of the global commodity chain which is where the lion's share of income from tourism is to be found (Lapeyre, 2010).

Ecotourism, politics and contested possibilities

A number of commentators have concerned themselves with the politics of how particular conceptualisations of ecotourism come to be favoured as a pathway for conservation and development, with a view to unpicking the implications for people and landscapes of how this process unfolds. Much of this analysis centres on pinpointing the politics inherent in whose visions of ecotourism determine, effectively, what it comes to mean in practice, and then considers who wins and who loses as a result.

Some have asked why ecotourism comes to be so dominant as a policy option in the first place. Again, the example of Costa Rica comes to the fore, in work which seeks to explain the politics lying behind the explicit decision to foreground ecotourism within the country's broader development strategy and publically declare the country an ecotourism destination. A number of scholars have argued that the exceptional importance of environmental issues to Costa Ricans politics owes chiefly to the influence of a small but well organised group of Costa Rican conservation scientists and environmentalists (Evans, 2010; Steinberg, 2001). In addition to their own elite background and close, often personal, ties to high-ranking politicians, this group had links with transnational actors, through which they gained access to funding from US universities and international conservation organisations (Steinberg, 2001; Wallace, 1992). Lynn Horton wrote a richly nuanced account of the kinds of ecotourism that resulted, on the Osa Peninsula, from this politics of 'environmental exceptionalism'. She traces the many-splintered trade-offs and payoffs, winners and losers that accompanied a situation (see Chapter 9) in which '"Global" environmental claims were privileged over local claims to land and natural resource use' (Horton, 2009: 96).

A recurring theme in this literature is that ecotourism, like other forms of conservation and development, is often better at achieving environmental objectives than it is at achieving development ones, and that this is no coincidence, but rather a reflection of a political dispensation which systematically set conservation above development. In the context of ecotourism, the work of Jim Butcher (2007, 2011) is perhaps archetypal. Butcher sees predominant ecotourism narratives as a way of justifying development in terms of basic needs, upholding a distinction made between developed people who can enjoy the fruits of nature as provided by the global economy, and traditional indigenous or similar people. The latter group must stick to using nature in ways which are compatible with conservation objectives, this usage then being defined as the only appropriate form of development available or justifiable. Because this limits pathways to development to marginal activities with no real potential for poverty reduction, he advocates a separation of conservation and development as a necessary step towards opening space for alternative visions of

Box 11.3

Ecotourism and neoliberal agendas in Madagascar

Duffy (2008) views ecotourism in Madagascar as one of the ways in which nature is being 'neoliberalised'. In other words, ecotourism is effectively being used to extend a way of thinking, making decisions and gaining control over the use of nature which revolve around its monetary value. Whilst there are claims that these processes are going to bring development to local communities and to involve them centrally in that decision-making process, it is less than clear that this is what will actually unfold. Duffy points to the networks of actors who are brought together by the prospect of the win-win scenario that ecotourism represents. These comprise a mix of international and local actors, private and state, conservation and development focused, who bring different interests and agendas which converge to some extent around the idea that only through giving nature sufficient economic value can conservation and development be achieved simultaneously. This coalescence has the effect of legitimising neoliberal policy prescriptions, such as the shrinking of the state, the greater involvement of non-state actors, and of course crucially a reliance on market mechanisms. There were innovative initiatives which promised conservation and development outcomes through ecotourism, such as a new, ostensibly people-friendly protected area category known as a '*site de conservation*', and a luxury eco-lodge that invests in development projects. However, Duffy argues that similar initiatives elsewhere have struggled to achieve conservation and development objectives (see Chapter 9). Whilst it was too early to gauge the outcomes of these initiatives, they were underpinned by the notion of ecotourism as a market mechanism to bring about conservation and development and accepted by a wide range of actors, from the World Bank to WWF. The danger, for Duffy, was that as in many other initiatives which claimed to be community-based, it would be a narrow elite of people who would most benefit, and whose gains would be obscured by a discourse of participation, conservation and development.

development, which need not be environmentally destructive but hold open the prospect of more fundamental transformation and prosperity

Other commentators have focused on the extent to which ecotourism, whilst positing itself as an alternative to mass tourism, is just as concerned with capital accumulation and plugs neatly into the current global political and economic status quo (Mowforth and Munt, 2009) As such, it is fundamentally bound up with unequal distributional impacts, and underpinned by an ideological commitment to competition, individualism, consumption and commodification of nature (ibid). It is much less interested in challenging prevailing power structures or addressing inequality, but instead may leave poor people disempowered, fragmented and subject to unsolicited and unwelcome forms of socio-cultural change (Duffy, 2002; Mowforth and Munt, 2009). Work by Rosaleen Duffy in Madagascar (2008) explores the extent to which ecotourism is another means through which nature is being 'neoliberalised', that is, where the tenets of neoliberal thought increasingly regulate human interactions with nature, through the medium of the market. (See Box 11.3).

Summary

- Both conservation and development constituencies share a strong interest in ecotourism, and its origins lie in the work of both conservation and development policymakers and practitioners.

- Ecotourism is a term with many different definitions, some more stringent than others. Broadly speaking, ecotourism initiatives must at the very least improve the conservation status of an area and deliver benefits to local people.

- Ecotourism started off, like many integrated conservation and development projects, being vaunted as a panacea. The sheen, however, has long since worn off.

- Like much of the rest of the ICDP discourse, ecotourism is very explicitly about using market mechanisms to achieve conservation and development objectives.

- Whilst ecotourism in many ways is one of the most lucrative examples of market based conservation and development, it is also constantly at risk of providing cover for 'greenwash'; although the advent of ecotourism certification schemes have sought to tackle this problem.

- Because of its commitment to free markets and trade, some commentators see ecotourism as a way of 'neoliberalising nature', that is, of bringing nature more and more under the control of market mechanisms.

- As long as ecotourism remains 'the road less travelled', it risks co-option by a broader tourism industry which is very far from internalising environmental objectives into its core model, or to contributing to reducing poverty and inequality.

Discussion questions

1. How much hope should those who wish to reconcile conservation and development objectives place in ecotourism?

2. Who benefits most from ecotourism?

3. How stringently and precisely should we define ecotourism?

4. Has ecotourism been 'eaten up' by a broader tourism sector driven by the workings of a neo-liberal, capitalist political and economic order?

Notes

1. http://data.iucn.org/dbtw-wpd/html/tourism/section14.html (accessed 19.2.2014)
2. https://www.ecotourism.org/book/ecotourism-definition (Accessed 20.2.2014)
3. http://sdt.unwto.org/en/content/global-sustainable-tourism-criteria-and-council-gstc (accessed 13.04.2014)
4. Strasdas, Wolfgang (2010), 'Ecotourism and climate change – how to adapt to a new challenge?' Presentation given at the European Ecotourism Conference 2010, http://www.slideshare.net/fullscreen/Ruukel/euro-eco-2010-wolfgang-strasdas/1 (accessed 10.03.2014)
5. http://www.world-tourism.org/sustainable/IYE/quebec/cd/stat%20mnts/pdfs/sobrae.pdf#search=%22proecotur%22 (accessed 28.2.2014)
6. http://ens-newswire.com/ens/%20jan2003/2003-01-09-02.asp (Accessed 28.2.2014)

Further reading

Honey, M. (2008) *Ecotourism and Sustainable Development: Who Owns Paradise?* (second edition). Island Press.

Kirkby, C.A., Giudice-Granados, R., Day, B., Turner, K., Velarde-Andrade, L.M., Duenas-Duenas, A., Lara-Rivas, J.C., and Yu, D.W., 2010. The Market Triumph of Ecotourism: An Economic Investigation of the Private and Social Benefits of Competing Land Uses in the Peruvian Amazon. *Plos One 5.*

Kiss, A. (2004) 'Is community-based ecotourism a good use of biodiversity conservation funds?', *Trends in Ecology & Evolution*, 19(5): 232–237.

Marzouki, M., Froger, G., and Ballet, J., (2012) Ecotourism versus Mass Tourism. A Comparison of Environmental Impacts Based on Ecological Footprint Analysis. *Sustainability* 4: 123–140.

Mowforth, M., and Munt, I. (2009) *Tourism and Sustainability: Development, Globalisation and New Tourism in the Third World*, 3rd edition, Taylor & Francis.

Online resources

The International Ecotourism Society – http://www.ecotourism.org/ (accessed 03.04.2015)

Centre for Responsible Travel http://www.responsibletravel.org/home/index.html (accessed 03.04.2015)

United Nations World Tourism Organisation http://www2.unwto.org/ (accessed 03.04.2015)

The International Centre for Responsible Tourism http://www.icrtourism.org/ (accessed 03.04.2015)

International Centre for Ecotourism Research, Griffith University (Australia) http://www.griffith.edu.au/environment-planning-architecture/international-centre-ecotourism-research (accessed 03.04.2015)

Global Sustainable Tourism Council http://www.gstcouncil.org/ (accessed 03.04.2015)

12 Conservation and development in the context of religious, spiritual and cultural values[1]

- Linking conservation and development with religion
- Can faith groups make an important contribution to conservation and development?

Introduction

This chapter examines the challenges and opportunities in applying religious, spiritual and cultural values to conservation and development. It looks at the problems and prospects of faith-based conservation and development. A variety of individuals and groups affiliated with what are conventionally called 'world religions' have provided moral inspiration for historically notable partnerships focusing on sustainable development and environmental conservation. Examples include: Oxfam (Oxford Committee for Famine Relief), which was initiated in 1942 by the Oxford University Vicar Canon Milford, now works in over 100 countries across the developing world and has an annual budget of over US$300 million (Oxfam, 2010); the Aga Khan Development Network, which was started by a spiritual leader of the Shia Ismaili Muslims, and with an annual budget of over US$450 million, helps to bring together a number of development agencies, institutions, and programmes that work primarily in the poorest parts of Asia and Africa (AKDN, 2010); the WWF (World Wide Fund for Nature), which in 1986 initiated a dialogue with five of the world's most prominent religions – Buddhism, Christianity, Hinduism, Islam and Judaism – about how the tenets of their faiths could help environmental conservation; the Alliance of Religions and Conservation was formed in 1995 to enhance links between secular organisations engaged in conservation and what its leaders considered to be the world's major faiths (ARC, 2010). The efforts of these non-governmental organisations to link religion to conservation and development were further spurred by an important initiative undertaken in 1997 by James D. Wolfensohn, then the President of the World Bank, and George Carey, the Archbishop of Canterbury (United Kingdom), who organised a meeting that brought together senior leaders from major world faiths. This meeting forged greater links between conservation and development organisations and faith groups under the auspices of the World Bank. This collaboration has facilitated discussion of ways in which religious groups could help conservation and development (WFDD, 2009). The World Bank has since been working with faith-based groups to reduce poverty and to promote conservation in the developing world.

A growing body of literature has suggested that conservation and development are both driven by ethical or moral values and can earn legitimacy through cultural acceptance, public engagement and mass support (Votrin, 2005; Orrnert, 2006; Van Houtan, 2006; Child, 2009; de Groot and Steg 2009; De Cordier 2009). Many of the world's predominant religions promote moral codes of conduct and have played a role in supporting environmental conservation (Boyd, 1984; Kinsley, 1994; Palmer and Finlay, 2003) and sustainable development (Taylor, 1995; White and Tiongco, 1997; Belshaw et al., 2001). Can this also make religious groups potential partners in future conservation and development programmes? Religions are internally diverse, with competing versions and perspectives, some forms of which are more amenable to environmentalist concern than others (Beyer, 1992; Taylor, 2005; Taylor, 2010). This chapter critically examines whether partnerships with institutional religions are worthwhile and whether they are likely to benefit outreach and action by secular conservation and development organisations.

The chapter starts off by examining the challenges in forming partnerships with faith groups. Although there is literature suggesting that religion has played a role in conservation and development (Selinger, 2004; Haar and Ellis, 2006; Gottlieb, 2007; Dudley et al., 2009), problems in forming partnerships with faith groups have also been acknowledged (Clarke, 2007; Foltz and Saadi-nejad, 2007; Peterson and Liu, 2008; Winkler, 2008). Second, while the probability of achieving objectives associated with conservation and development is greater if the two are approached synergistically (Becker, 2003; Adams et al., 2004; Johannesen and Skonhoft, 2005; Fisher and Christopher, 2007), it remains to be seen whether religion can provide a common moral framework for action and why it might be critically important to

Figure 12.1 *Morning prayer in Bishkek marks the start the Orozo Ait (Eid ul-Fitr) celebrations; the end of Ramadan. A large proportion of the world's population subscribes to one of eleven mainstream faiths in the world.*

Source: Photographer Evgeni Zotov; used under Creative Commons licence

do so in comparison to non-religious organisations. This chapter therefore examines the possible role of religion in enhancing linkages between conservation and development. Third, this chapter reviews examples of collaboration between conservation and development organisations and faith groups, and evaluates future opportunities for enhancing such partnerships.

Linking conservation and development with religion

As Chapter 10 explores in depth, efforts to achieve conservation and development simultaneously, through the vehicle of integrated conservation and development projects, led to some successes at best, mixed results in the round, and at worst, disillusion. Moreover, while the concepts of 'conservation' and 'development' both are highly contested (e.g. Roe and Elliot, 2010), in this chapter conservation is taken to mean protection of the living environment, and development is taken to mean the improvement of the lives of the poor by way of improving the environment in which they live. The key question addressed in this chapter is whether religion can provide an overarching moral framework for action that intersects environmental conservation and sustainable development. Following Adams et al. (2004), Johannesen and Skonhoft (2005), Fisher and Christopher (2007) and, as this book has argued thus far, this chapter assumes that there are advantages in approaching conservation and development synergistically. It seeks to examine challenges, as well as opportunities, in linking conservation and development with religion.

There are eleven main religious faiths in the world, namely Bahai, Buddhism, Christianity, Daoism, Hinduism, Islam, Jainism, Judaism, Shinto, Sikhism and Zoroastrianism (ARC, 2010), although some nature religions such as Animism or Paganism are often viewed as promoting ways of living in harmony with their natural environment (Taylor, 2005). The reason this chapter focused on the faiths identified above is because some estimates suggest that over 80% of the world's population subscribes to one or another of them, and even if many of those affiliated with these religions do not strictly adhere to them, they provide at least a background ethical perspective to over four billion people in the world (O'Brien and Palmer, 2007).

Conflict of religion with conservation and development

Despite enthusiasm for linking conservation and development with religion, it must be acknowledged that partnerships between faith groups and conservation-development organisations are not always easy. Even though religions advocate ethical and moral values, some scholars (Foltz and Saadi-nejad, 2007; Winkler, 2008; Hall et al., 2009; Plant, 2009) have argued that these may not necessarily translate into action, and the reasons underlying attitudes of different religions toward conservation can vary widely. For example, an anthropocentric mainstream Muslim position is that humanity has responsibility for Allah's creation, which has been given to humans as a gift. At the other end of the spectrum, the biocentric Jain belief advocates that every being – animal, plant or human – has a soul and should be treated with respect (Hall et al., 2009). Foltz and Saadi-nejad (2007) provide an example of Zoroastrianism, a religion that respects and protects many aspects of nature including certain species (e.g. cows and dogs are considered sacred). However, Zoroastrianism also posits an ongoing struggle between the forces of good and evil; certain species groups (e.g. ants, snakes) are seen as evil and are thus to be destroyed whenever possible. Religious affiliation therefore does not always promote pro-environmental behaviour, as defined by Western conservation traditions.

Similarly, religions may not always support development activities that focus on poverty alleviation – faith traditions are different from one another in this respect (Plant, 2009). Dinham and Lowndes (2008) suggest that because beliefs and practices vary within and between different faith communities, the success of religion– development partnerships will depend on the socio-political context within which faiths are situated. Haar and Ellis (2006), for example, argue that religious ideas are often relevant to development thinking and therefore religion can play an important role in sustainable development. Winkler (2008), on the other hand, found no evidence of poverty reduction despite community development efforts in an area with widespread presence of churches of various denominations in an inner-city neighbourhood in South Africa. Flying in the face of enthusiasm for religion–state partnership in social welfare and development in the wider South African context, the burgeoning faith sector in this neighbourhood was marred by competing faith identities, exclusionary ideologies and instability, and consequently, it was unable to implement a coherent program of poverty reduction (Winkler, 2008). Religious institutions, therefore, may not always help in achieving development objectives.

For religious institutions, promoting the practice of religion does not often include promoting environmental concerns (e.g. Tomalin, 2002, 2009; Foltz and Saadi-nejad, 2007; Winkler, 2008). Scholarly literature even suggests that religion might be incompatible with conservation and development on various accounts, which can be broadly grouped into three categories:

1   differences in worldviews
2   conflict between identities
3   divergent attitudes and behaviour.

There are different societal scales at which these three incompatibilities operate. While all three are somewhat linked with each other, differences in worldviews pertain mainly to the presence or absence of God; conflict between identities arise because of adherence to different Gods; while attitudes and behaviour are driven both by the worldview and the identity of a religious group or individual. Therefore, differences in worldviews might affect interaction between religious and secular groups within society, while conflict between identities might affect interaction between two religious groups. Attitudes and behaviour of individuals often determine whether religious and secular values and practices are compatible. These incompatibilities together influence the dynamics between individuals, between religious groups and within society.

Differences in worldviews: Different worldviews of religious groups and conservationists often make religion and conservation incompatible (Ruse, 2005). Darwin's theory of evolution (Darwin, 1859), for example, was interpreted by some as a challenge to religious beliefs about the creation of the world by God and the superiority of humans over other forms of life (Dunlap, 1988). However, the differences in worldview arise not just between religious versus secular groups, but also between religions. For example, many religions consider the sacred to be above or beyond the earth, which leads to differences between religions that consider the earth sacred in some way, and those that do not (Taylor, 2010). Similarly, religion is often considered incompatible with certain development strategies. Selinger (2004), for example, argues that the modernisation paradigm of the twentieth century, which assumed that with assistance poor countries can be developed in the same manner as the rich countries, has been a key development strategy. Most development interventions aim to promote economic growth whilst side-lining cultural values in society. Furthermore, many secular development organisations do not want to be seen working with a religious organisation no matter how much influence faith groups have on local culture. As a result of this, religion – although an important force within many cultures – is marginalized. According to Ver Beek (2002), the reason for this marginalisation is the dichotomisation between sacred and secular in socio-political, and often religious, thought.

Conflict between identities: Although there is substantial overlap in ethical and moral values advocated by different religions, there is also a long history of conflict between religions (Kaplan, 2007). It has been suggested that these conflicts originate in strong religious identities, which when threatened, result in inter-group hostilities (e.g. Ysseldyk et al., 2010). Such strong identities might therefore be counter-productive to communicating a conservation message to mixed audiences consisting of members drawn from different faiths. Similarly, strong religious identities can be counterproductive to successful implementation of development programmes. Winkler's (2008) example of an inner-city neighbourhood in South Africa, where competing religious identities were detrimental to neighbourhood-wide anti-poverty campaigns, is illustrative. Clarke (2007) argues that involvement of faith-based organisations in development raises the concerns of some donors, who do not want to support organisations that actively proselytise or denigrate other faiths. Such actions can lead to faith-based tensions in some of the poorest and most culturally sensitive countries in the world (e.g. Vilaca and Wright, 2009).

Divergent attitudes and behaviour: Although most religions advocate ethical and moral values, these values may not always promote pro-environment behaviour. Peterson and Liu (2008) examined environmental worldviews of various social groups in the Teton Valley of Idaho and Wyoming, USA. After controlling for demographic factors, they found that environmental behaviour is not positively correlated to religiosity and those not affiliated with organized religion were the most environmentally concerned and active (Peterson and Liu, 2008). Tomalin (2004) provides an example of sacred grove conservation discourse in India and argues that while protection of sacred groves has been touted as an example of Hindu religious environmentalism, the attitudes and behaviour of a majority of middle-class elite Hindus do not demonstrate awareness of the environmental ethic of using natural resources responsibly.

Similarly, religious attitudes and behaviour are not always complementary to certain paradigms of sustainable development. In 2002, for example, the evangelical Christian President of the USA, George W. Bush, launched Faith-Based and Community Initiatives with a view to provide federal funding to faith-based organisations involved in combating social problems such as homelessness in the USA. In the same year, however, the administration withdrew international aid to US agencies either supporting or providing abortion services, casting a serious blow to population control and poverty reduction efforts in many developing countries (Crane and Dussenberry, 2004). Critics have therefore argued that such interference of religious attitudes and behaviour in aid agencies' work can be detrimental to sustainable development efforts.

Figure 12.2 *Mongol shamanic sacred mountain near Ulan Bator, Mongolia. Many faiths consider nature sacred and certain places are designated as sacred sites, often in natural surroundings.*

Source: Martin Vorel, Freepix.eu; used under Creative Commons licence

These incompatibilities might suggest that faith groups might not be good partners in conservation and development programmes. However, a wide variety of secular conservation and development organisations have shown increasing interest in forming partnerships with faith groups (CII 2005, WWF 2009, DDVE 2010, WFDD 2009). The following sections suggest that, in order to work effectively with faith groups, secular organisations will need to recognise that there are important intersections in the ethical and moral values undergirding conservation, development and religion.

Overlap of ethical values in conservation, development and religion

Although the world's longstanding and predominant religions are in decline in many areas of the world (most dramatically in Northern and Western Europe), Taylor (2010) argues that there is still a desire among those who eschew such religions for a spiritually meaningful understanding of the world, or even the universe, and the human place in it. Some within the world's predominant religions have demonstrated an increasing affinity towards some forms of nature spirituality and are engaging in practices that are understood as 'green' (Taylor 2005, 2010). Influential thinkers have suggested that ethical or moral values can play an important role in developing compassion toward non-human species (e.g. Wilson, 1984) and towards fellow human beings (e.g. Sen, 1999). Such values also therefore underpin both conservation (compassion towards non-human species) and development (compassion towards fellow human beings). Although the histories of modern-day conservation and development are complex, brevity requires that in the following examination of the origins of the ethical and moral values, only key historical watersheds are identified.

In Chapter 1, we outline how the contemporary Western conservation ethic arose in part with the emergence, during the late nineteenth century, of the wilderness protection movement in the USA. Although in the seventeenth century, some Puritans viewed nature with foreboding, by the nineteenth century, figures such as Ralph Waldo Emerson, Henry David Thoreau and John Muir transformed the American view of wilderness into something of great spiritual value (Schmitt, 1969; Nash, 2001; Callicott, 2008; Taylor, 2010). This ethic, articulated most forcefully by Thoreau and Muir in the nineteenth century, acknowledges the moral right of non-human species to inhabit the Earth and aims to set aside areas for their preservation. During the twentieth century, many other figures, most notably Gifford Pinchot, Aldo Leopold and Rachel Carson, effectively championed environmental conservation, and in diverse ways, linked their cause to spiritual or religious perceptions (Schmitt, 1969; Fox, 1985; Oelschlaeger, 1991; Nash, 2001; Gatta, 2004; Taylor, 2010). Although some indigenous and native communities around the world developed environmentally sustainable ways of living, such cultures have often come into conflict with the modern Western conservation apparatuses – as chapters 9 and 10 document. Yet increasingly, traditional methods of conservation and involvement by local communities are valued by conservationists as key elements of conservation strategies (Wild and McLeod, 2009).

The contemporary approach to development acknowledges the moral right of all human beings to social justice and a belief in the capacity of human beings to shape their own progress, instead of assuming that poor countries should engage in development in the same fashion as affluent ones. The twenty-first-century approach to development, particularly that advocated by the United Nations, asserts that 'only with a universal morality of justice is there a future for humanity' (Schuurman, 2000: 19). The challenge for sustainable development in less developed countries, however, is to reduce poverty without increasing the environmental impact. Rees (2002: 15) argues that 'Sustainability with social justice can be achieved only through an unprecedented level of international cooperation rooted in a sense of compassion for both other peoples and other species'. This contemporary approach to development resonates with the ethical positions that are assumed by some religions.

Although religions vary in their ethical positions (Morgan and Lawton, 2007) there are some themes found within the world's major religions that can be understood in ways that cohere with the most common, global, understandings of the ethics of conservation and development. Given the wide variety of values found in the world's diverse religions, a thorough examination of their ethical positions is beyond the scope of the current analysis. Instead, this chapter postulates values that can provide common ground between people variously involved in religion, conservation and development. It focuses only on a few elements common in religious ethics that are relevant to environmental conservation and sustainable development: stewardship, reverence for nature, altruism and compassion.

Ecological advocates within many religions adhere to the principle of stewardship of nature or the idea that nature should be revered. The Abrahamic traditions – Christianity, Judaism, and Islam – all tend to centre their environmental ethic around the concept of stewardship, and therefore, the upkeep and management of nature as the responsibility of humans. Religions originating in Asia, and some indigenous religions, by way of contrast, often tend to emphasize the divinity in nature, its forces, or its creatures. All of these religions, in various ways tend to value altruistic behaviours and consider compassion a virtue, sometimes even compassion towards non-human beings

Figure 12.3 *Prince Karim Aga Khan. Many faith-based institutions such as the Aga Khan Foundation have made significant contributions towards addressing poverty in the world*

Source: US Department of State; used under Creative Commons licence

(Schuurman, 2000). These sorts of themes can inspire environmental concern and action (e.g. Palmer and Finlay, 2003; Tucker and Grim, 2004; Gottlieb, 2007; Colwell et al., 2009).

Can faith groups make an important contribution to conservation and development?

It is clear that there are ethical values that faith groups can bring to conservation and development efforts, even if challenges and difficulties will also be encountered. The chapter now turns to examine the links between ethical or moral values, and activities of faith groups. The values of compassion toward others, and the stewardship of nature, that are often found in religions, have motivated and continue to motivate many faith-based organisations to engage in environmental conservation and sustainable development. At the same time, prominent conservation organisations (CII, 2005; WWF, 2009) have mutual interest in forming partnerships with faith-based groups. Similarly, the World Bank's Development Dialogue on Values and Ethics (DDVE, 2010) has successfully enlisted support and involvement of religious leaders and organisations in starting to address the

Millennium Development Goals in some of the poorest countries in the world. Many examples suggest that religious groups are attempting to address perceived incompatibilities between their values and practices and those of secular conservation and development organisations.

Differences in worldviews: Despite differences in worldviews, in some regions religious organisations are working co-operatively. For example, Christian organisations such as Plant With Purpose have taken direct conservation action by planting trees in a wide range of developing countries across the world (Floresta, 2008), even though Christianity is a minority religion in many of these countries. Members of other religions including Buddhism, Daoism, Hinduism and Islam have put their differences in worldviews aside and found a common moral ground, and are now cooperating in environmental conservation and sustainable development projects (Vigne and Martin, 2000; ARC, 2004, 2006; CII, 2005; WWF, 2009; Basil, 2010). The World Council of Churches includes churches with over 560 million members (WCC, 2010). Despite denominational differences within it, this organisation is engaged in various development activities. The Women, Faith and Development Alliance as well as the World Conference of Religions for Peace, both have institutional members from different faith groups working together (WFDA, 2010; WCRP, 2010).

Conflict between identities: Many faith groups have attempted to leave aside conflict between identities and have jointly contributed to ethical investments. The International Interfaith Investment Group (3iG, 2010), for example, has been instrumental in encouraging substantial investments from religious organisations in environmentally responsible and ethical projects (Dudley et al., 2009). Although inter-religious conflicts do arise due to the sectarianism and fundamentalism found in most religions, consolidated actions such as 3iG suggest that some faith groups are able to overcome their strong identities and faith-related tensions and are able to co-operate. The Community of Sant'Egidio, a movement which has its origins in the Catholic Church, is actively engaged in brokering peace in situations of inter-religious conflict (Sant'Egidio, 2010). The annual Prayer for Peace organised by this community has become a prominent global interfaith meeting where many groups work co-operatively.

Divergent attitudes and behaviour: Many faith groups are making efforts to change the attitudes and behaviours of their members. Conservation-focused education programmes undertaken by Islamic religious leaders in Pakistan or Malaysia are illustrative (Sheikh, 2006; WWF, 2009). Similarly, active forest protection efforts by individuals from faiths including Buddhism, Daoism and Hinduism (ARC, 2004, 2006; CII, 2005; Bhagwat and Rutte, 2006; Chimedsen-gee et al., 2009; Ormsby and Bhagwat, 2010) also suggest stronger pro-environment attitudes and behaviours are emerging. Initiatives such as the World Faiths Development Dialogue attempt to bridge faith-based and mainstream development thereby changing attitudes of secular development organisations towards faith groups and vice versa (WFDD, 2009).

Examples of faith groups involved in organic agriculture

One sector where faith groups have been successful in promoting conservation and development is organic agriculture. The present model of industrialised, export-oriented agriculture has increased world food production, but has not always benefited rural communities, particularly in the developing world (Pretty, 2002). Furthermore, intensive agricultural practices alongside heavy use of artificial chemical fertilisers and pesticides have resulted in ecosystem degradation and biodiversity loss (Carson, 1962; Tilman et al., 2002). A search for sustainable agricultural practices has often pointed to small-scale organic methods, which are less efficient than industrialised agriculture, but more suitable for small-scale farming particularly in developing countries (Pretty, 2007). In addition, small-scale agriculture is considered a way of making rural communities self-sufficient, thereby enhancing their food security (Scialabba, 2007).

The International Federation of Organic Agriculture Movements (IFOAM) suggests that this form of agriculture relies on ecological processes, biodiversity and crop cycles adapted to local conditions rather than using artificial inputs that can have adverse effects on ecological processes. According to IFOAM, this form of agriculture sustains the health of soils, ecosystems and people by combining tradition, innovation and science to benefit the environment and by promoting fair trade to improve the quality of life of the rural poor. Organic agriculture worldwide is developing rapidly with 35 million hectares now managed organically, representing approximately 0.8% of total world farmland (IFOAM, 2010). When viewed as an alternative to modern, intensive agriculture systems, organic agriculture invites criticism because of the concerns over meeting food demands of the world's rapidly growing population. However, there is increasing evidence from Africa and elsewhere in the developing world that organic agriculture can play its part in feeding the world and in meeting various sustainability goals, from water conservation and improved soil quality to delivering higher levels of employment and conservation of biodiversity (FAO-UNEP, 2008; Jhamtani, 2010). Furthermore, the International Labour Organization recognises the benefits of organic farming for secure and sustainable employment and conditions of work (ILO, 2010); and the United Nations Environment Program is currently examining the potential economic, employment and environmental benefits of greater investment in sustainable agriculture in Eastern Europe, the Caucasus and Central Asia (UNEP, 2010).

Various faith communities around the world have started to promote organic farming through public education programmes or by managing their own agricultural lands organically. These include faith groups in the developed and the developing world. The Orthodox women's monastery of Ormylia, Greece, for example, introduced organic farming in 1991 on the monastery's existing agricultural lands, based on the need to 'celebrate and preserve nature while engaging in sound ecological practices designed to protect that very environment' (Goarch, 2010). The goal of Project Ormylia is to act as a focus and catalyst, initially for nearby monasteries (e.g. Simonopetra Monastery on Mount Athos) and subsequently for the surrounding community. The Christian Andechs Monastery in Bavaria, Germany, holds 110 hectares of cultivated land that has been managed organically since 1995. The motivation behind the shift in agricultural practice to an approach that calls for extra care and restraint in dealing with nature is based on Andechs' aim of 'doing justice to the task of preserving God's creation' (Andechs, 2010).

More importantly, from the conservation and development perspectives, there are examples of faith groups promoting organic agriculture in the developing world. In 2005, an organic farming training programme run by the Orthodox Church in Assela, Ethiopia, introduced farmers, clergy and community leaders to alternative and improved methods of agriculture. The Ethiopian church, pointing to the Biblical mandate of concern and care for the environment, states that it sees caring for nature as part of its sacred mission and duty (ARC, 2005). In 2006, the Hindu community's International Society for Krishna Consciousness (ISKON) founded a training centre for organic agriculture on the outskirts of Mysore, Karnataka, India. Named Basil Academy, this organisation runs 'Eco Agri Research Foundation', which aims to 'make the farmer a responsible, respected member of society who plays a significant role in protecting and sustaining our planet and its people' (Basil, 2010). Such initiatives in developing countries are contributing to environmental conservation and economic development for the rural poor.

Although no data are available on the land mass cultivated organically by religious organisations, considering that nearly 15% of Earth's land surface is under the influence of religious or sacred institutions (Bhagwat and Palmer, 2009), faith communities can play an important role in conservation and development if they manage their farmland organically. The examples discussed above also suggest that the motivations for many of these practices arise from ethical and moral values. These examples also indicate that tangible opportunities exist for religious and secular groups to collaborate. IFOAM's (2010) advocacy campaigns on food security, climate change or biodiversity, for example, could benefit from partnerships with faith groups that are actively working in these areas. Similarly, ILO's (2010) or UNEP's (2010) campaigns to generate economic, employment and environmental benefits through sustainable agricultural practices, will benefit from partnerships with faith groups.

Strengthening linkages with faith groups

While environmental conservation and sustainable development are in need of greater public support, which faith groups might well be able to provide, there are a number of things that secular conservation and development organisations can in turn do to enhance linkages with religious organisations. This can be brought about by establishing strong working relationships with faith groups, leaders and adherents. To this effect, focusing on the common ground between ethical or moral values advocated by religious groups and conservation and development organisations might be necessary, rather than dwelling on incompatibilities. 'Planetary stewardship' as a framework for science and society to rapidly reduce anthropogenic damage to the biosphere (Sensu Power and Chapin, 2009), for example, has a shared subtext with the Abrahamic religions' ideology of 'stewardship of nature'. Identification of synergies between religious programmes and conservation and/or development initiatives is likely to help establish a common ground between faith groups and secular organisations. There are two areas in which such synergies can be explored:

1 outreach and education
2 action.

Outreach and education: Religious leaders can be important agents in effectively communicating issues in conservation and development, and their support is often necessary for outreach activities. Faith group leaders are conversant with their group's vocabulary and are most effective in communication, knowing what is evocative to their religious audience. For example, phrases such as 'the creation' and 'creation care' are helpful to communicate a conservation message when working with Abrahamic religions (Wilson, 2006). In addition to faith group leaders, involvement of institutions such as religious colleges, universities and training centres in conservation and development activities might be an effective way of reaching out to the target audience. It is also vitally important to develop training material in environmental conservation and sustainable development for faith groups. Each group in the US-based National Religious Partnership for the Environment, for example, has resource kits enabling congregations to integrate environmental concerns (Gorman, 2009). Such material is not yet available for other religions and in other parts of the world, but the development of this type of material might be an effective way for conservation and development organisations to reach out to religious groups.

Action: While religious audiences may be able to identify with ethical or moral values of environmental conservation and development, their own faith is often their primary concern (e.g. Tomalin 2002). Furthermore, religious groups might place a high value on maintaining a separate cultural identity. When planning activities with faith groups, it is essential to recognise this and to respect social and cultural boundaries in order to initiate effective partnerships. However, it might be important to encourage religious groups to create opportunities for dialogue across faiths and with secular organisations so as to dissolve the social and cultural boundaries. The Alliance of Religions and Conservation (ARC, 2010), a secular body that helps the major religions of the world to develop their own environmental programmes, based on their own core teachings, is notable in this respect. This organisation actively creates opportunities for predominant religious faiths around the world to promote environmental conservation and sustainable development, and occasionally to encounter one another. Effective action for helping conservation and development in most cases, therefore, seems to demand changes in attitudes and behaviours from religious adherents and secular organisations alike. Such changes can be brought about by highlighting success stories where adherents are shown examples where people of their own faith are engaged in conservation and development activities (e.g. Kula, 2001). This way of encouraging participation is often effective because members of the same faith group share the strongest cultural associations with each other. Similarly, there is a need for empathy and an open mind towards faith groups on the part of conservation and development volunteers, professionals, NGOs and donors. Finally, it is important to help religious leaders to establish common ground between their practices and those that help environmental conservation and sustainable development. Although this might often require redefining conservation and development, it is also likely to ensure that collaborative actions with faith groups are most effective.

Summary

- The examples discussed in this chapter suggest that ethical or moral values related to conservation and development may at least sometimes play a role in the faith groups' choice of ideas and practices.

- Despite these examples, it should be acknowledged that there is no direct causal effect of faith on conservation and development at least that has not yet been demonstrated by social scientists. Nor is there concrete evidence that individuals or organisations affiliated with religions are more likely to be concerned about the environment than those who are not.

- Many non-governmental organisations as well as prominent donors such as the World Bank concur that faith groups are important, potential partners in conservation and development. Partnerships with faith groups might be valuable because these groups can enhance public support for conservation and development.

- While there are challenges in forming partnerships – including differences in worldviews, conflict between identities, and divergent attitudes and behaviour – it is possible for conservation and development practitioners to work with religious individuals and groups to promote environmental values and sustainable development.

Discussion questions

1. In what ways have the world's religions made a contribution to environmental conservation and sustainable development?

2. What are the challenges in making the partnerships between faith-based organisations, conservation NGOs and development NGOs work?

3. What are the opportunities in partnerships with faith groups for the practitioners of conservation and development?

Note

1. Adapted from: Bhagwat, S.A., Ormsby, A.A. and Rutte, C. (2011) The role of religion in linking conservation and development: challenges and opportunities. *Journal for the Study of Religion, Nature and Culture*, 5(1): 39–60.

Further reading

Adams, W.M., Aveling, R., Brockington, D., Dickson, B., Elliott, J., Hutton, J., Roe, D., Vira, B., and Wolmer, W. (2004) 'Biodiversity Conservation and the Eradication of Poverty.' *Science* 306: 1146–1149.

Bhagwat, S.A., Dudley, N., and Harrop S.R. (2011) 'Religious Following in Biodiversity Hotspots: Challenges and Opportunities for Conservation and Development.' *Conservation Letters*.

Belshaw, D., Calderisi, R., and Sugden, C. (2001) 'Faith in Development: Partnership Between the World Bank and the Churches of Africa' (Washington DC, USA: World Bank).

Dudley, N., Higgins-Zogib, L., and Mansourian, S. (2009) 'The Links Between Protected Areas, Faiths, and Sacred Natural Sites.' *Conservation Biology* 23: 568–577.

Palmer, M., and Finlay, V. (2003) Faith in Conservation: New Approaches to Religions and the Environment. (Washington, DC: The World Bank).

Van Houtan, K.S. (2006) 'Conservation as Virtue: a Scientific and Social Process for Conservation Ethics.' *Conservation Biology* 20: 1367–1372.

Online resources

ARC (Alliance of Religions and Conservation) 2010. http://www.arcworld.org (accessed December 2010).

Tucker, M.E., and Grim, J. 2004. Overview of World Religions and Ecology. Forum on Religion and Ecology http://fore.research.yale.edu/religion/ (accessed December 2010).

WFDD (World Faiths Development Dialogue) 2009. History and Objectives of the World Faiths Development Dialogue http://berkleycenter.georgetown.edu/wfdd/about (accessed December 2010).

⑬ Payments for ecosystem services in the context of material and cultural values

- **Using PES to achieve conservation and poverty reduction simultaneously**
- **Towards non-material values in nature conservation**

Introduction

In previous chapters (4, 5 and 9), we have explored – and in some instances critiqued – the theory of payments for ecosystem services. This chapter looks at payments for ecosystem services in practice, first in terms of the more common conception of ecosystem services in terms of material (monetary) values, and second, in terms of cultural values. At the core of the chapter is a concern with the juxtaposition between 'material' values enshrined in the ecosystem services concept (conservation of nature *because* it is economically useful to people) and 'non-material' values enshrined in the conceptualisation of nature as sacred (conservation of nature because it is sacred). This concern reflects increasing scepticism about the extent to which offering material rewards for ecosystem services delivers either conservation or poverty reduction objectives, separately or in tandem, and about whether market mechanisms lend themselves to the conservation of ecosystem services and functions. After a brief review of the conceptual origins of payments for ecosystem services, the chapter considers the available evidence on the conservation and poverty-reducing outcomes of existing payments for ecosystem services initiatives grounded in 'material' values. It goes on to consider ecosystem services in the context of cultural values, using 'sacred natural sites' as a case. The chapter then takes one step further by attempting to find synergies between material and non-material values. It looks at a practical example of 'elemental equity' initiative which translates cultural values of nature to instigate the investment in its conservation.

Using PES to achieve conservation and poverty reduction simultaneously

PES, conservation and poverty reduction in the context of material values

Ecosystem services are defined as processes by which the environment produces resources utilised by humans – such as clean air, water, food and materials (Defra, 2006). Early references to the idea of ecosystem services go back to the mid-1960s and early 1970s (De Groot et al., 2002). However, the tendency to produce valuations of benefits of natural ecosystems to human society was accelerated by Daily (1997) in her book *Nature's Services*. In the same year, Costanza and a number of co-authors published an influential paper in Nature valuing the world's ecosystem services at US$16–54 trillion per year – as much as three times global gross national product at the time (Costanza et al., 1997). The exponential growth in literature on ecosystem services is evident in the number of citations (totalling 5,297 as of April 2015, across the Web of Science databases) that Costanza et al.'s paper has received. Two decades later, Costanza and colleagues (Costanza et al., 2014) revisited the 1997 paper, repeating the exercise and method with more recent data. For 2011, they estimated that the total global ecosystem services in 2011 had reached $125 trillion/yr (assuming updated unit values and changes to biome areas) and $145 trillion/yr (assuming only unit values changed), both in 2007 $US.

In this work related to ecological economics the terms 'ecosystem functions', 'ecosystem services' and 'ecosystem goods and services' are commonly used. De Groot (1992) uses the term 'ecosystem functions' and defines those as 'the capacity of natural processes and components to provide goods and services that satisfy human needs, directly or indirectly'. Daily (1997) defines 'ecosystem services' as 'conditions or processes through which natural ecosystems and the species that make them up, sustain and fulfil human life'. Costanza et al. (1997) distinguish between 'ecosystem functions' and 'ecosystems goods and services', but demonstrate linkages between the two.

Figure 13.1 *Injecting a little humour into the valuation of ecosystem services in monetary terms*

Source: http://naturalcapital.us/about/ (accessed 17 April 2015)

They suggest that 'ecosystem functions' refer to 'habitat, biological or system properties or processes of ecosystems' whereas 'ecosystem goods' and 'ecosystem services' 'represent the benefits human populations derive, directly or indirectly, from ecosystem functions'. Much has also been written in recent years about the typology of ecosystem services in order to standardise their assessment (De Groot et al., 2002; Boyd and Banzhaf, 2007; Egoh et al., 2007; Wallace, 2007; Costanza, 2008; Fisher and Turner, 2008; Wallace, 2008). Despite a wide variety of interpretations, a common theme that runs across all definitions and typologies is that they focus on the 'material' benefits that humans can derive by putting value on natural resources. Sahlins et al. (1996) argue that this approach originates from the Judaeo-Christian roots of science and its influence on Western economic behaviour. Whilst the continental European approach uses 'ecosystem functions' for such valuation, the Anglo-Saxon approach focuses on 'ecosystem services' (Ansink et al., 2008).

High-profile global conservation initiatives, such as the Millennium Ecosystem Assessment (MEA), have adopted the 'ecosystem services' framework and this has promoted the idea of valuing nature in monetary terms, which has become increasingly popular among ecologists and conservation biologists (Spash, 2008a). The magnitude and importance of the environmental problems that we face make the popularity of PES approaches understandable. For instance, estimates for the contribution of twenty-first-century deforestation rates to global annual carbon emissions range from 10–17% (Samii et al., 2014). We know very well the importance of forests as 'carbon sinks', both in terms of storing carbon dioxide and removing it from the atmosphere. Therefore, increasing rather than reducing forest cover could make a fundamental contribution to climate change mitigation. In the words of one influential publication, 'current annual rates of tropical deforestation from Brazil and Indonesia alone would equal four-fifths of the emissions reductions gained by implementing the Kyoto Protocol in its first commitment period [i.e. 2008–2012]' (Santilli et al., 2005: p. 267). It has also been argued that forest protection may in fact be cheaper, per unit cost, than other ways of attempting to reduce carbon emissions (Gullison et al., 2007).

From this standpoint, it could be argued that emphasising the economic value of forests, as well as the cost efficiency of afforestation as a mitigation strategy, is ultimately the only way to save them. In this respect, payments for ecosystem services are merely the most recent reincarnation of the 'use it or lose it' philosophy that underpinned integrated conservation and development projects (ICDPs), covered in Chapter 10. Indeed, early proponents of payments for ecosystem services saw them as in one sense the successors to ICDPs by virtue of improving on them. For instance, Ferraro and Simpson (2002) argued that one of the biggest problems with ICDPs was that, because of the way projects were set up, sometimes people were given payments for participating in projects regardless of the conservation outcomes (see also Brandon et al., 1998; Newmark and Hough, 2000). If, however, a more direct relationship could be established in which those who received the benefits of ecosystem services and functions directly paid those who 'provided or maintained them', integrated conservation and development projects might become more effective (Ferraro and Simpson, 2002). PES arrangements could also allow ICDPs to 'break free' from the project cycle (Ferraro and Simpson, 2002). In this way, it is precisely the focus on the economic value of nature which has led PES to become a vehicle for poverty reduction as well as conservation objectives. Perhaps the best known example of this rationale is the global initiative now known as REDD+ (reducing emissions through avoided deforestation and degradation), an international programme coordinated through inter-governmental agreements since 2007. Payments for ecosystem services are viewed as the key mechanism for delivering both forest protection/mitigation and poverty reduction outcomes. As such, REDD+ is one of the more common examples of the stronger focus on the poverty reduction prospects for payments for ecosystem services.

There are two overarching positions for and against utilising PES to achieve conservation and poverty reduction objectives simultaneously. In the following sections, we look at the use of PES to maintain forests, because the bulk of the evidence relates to this sector; but the lessons from the experience with forests also have relevance to PES in other ecosystem settings. The first position contends that, in the context of deforestation, it is better to try to do conservation and poverty reduction separately. In short, it is argued, this is because forest conservation and poverty reduction are orthogonal or even conflicting processes, requiring different instruments and interventions which would be better aimed at different groups of people. Pagiola et al. (2005) argue that it is not wise to assume that poor people, for instance living in or at the edge of forests, are responsible for deforestation, even though they might benefit from poverty reduction initiatives. Their contribution to deforestation, in fact, is miniscule in comparison with intensive logging and farming activities, both legally and illegally. Following this logic, one of the better-known commentators, Sven Wunder (2005), has argued that targeting poor people will not aid conservation objectives if, in consequence, it moves away from targeting those who are much more responsible for deforestation. As well as its inefficacy as a conservation tool, whether PES would serve as the optimal poverty reduction tool in such circumstances is also moot.

The second position, whilst it is not always diametrically opposed to the first, takes its cue from moral and practical considerations. If conservation and poverty alleviation are not coordinated, at the very least, the argument goes, the risk is that the livelihoods of poor people are adversely impacted by conservation efforts such as forest protection (cf. Agrawal and Benson, 2011). Tropical forests provide a compelling example of this difficulty: not only are they very well suited to carbon capture and storage, communities living in and/or adjacent to them are often very poor (Deveny et al., 2009; Sunderlin et al., 2005). This being the case, commentators such as Agrawal and Benson (2011) maintain that it is impractical and potentially counter-productive to implement forest conservation in ways which do not take poverty reduction into account, because they will inevitably take place, at least in some cases, in the same physical space. Turner et al. (2013), in a global study, found that the land containing biodiversity most valued for conservation also provides 56% of the essential ecosystem services that are of use to the world's poorest people; although this study may make controversial assumptions about the distribution of poverty between rural and urban areas. On this reading, conservation and poverty alleviation are almost impossible to disentangle.

To make, therefore, better decisions about the design of PES mechanisms and what objectives they should be designed to serve, we need to be clear on how the relationship between conservation and poverty reduction objectives can or cannot be configured. But beyond these broad brush positions, how effective has PES been either at conservation or poverty reduction, or indeed both in tandem? It is these questions to which we now turn our attention, although with more of a focus on conservation or poverty reduction benefits separately, because there is very little written as yet which evaluates the results of initiatives which attempt to achieve both simultaneously.

The efficiency of PES as a conservation measure

A recent systematic review by Cyrus Samii and colleagues (2014) has attempted to evaluate the efficacy of forest-related payments for eco-system services both in terms of conservation and poverty outcomes. Sadly, their systematic review is framed in terms of the weakness and incompleteness of the evidence base currently available. Albeit steeped in caveats, they do offer tentative conclusions about the efficacy of PES as a means of reducing deforestation rates. The criteria of their systematic review left them with a small sample, covering examples from Africa, Asia and Latin America. From this, they suggest that on average, a reduction in annual deforestation rates in the region of 0.25% might be attributable to PES schemes. Insignificant though this figure may seem, they contend that it may represent an effect which is more important than it looks, for a number of reasons. First, this percentage does not substantially vary between the different continental contexts that they identify and might therefore hold statistical significance. It also suggests that PES can increase forest cover in addition to conserving existing forest reserves. Figure 13.2 displays the range of results reported in various studies on the contribution of PES interventions to forest cover levels.

In spite of this modest contribution, however, PES for forest conservation has more commonly been viewed as an inefficient approach (Alix-Garcia et al., 2012; Robalino et al., 2008; Robalino and Pfaff, 2013). One reason for this inefficiency is the high fixed costs of setting up and managing PES forest projects: demarcating and measuring a parcel of land, first and, second, setting up monitoring and evaluating systems does not come cheap (Honey-Rosés et al., 2009). Such high costs and logistical complexity underlie Samii et al.'s (2014) finding that the 'additionality' of PES in a forest context – that is, the extra forest that now exists, but which would not have in the absence of a PES intervention – amounts to a 0.3% reduction in the annual deforestation rate. Extrapolating from this finding, they contend that, over a ten-year period, 97% of the forest in existence would probably have continued to be there regardless of whether there had been a PES programme or not. By this logic, PES mechanisms would appear to contribute negligently to forest cover. Thus, Samii et al. conclude that 'the vast majority of parcels on which payments are issued are made "for nothing" from an *ex post* perspective' (2014: 46).

Figure 13.2 *Effects of PES on forest cover due to deforestation*

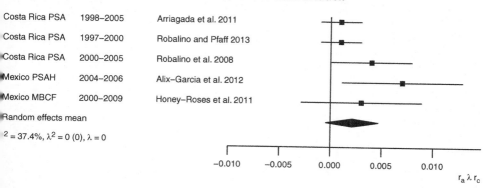

Source: Reproduced from Samii et al., 2014.

This finding also implies that the PES mechanisms examined do not necessarily change the behaviour of people who are paid to provide ecosystem services. Arriagada et al. (2009), in the context of Costa Rica, reported a tendency of landowners to nominate for inclusion in the country's national PES programme those parcels of land which they had no intention of deforesting. In order for payments for ecosystem services to contribute more substantially to reducing deforestation rates, even in countries feted as success stories (like Costa Rica), they need to change the behaviour of people who are deforesting in a big way. Conversely, offering people compensation to keep forest on land that they had no plans to cut down fails to address the underlying dynamics driving deforestation. This finding led Arriagada et al. (2012) to attempt to gauge the effect of PES schemes in a more targeted manner, by investigating their efficacy in the Sarapiqui region of Costa Rica where deforestation rates were higher. In this area, they found that PES schemes brought about between 0.7–1.7% of a reduction in the annual deforestation rate.

Figure 13.3 *A hanging bridge in the Monteverde Cloud Forest Reserve, Costa Rica*

This is considerably higher than Samii *et al*'s (2014) 0.3% average, but the implication remains that, after 10 years, 88% of PES payments would have been made upon forests that would likely have continued to exist regardless.

Payments for ecosystem services and poverty reduction

As with the conservation outcomes of payments for ecosystem services, the evidence base covering the impacts of PES on poverty reduction is insubstantial (Samii et al., 2014). This echoes Vira and Kontoleon's (2013) finding that our overall grasp of the relationship between conservation and poverty is under-researched. Nevertheless, an emerging literature is starting to chart the extent to which payments for ecosystem services interventions have contributed to poverty reduction objectives. An early headline from this literature is that PES mechanisms have provided significant sources of revenue for some people, but it remains unclear whether these benefits are available to the poorest.

Samii et al. (2014) found only two quantitative studies on the effects of PES initiatives on poverty. One of these, by Liu et al. (2010), looked at the Sloping Land Conversion Programme in China (SLCP); the other explored the Nhambita Community Carbon Livelihoods (NCCL) programme in Mozambique (Hegde and Bull, 2011). In the Chinese example, Liu et al. (2010) estimate that, on average, the SLCP increased household income by approximately 14%. In the NCCL, Hegde and Bull reported an average increase of 4% in household income. These studies appear, in the first instance, to offer evidence of significant poverty impacts from PES initiatives. However, other factors, which qualitative studies have uncovered, suggest a more nuanced picture behind the headline figures, and point to issues that will need to be addressed, if PES is to deliver on its poverty-reduction potential. The first factor for this is the difficulty faced by poor people in participating in payments for ecosystem services initiatives, often leading to their exclusion (Pagiola, 2008; Samii et al., 2014; van Noordwijk et al., 2012; Zbinden and Lee, 2005).

The key barriers to participation stem from the prevalence with which payments flow to beneficiaries who own property and who are able to demonstrate ownership. People who have land titles over land which provides ecosystem services are therefore well-placed to receive payments for ecosystem services. Conversely, people who do not own land, cannot prove legal ownership or whose claim to land ownership is disputed, are much likelier to be excluded from PES schemes. Pagiola (2008) argues that this barrier is common in Costa Rica, which often features as a 'paradigm case' of good practice in PES initiatives. Eligibility for PES schemes in Costa Rica include land title, proof of tax compliance, absence of debts or fines, production of a notarised land map and other high transaction costs. As a result, he maintains, many poor people may have been excluded from receiving payments.

Non-participation in PES schemes can also occur even when candidates are eligible for inclusion. Also in Costa Rica, Arriagada et al. (2009) found that many eligible farmers were not participating in PES initiatives because of a lack of understanding of the programme. Zbinden and Lee (2005) have suggested that not only do poorer Costa Rican farmers often not know of available PES schemes for which they are eligible, they also lack knowledge and skills necessary for taking part and benefiting. In the context of Mexico, Pagiola concludes that 'the bulk of program benefits tend to go to larger and relatively better off farmers' (2008: 721).

Pagiola proceeds to argue that these exclusionary effects are in fact a result of the way in which PES in Mexico has been designed primarily in order to further conservation objectives. If accepted, this argument bodes ill for the prospects for reconciling conservation and development objectives. Here, we would seem to have an example of a trade-off between conservation and poverty outcomes, rather than a reconciliation of them.

However, the opposite effect has also been observed, also in a Mexican context, in relation to the PSAH (Payments for Hydrological Environmental Services) programme. Wunder et al. (2008) reported that the involvement and influence of government agencies focussed on poverty reduction, led the design of the programme to privilege poverty reduction even at the expense of conservation objectives, because payments were targeted at poor people regardless of whether they were deforesting or not. To achieve its conservation objectives, the programme would have been better designed if it targeted those implicated heavily in deforestation, regardless of whether they were poor. It is not the intention of Wunder et al. to question the legitimacy of poverty reduction as such. Rather, they try to show how poverty reduction measures are as likely to conflict with as they are to serve conservation objectives.

Even when poor people qualify for, know about and have the skills to make the most of PES schemes, they may actively decline to participate; chiefly in two circumstances. First, if they lack capital and/or are unable to access affordable credit, poor people may not be convinced they can forgo the opportunity costs of not clearing forests. The benefits from the sale, for instance, of agricultural produce (and timber) in a much quicker timeframe may outweigh those of a PES scheme, especially where these will take some years to deliver a return upon conserved land (Alix-García, 2012). Second, returns on the land are not always greater if it is conserved, rather than deforested. In sum, poor people may not find it either desirable or practicable to consider a reduction in income to serve conservation objectives which do not adequately compensate them, and which may in fact deliver greater benefits to other actors (conservation agencies, tour operators and tourists, etc.).

In spite of these difficulties, poor people do participate in PES initiatives; but how do they fare, relative to better-off participants in the same initiative? Hegde and Bull (2011) attempted to answer this question in research which explored an agroforestry programme in Mozambique, finding that the levels of benefits were lower in poorer households relative to less poor participants.

Towards non-material values in nature conservation[1]

The rationale for ecosystem services and non-material values

We have seen, then, that on the face of it, the idea of 'use it or lose it' focused on the economic value of nature looks persuasive. However, the evidence on using leveraging the economic value of ecosystem services to achieve conservation and poverty reduction is less convincing, at least in relation to forests. It is consistent, conversely, with the argument that ecosystems services and functions are not always easily translated into marketable commodities (Gustafsson, 1998; van Noordwijk et al., 2012). A thought-provoking commentary, written by McCauley (2006) in *Nature* makes this argument across the spectrum of global ecosystems. McCauley argues that there is very little evidence for the effectiveness of market-based conservation; and therefore nature should be protected for nature's sake. Recent evidence has also started to suggest that market-based environmentalism in general is struggling in the face of the global economic downturn. For example, the value of carbon credits in some voluntary markets saw a fall of 40 per cent between December 2008 and March 2009 (New Carbon Finance, 2009). The recent global economic downturn, therefore, provides further support to McCauley's argument and calls into question market-driven mechanisms for conservation devised to promote 'material' benefits from nature.

Here, we look at sacred natural sites as a potential counterpoint to economic valuations of nature. These are places of 'non-material', for example spiritual, significance to people (IUCN, 2008). The examples include iconic sacred sites such as Machu Picchu in Peru or Uluru (Ayers Rock) in Australia, but also lesser-known sites such as sacred groves in Ghana or sacred lakes in India. A wide variety of informal institutions have traditionally governed such sites. To a large extent, sacred natural sites have remained relatively unaffected by strong market forces because of many indigenous peoples' active struggle to protect these sites (Verschuuren, 2007). The rest of this section addresses three key questions in relation to these sites: (1) Do sacred natural sites provide 'ecosystem services' as defined by ecological economists? (2) Does the 'ecosystem services' framework accurately represent values indigenous people attribute to sacred natural sites? (3) Can material values of ecosystem services and non-material values of sacred natural sites be effectively reconciled?

Ecosystem services from sacred natural sites?

Five main categories of ecosystem services are variously recognised by ecologists working within the frame of ecological economics (Costanza et al., 1997; Daily, 2000; De Groot et al., 2002) These include: (1) preserving; (2) supporting; (3) provisioning; (4) regulating; and (5) cultural. MEA (2005) recognises four of these five services – supporting, provisioning, regulating and cultural. The preserving services include maintenance of genetic and species diversity for their evolutionary potential. The supporting services include purification of air and water, pollination of crops and dispersal of seeds. The provisioning services include provision of foods, herbal medicines and sources of energy such as hydropower or fuel wood. The regulating services include carbon sequestration or climate regulation, waste decomposition or nutrient dispersal. The cultural services include recreational experience or intellectual inspiration. The examples below illustrate these services with reference to sacred natural sites.

Preserving: Sacred groves are patches of forest in otherwise open landscapes (Bhagwat and Rutte, 2006). Such patches provide habitat for forest-dwelling species within agricultural landscape and permeable landscape matrix for species to move between reserves – thereby preserving species diversity (Bhagwat et al., 2005). Such patches also provide refuges for populations of many species outside formal reserves. While isolated populations in reserves are at risk of genetic isolation, a network of patches across a landscape preserves genetic diversity.

Supporting: Many sacred lakes span entire watersheds, supporting all forms of life within those watersheds. For example, Lake Titicaca on the border between Bolivia and Peru is among the highest and deepest lakes in the world. Considered sacred by the Incas, this lake supports a large watershed of over 8000 sq. km, giving protection to forests upstream and recharging aquifers downstream (Salles-Reese, 1997).

Regulating: Large tracts of forest are important for regulating atmospheric cycles such as carbon, nutrients and water (Nunez et al., 2006). For example, Mount Athos is an Orthodox Christian monastery – a sacred site covering an entire mountainous peninsula in northern Greece. This peninsula covers an area of 336 square kilometres with its steep, densely forested slopes reaching over 2000 metres (Mount Athos, 2008). The forests on this mountain remain largely untouched because of inaccessibility, and play a role in regulating the local atmosphere such as through carbon sequestration, nutrient cycling and water storage.

Provisioning: In many parts of the world livelihoods of people still depend on natural resources. For example, sacred mountains in Tibet (Menri) are also where local people harvest plants commonly used in traditional medicine. These medicinal plants are harvested in such a way that plant populations do not deplete (Anderson et al., 2005). The Tibetan mountains have supported a thriving tradition in herbal medicine for centuries.

Cultural: All sacred natural sites bear cultural significance to many indigenous communities who maintain them (IUCN, 2008). In these sites, annual festivals are held, providing opportunities for community gatherings. In addition to promoting cultural integrity, such gatherings also provide recreational experience for people.

Thus, sacred natural sites provide a wide range of 'ecosystem services', as ecologists define them. However, protection of these sites is not driven by material benefits, but by cultural traditions of indigenous people that have been handed down through generations.

The literature on ecosystem services borrows the language and concepts of economics and refers to ecosystems, and resources derived from them, as 'natural capital' (Costanza et al., 1997; Turner and Daily, 2008). While economics concerns itself with the efficient allocation of scarce resources as a means of satisfying human wants or desires (Tietenberg, 1991), it has been contended that the framework of economics is:

a) utilitarian, because things count to the extent that people are willing to pay for them;
b) anthropocentric, because humans assign monetary value; and
c) instrumentalist, because various components of the natural world are regarded as instruments for human satisfaction (Randall, 1988
d) Therefore, an entity is considered to have economic value only if people consider it desirable and are willing to pay for it (Chee, 2004).

The economists' framework for valuing nature suffers from a number of limitations (Kumar and Kumar, 2008):

a) Individuals are not always driven by a value-for-money approach (Simon, 1957). For example, individuals may make a conscious decision to pay a premium for power generated from clean technologies even if power produced using fossil fuels is cheaper.
b) Individuals are not always utility maximisers (Gowdy and Mayumi 2001). For example, people routinely take non-utilitarian ethical standpoints when it comes to conservation of endangered species the endangered species in question do not then have use value per se (Spash, 2000, 2006).
c) Individuals' preferences are often determined by established cultural practices (Tversky and Kahneman, 1991). For example, many modern-day societies follow conservation practices that their forefathers started despite significant changes in the way of life.

Attempts to give market values to ecosystem services, and more generally environmental change, through contingent valuation are increasingly proving problematic (Spash, 2008b). Spash (2008b) and Sagoff (2008) specifically criticise the ecosystems valuation literature. Sagoff (2008) contends that attributing a market value to ecosystem services is ineffective because these services are either too widely accessible to be priced, priced competitively, or too cheap to meter. These various limitations suggest the contingent valuation of 'ecosystem services' cannot accurately represent values which traditional societies attribute to sacred natural sites. Gowdy (1997) has argued that economists need to broaden their concept of value beyond that determined by market exchange. We need to define ecosystem services in a new context – a context that treats nature as such, rather than from utilitarian, anthropocentric or instrumentalist perspectives. There is then a question as to how far we can reconcile the contrasting values that financial markets and traditional societies would attribute to nature conservation.

In ancient classical traditions earth, water, fire and air are recognised as the elements that form all living beings and non-living objects. While these four elements can be seen and experienced there is a fifth element, often referred to as ether or aether, which according to many traditions, is invisible yet all-pervasive. Aristotle's view was that the aether was only found in space, and was the element responsible for the movement of the stars and planets. Followers of Plato, however, held that the aether was a substance like the other elements, and that it was the element from which the soul fashions rational creatures (Karamanolis, 2006, 104). Thus, according to ancient Greek philosophy, the five elements are essential ingredients of living and non-living objects in nature. According to the Sankhya philosophy, a pre-classical school of thought that developed in India between 1000 and 100 BCE, Pancha-Maha-Bhootas, the five great elements, are represented in the human body itself (Feurstein, 2001). This implies that spiritual progress of humans is closely linked to the five elements of nature within. For example, Shvetashvatara Upanishad (2.12) says: 'When the [aspirant of spiritual progress] has full power over his body composed of the elements of earth, water, fire, air and ether, then he obtains a new body of spiritual fire which is beyond illness, old age, and death' (Easwaran, 2005). The worldview of ancient philosophies is reflected in the reverence for nature which many traditional societies have even today. The conservation of sacred natural sites, therefore, is rooted in the belief that humans are an integral part of nature; not separate from it.

Each of the five elements of nature defined here – earth, water, fire, air, and ether – signifies one aspect of nature and natural processes. Earth represents objects in solid state such as rocks which form soil and contain minerals. Water represents objects in liquid state – the hydrological cycle containing water in all forms. Air represents objects in gaseous state, which include all atmospheric gases. Fire represents the fourth state of matter, plasma, relatively new to science. Ether represents something that is of non-material nature, interpreted as spiritual.

Furthermore, each of the five elements is related to one of Earth's spheres – lithosphere, hydrosphere, atmosphere, biosphere and space (Table 13.1). The earth element is related to the lithosphere made up of rocks; the water element to hydrosphere made up of water in all forms and the air element to atmosphere made up of gases. The fire element is related to biosphere because it represents all forms of energy, which living beings need for sustenance. The ether element relates to space, which is present everywhere, but cannot be seen or felt physically.

The linkages between the elements of nature, the Earth's spheres and the ecosystem services are also evident. Earth is considered to represent stability; water, purity; air, pervasiveness; fire, energy and power and space, creativity and dynamism (Feurstein, 2001). These qualities define ecosystem services derived from nature (Table 13.1). The element earth, which represents stability, delivers the service of preservation – for example maintenance of genetic and species diversity. The element water, which represents purity, delivers the supporting service – supporting all living beings that depend on clean fresh, uncontaminated water. The element air, which represents pervasiveness, delivers the regulating service – including atmospheric regulation. The element fire, which represents energy and power, delivers the provisioning service – providing energy in the form of

Table 13.1 *Reconciling 'ecosystem services' with 'elements of nature' framework*

The ancient element of nature	Corresponding sphere where ecosystem function originates	Service delivered (cf. ecologists' framework of ecosystem services)
Earth	Lithosphere	Preserving
Water	Hydrosphere	Supporting
Air	Atmosphere	Regulating
Fire	Biosphere	Provisioning
Ether	Space	Cultural

food to all living beings. The element ether, which represents creativity and dynamism, delivers the cultural service – including recreational experience or intellectual inspiration.

Are these ancient elements of nature and their perceived properties in harmony with the conceptualisation of ecosystems as services? While ecosystem services are rooted in material values, the 'elements of nature' framework can be considered as a non-material standpoint for the same – reconciling the two seemingly opposite ideologies. Yet there are good reasons why a non-material standpoint matters. Nature conservation always has to find a delicate balance – it has to appeal as much to Western science as to traditional societies. In many traditional societies, nature conservation is seen as something that has been 'imposed' by the Western world. There are numerous examples where indigenous communities displaced from conservation areas feel alienated from the Western world and its values (West et al., 2006).

For conservation to be successful, conservationists need to make allies. Therefore, a conservation approach should be such that it is acceptable to Western science whilst being attractive to traditional societies. The imposition of Western values on traditional societies is dangerous because their alienation will mean failure for the conservation of critical ecosystems, habitats and species. Experience has shown that conservation initiatives fail when people are alienated (Pyle, 2003). Nature conservation needs win-win solutions – a reason why the non-material 'elements of nature' framework is likely to be more successful for the conservation of sacred natural sites.

Linking conservation and development through an 'elements of nature' framework

Sacred natural sites enshrine non-material values of nature conservation. In order to ensure their continued protection, a conservation approach that is sensitive to the plural values people attribute to these sites is essential. The 'elements of nature' framework proposed in this chapter can be sensitive to people's beliefs. While sacred natural sites provide all the ecosystem services, as defined in the literature, these services are a by-product of traditional protection of these sites. Beyond sacred natural sites, there are already successful examples where the 'elements of nature' framework has been translated into payments for ecosystem services. Referred to as 'elemental equity' this initiative accepts payments for restoration of the elements of nature – earth, water air and fire – in a South African landscape (see the online resources section for a weblink). The fifth element, according to the elemental equity initiative, is people, because the main goal of this initiative is 'helping hands for healing lands'. Such initiatives may provide a bridge between academic debates on material vs non-material values of nature and the practice of natural resources conservation.

Summary

- Ecosystems services are provisions that humans derive from nature. Ecologists trying to value ecosystems have proposed five categories of these services: preserving, supporting, provisioning, regulating and cultural.

- To date, this ecosystem services framework has been deployed in ways which focus on the 'material' or economic value of nature, leading to PES being positioned as a kind of 'successor' approach to integrated conservation and development projects (ICDPs – see Chapter 10).

- Whilst a recent valuation exercise set the economic value of global ecosystem services in the region of US$125–US$145 trillion, the (limited) evidence on the extent to which this value is being leveraged to achieve conservation and poverty reduction objectives simultaneously does not yet paint a promising picture.

- Some commentators worry that PES schemes do not realise their potential to achieve conservation objectives because, owing also to the expectation of doing poverty reduction simultaneously, they are not targeted at the people whose behaviour is most indicated in environmental degradation.

- Conversely, other commentators raise the concern that poor people – and especially the poorest – benefit little or not at all. This is because a) they are not always eligible for inclusion in PES schemes, or b) even when they are eligible, are still not in a position to forgo shorter term economic returns required by conserving forest rather than converting it to other uses, even when conservation would in the longer term deliver higher returns.

- Yet the predominant ecosystem services can also be recognised with cultural, 'non-material' values. Sacred natural sites, spaces protected for their cultural or religious significance, are areas of 'non-material' spiritual significance to people.

- Ancient classical traditions recognise five elements of nature: earth, water, air, fire and ether. The perceived properties of these elements correspond with the ecosystem services framework.

- Can material and non-material values be reconciled through an 'elements of nature' framework? We argue that if the two can be reconciled, the 'elements of nature' framework is more suitable to make a case for conservation of sacred natural sites because it can be attractive to traditional societies whilst being acceptable to Western science.

Discussion questions

1. What are the caveats in the ecosystem services approach to the conservation of nature and the alleviation of poverty?

2. How can sacred natural sites support a range of ecosystem services whilst following a non-material approach to the conservation of nature?

3. How might reconciling material and non-material values of ecosystem services help towards the conservation of nature and the alleviation of poverty?

Note

1. The following sections have been adapted from: Bhagwat, S.A. (2009) Ecosystem Services and Sacred Natural Sites: Reconciling Material and Non-material Values in Nature Conservation. *Environmental Values,* 18: 417–427.

Further reading

Bhagwat, S.A., and C. Rutte. (2006) 'Sacred groves: potential for biodiversity management'. *Frontiers, Ecology and the Environment* 4: 519–524.

Daily, G.C. (1997) *Nature's Services: Societal Dependence on Natural Ecosystems.* Washington, DC: Island Press.

Costanza, R., R. d'Arge, R. deGroot, S. Farber, M. Grasso, B. Hannon, K. Limburg, S. Naeem, R.V. O'Neill, J. Paruelo, R.G. Raskin, P. Sutton, and M. van den Belt. (1997) The value of the world's ecosystem services and natural capital. *Nature* 387: 253–260.

Egoh, B., M. Rouget, B. Reyers, A.T. Knight, R.M. Cowling, A.S. van Jaarsveld, and A. Welz. (2007) Integrating ecosystem services into conservation assessments: a review. *Ecological Economics* 63: 714–721.

McCauley, D.J. (2006) Selling out on nature. *Nature* 443: 27–28.

Samii, C., Lisiecki, M., Kulkarni, P., Paler, L., Chavis, L., (2014) Effects of Payment for Environmental Services (PES) on Deforestation and Poverty in Low and Middle Income Countries: A Systematic Review. *Campbell Systematic Reviews* 11.

Turner, R.K. and G.C. Daily. (2008) The ecosystem services framework and natural capital conservation. *Environmental and Resource Economics* 39: 25–35.

Online resources

Defra. 2006. Ecosystem Services: Living with Environmental Limits http://www.ecosystemservices.org.uk/ecoserv.htm (accessed 15.01.2015).

IUCN. 2008. Protecting the Sacred Natural Sites of the World. http://cms.iucn.org/about/work/programmes/social_policy/sp_themes/sp_themes_sns/index.cfm (accessed 15.01.2015).

Elemental Equity 2015. http://livinglands.co.za/portfolio/elemental-equity/# (accessed 15.01.2015)

14 Conclusion: conservation and development futures

- Conservation and conflict
- Resilience as the bridge between conservation and development?
- Missing the wood for the trees – alternative relationships between conservation and development?

Introduction

Up until now, this book has been focused on efforts to reconcile conservation and development to date. Rather than summarising the key arguments presented in the chapters of this book so far, the conclusions make an attempt at 'horizon scanning' to identify key issues that are re-shaping the relationship between conservation and development. It picks up on three trends which are either receiving substantial attention within conservation and/or development circles, or which will have fundamental ramifications for the future of the relationship(s) between conservation and development. The first of these trends is the increased focus on conservation and conflict, mainly clustered around sudden spikes in the twenty-first century in the levels of international trade in illegal wildlife products. The second trend picks up on the increased influence of thinking around resilience, a term which has risen to prominence in the development arena rather more slowly than it did in conservation circles. Now that it is prevalent in both, and to some extent seen as the successor to the concept of sustainability, can it provide the bridge between conservation and development which, to date, has been lacking? The third trend is the tendency towards concern that efforts to achieve conservation and development simultaneously do not really engage either with the drivers of biodiversity or with the productions/ reproduction of poverty, in large measure because of a disproportionate focus on rural areas high in biodiversity. What might an alternative consideration of the relationship between conservation and development look like?

The three key trends picked up in the conclusions emphasise our approach laid out at the beginning of this book that conservation and development need to be looked at in the wider context, rather than the specific meanings associated with 'conservation' and 'development' alone.

Conservation and conflict

The drivers and extent of illegal wildlife hunting and conflict

Little work to date has explicitly considered the relationship between conflict, conservation and development. However, it is starting to receive attention, and in ways which already demonstrate how useful it is for conservation and development practitioners and scholars to apply a broader conceptualisation of conflict to their own work. As Duffy (2014) notes, if conflict is mentioned at all, then it is in relation to human–wildlife conflict. This is perhaps at least in part because of the tendency to do neither conservation nor (long-term) development work in conflict zones. What have been documented to some extent are the direct and indirect consequences of conflict for wildlife populations (Beyers et al., 2011; Dudley et al., 2002; Shambaugh et al., 2001). Recently, there have been calls to use economic valuations of ecosystem services as a means through which to calculate more precisely reparations that should be paid as a result of environmental harm caused by conflict (Francis and Krishnamurthy, 2014). There is, moreover, clearly a geographical dimension, in the sense that 80% of all conflict between 1950 and 2000 occurred in countries which are rich in biodiversity (Hanson et al., 2009). This brings into sharp relief the question of whether conservation can and should be used more in conflict situations, and the extent to which drawing what some see as our 'false dichotomy' between conservation and development hinders progress in this area (Milburn, 2012). Bringing a broader conflict lens to our conservation and development thinking reveals two sets of issues/trends worthy of attention:

1   the role attributed (somewhat problematically) to the illegal wildlife trade in funding and provoking conflict and/or destabilising national security;

2   renewed efforts at militarising conservation interventions in response and contributing to heightened risk of armed conflict over wildlife resources.

Douglas and Alie (2014), drawing upon works which frame conflict in terms of greed and grievance (see, for instance, Collier, 2007; De Soysa, 2013; le Billon, 2005), argue that wildlife can be seen as a 'high-value resource', akin to oil, diamonds or timber and implicated in generating conflict in sometimes identical ways. Certainly, wildlife trade is big business: Engler (2008) estimates that declared import revenues from wildlife, fisheries and wild timber combined amount to US$332 billion per annum. Illegal trade in animals has been put at US$10–US$20 billion per year (Bliss, 2009), with the standard caveats about how difficult estimation of this sum is, owing to the clandestine nature of this activity. Table 14.1 gives a flavour of the kinds of wildlife products that are sold illegally, the uses to which they are put and the prices they fetch.

Table 14.1 *Illegal wildlife trade – products, uses and prices (adapted from Douglas and Alie, 2014)*

Species or animal part	Uses	Reported retail price: range or max in US$
Mammals		
Lion	Trophy	$10,000–$50,000 ea.
Lion: bones	Tiger bone substitute	$165 per kilo
Orangutans	Exotic pet; tourism entertainment; zoo exhibit	$1,000 ea.
Rhino: horn	Traditional medicine	$30,000–$65,000 per kilo. One horn sells for up to $250,000
Tiger: skin, pelt,	Decorative; clothing	Up to $20,000 ea.
Snow leopard	Pelt	Up to $20,000 ea.
Elephant: raw ivory	Decorative/various; trophy	$6,500 per kilo
Polar bears	Trophy	$20,000 ea.
Cubs of lions, hyenas and leopards	Exotic pet for wealth elite of the Gulf states of Yemen, Saudi	Up to $12,700 ea.
Pangolins	Luxury food item; traditional medicine	Up to $7,000 ea.
Amphibians		
Tokay gecko	Exotic pet; traditional medicine	Up to $2,330 ea.
Reptiles		
Madagascan ploughshare tortoise	Exotic pet	$30,000 ea.
Angolan python	Exotic pet	$65,000 ea.
Komodo dragon	Exotic pet	$30,000 ea.
Birds		
Pakistani falcons	Sports hunting	$10,000 to $100,000 ea.
Insects		
Colophon beetles	Decorative display	$15,000 ea.

Not only is the illegal wildlife trade big business, but the revenues it generates are sometimes implicated in conflict scenarios for reasons of 'greed' (Douglas and Alie, 2014). Wyler and Sheikh (2008) argue that Al Qaeda operatives received funds from the trade of rhino horn and ivory between Somali 'warlords' and a variety of Indian Islamic extremist groups. Sudanese militia have been accused of entering Cameroon, Chad and Kenya in search of ivory (WWF, 2012), which is also said to be a source of revenue for the Lord's Resistance Army in Uganda in its long-standing conflict with the state (UNEP et al., 2013). However, the evidence base for any of these claims is very thin and they remain very hard to substantiate (Duffy, in review). There are numerous reasons why wildlife is an attractive source of revenue for actors involved in conflict. It is quite frequently easy to acquire, especially for well-armed groups, in comparison to the costs of extraction of other high-value resources, such as oil or diamonds (Douglas and Alie, 2014). Moreover, its value-to-weight ratio, in combination with the difficulties of distinguishing legally from illegally hunted animals at the point of purchase, give wildlife the characteristics of what

Figure 14.1 *Black rhino in the Ngorongongo crater, Tanzania*

Source: http://en.wikipedia.org/wiki/File:Ngorongoro_Spitzmaulnashorn_edit1.jpg.

conflict analysts refer to as a 'lootable resource' (Lujala and Rustad, 2011).

Wildlife is also implicated in conflict associated with grievance. For instance, Mai Mai rebels are reported to have slaughtered hundreds of hippopotamuses and threatened to do the same to gorillas in Virunga national Park, in the Democratic Republic of Congo, in order to bring international attention to their cause (Douglas and Alie, 2014). When Maoist rebels occupied protected areas in Nepal, appropriating revenues from ecotourism, they were viewed sympathetically by some at the local level for resisting unwelcome external influence, of which lucrative forms of ecotourism had become symbolic (Seeland, 2000).

Militarising the hunting and protection of wildlife: effects and consequences

Whatever the motivations for illegal wildlife trafficking, the result has been an increasing capacity and willingness to use violent, sometimes lethal force against humans. This tendency is manifest both on the part of some of those engaged in illegal hunting and on the part of conservation agencies and authorities seeking to respond to substantial increases in the levels of illegal hunting in recent years. The most attention grabbing variety of illegal hunting is now being conducted by criminal groups who, Gettleman (2012) reports, have access to 'serious' military equipment including heat-seeking telescopes, GPS satellite receivers, semi-automatic weapons and even helicopters. Conservation groups have responded in kind, arming rangers and increasing the number of 'boots on the ground' (Brockington et al., 2006; Duffy, 2014), as well as extending their discretion to act, for instance through the existence of shoot-on-site policies in protected areas (Humphreys and Smith, 2011). Additionally, there has been investment in surveillance capacity through the deployment of aerial unmanned vehicles (more commonly known as drones).[1] One disturbing consequence of what Duffy (2014) refers to as the militarisation of conservation is that in Africa alone, hundreds of people die on an annual basis in clashes between illegal hunters and wildlife protection staff (Gettleman, 2012, Humphreys and Smith, 2011). See Box 14.1 for an illustration from South Africa of violent illegal hunting and a strong crackdown from conservation/ state authorities.

Box 14.1

Hunting the hunters in Kruger National Park, South Africa

The Huffington Post reports that an illegal hunter in South Africa was given 77 years in prison in a deliberate effort to deter the targeting of rhinos for their horns. The sentence was awarded to South African Mandla Chauke, in the same month that two Mozambicans were sent to prison for 16 years for killing a rhino in order to cut off its horn last year in Kruger National Park. According to the article, 370 rhinos had been killed by illegal hunters in the first half of 2014, whilst 62 people have faced arrest in connection with some of these incidents. These figures come on the back of reports of the illegal killing of 1004 rhinos in South Africa in 2013.

Mandla Chauke, along with two accomplices, were convicted of being present illegally in Kruger National Park, shooting three rhinos and engaging in a firefight with patrolling rangers. One suspect was killed in the fighting, whilst another fled, according to the prosecution. In addition to convictions for illegal hunting and possession of firearms, Chauke was additionally found guilty of the murder of the accomplice, after prosecutors successfully argued that it was he who should be held responsible, rather than the ranger who shot the accomplice. The judge did not accept Chauke's claim that his fellow accomplices coerced him into taking part in the illegal excursion.

See article at: http://www.huffingtonpost.com/2014/07/23/south-africa-rhino-poachers_n_5612381.html (accessed 24.07.14)

There is little dispute surrounding the reality of the increased incidence of illegal hunting, but there is less consensus around the extent to which the use of this increase justifies the kinds of responses we are seeing. Duffy (2014) contends that beefing up the military capacity of conservation authorities is too crude and simplistic a response to the complexities and plurality of motivations underlying the varied groups of people involved in illegal hunting. For a start, she argues, there is a real danger of conflating highly organised, well-armed criminal gangs with the 'flip-flop' wearing variety of illegal hunter more concerned with subsistence than with the prize of a rhino horn or an ivory tusk smuggled onto South Asian black markets. Shoot-on site policies, as other commentators have also pointed out, with their emphasis on the rapid response, run the risk of making it impossible for protection staff to make this distinction accurately (Brockington et al., 2006). Arguably, moreover, again in the African context, they reinforce a centuries-old tradition of criminalising the livelihood activities of precolonial populations. This has been done by casting their hunting practices as barbaric, in contrast with the sporting, civilised forms of hunting pursued by European explorers, colonial administrators and expatriates (Duffy 2014; see Chapter 1 for more details). Indeed, the term most associated with illegal hunters, 'poacher', is a framing which is heavily implicated in the historical criminalisation of hunting related livelihood activities (Adams, 2004).

The use of surveillance technologies such as drones raise similar questions. To date, conservation organisations appear to have two main uses for UAVs. First, they have been seen as a way of collecting data on animal movements and biodiversity change (Duffy, 2014). Second, they are used to monitor the movements of illegal hunters, sometimes to engage them and sometimes to ensure that wildlife protection staff avoid coming into conflict with well-armed hunters. Nevertheless, the use of this technology poses questions around the extent to which the human rights of communities within and adjacent to protected areas, as well as the rangers who work in them, are respected (Duffy, 2014). Additionally, there is abundant evidence of state involvement in large-scale illegal hunting of wildlife, sometimes with a view to causing instability in neighbouring states (as in the case of apartheid South African incursions into Angola and erstwhile South-West Africa to hunt game with a view to depriving governments and people of important and lucrative resources) (Ellis, 1994; Kumleben, 1996; Reeve and Ellis, 1995). Finally, militaristic attempts to crack down on illegal hunting do not address another underlying cause, namely international demand for illegal wildlife products (Duffy et al., 2013).

Above all, the greatest, and perhaps most obvious, risk of meeting violent force with more violent force is that of escalation, as illegal hunters and traffickers adapt to more militarised conservation interventions by increasing their own capacity to engage in violent conflict (much as Mexican drug gangs have done in response to crackdowns by the Mexican and US governments). Elsewhere, Duffy argues that the international trade in illegal wildlife is increasingly being linked, discursively, to the war on terror, turning 'poachers' into terrorists and, thereby, waging 'war, by conservation' (Duffy, in review). She notes that both Hillary Clinton and Barack Obama have explicitly identified a link between international trade in illegal wildlife products and global security, because of the alleged involvement of terrorist groups in illegal wildlife trafficking. These statements have subsequently been endorsed by important conservation actors such as the UNEP (2013) – even though, as noted above, the veracity of claims about terrorist involvement in illegal wildlife hunting remain far less substantiated than one might be led to believe by looking merely at high-profile announcements on the matter. The consequence, Duffy argues, is a conservation approach that goes beyond 'back to the barriers', becoming very much an offensive position rather than a defensive one. Even more concerning is that she raises the possibility that military operations which would not otherwise be politically feasible might be taking place under the cover of conservation related to illegal hunting (Duffy, in review).

The risk from the speed with which this discourse is gathering momentum is the use of conservation for the legitimation of violence to which many, beyond the supposed terrorists at which it is aimed, may become subjected. Attached to this is the distinct possibility that conservation approaches based on outreach and collaboration with people who bear the costs of conservation lose yet more ground to antagonistic interventions. Within conservation circles, there is increasing awareness and concern about the implications for community-based conservation of increasingly muscular approaches to enforcement in the face of illegal hunting. For instance, a recent symposium in South Africa looked at what could be done 'beyond enforcement' to address illegal hunting, and to attempt to ensure that the costs of enforcement did not fall too heavily on people living in or adjacent to protected areas (IUCN et al., 2015). Nevertheless, the greater the priority given to stopping illegal hunting, the likelier it is that attention is distracted away from the exercise of a more fundamental rethink of the relationship(s) between conservation and development.

Resilience as the bridge between conservation and development?

The extending reach of resilience thinking

Resilience has for some decades already been influential across a number of academic disciplines and areas of study, including ecology, environmental history, traditional environmental knowledge, political ecology and common pool resource theory and management.

As Chapter 3 explores in greater detail, whilst in the 1970s, it may have been common to speak of resilience of ecosystems, in the twenty-first century it is more common to refer to the resilience of social-ecological systems, as it is increasingly recognised that the resilience of ecological systems is affected by and has impact upon the resilience of social systems in ways that are difficult to disentangle. For a long time, therefore, resilience has been seen as central to understanding human-environment interactions.

In conservation spheres, there is little new in these statements, because the term is very much part of the conservation lexicon. In the development sphere, its influence is not so deeply rooted and, until recently, was all but exclusively in relation to parts of the development process where it is harder to get away from environmental phenomena when thinking about how to plan and implement interventions. For instance, disaster risk reduction practitioners and others working on the impacts of climate change upon the development process have been using the concept to organise much of their work (Klein et al., 2003; Moser et al., 2010). The use of the term has since exploded into the rest of the development arena (Fischer and Kothari, 2011). Perhaps more than anything else, the appropriation of the concept by bilateral donors, such as USAID and the UK Department for International Development (DFID), shows quite how influential and important it has become. For instance, whilst little mention could be found of 'resilience' in DFID documents dating back to 2006, by 2011 it had been acknowledged as an objective as central as poverty reduction and growth (DFID, 2011). USAID, similarly, launched a 'Resilience to Recurrent Crisis' Policy and Program Guidance Initiative (USAID, 2012). Finally, the fact that resilience was also at the core of both the World Development Report and the Human Development Report in 2014 demonstrates quite how dominant it has become within the mainstream development discourse (Béné et al., 2014).

One way of capturing how the engagement with resilience through different areas and disciplines has moved into the development arena is through the framework of Béné et al. (2014). They argue that resilience can be defined as the ability to deal with the impacts of adverse changes and shocks. This ability encompasses 'buffering impacts', 'returning to pre-shock situation' or 'bouncing back', but also 'shock absorbing', 'evolving and adapting' or even 'transforming' (Berkes et al., 2008; Walker et al., 2002).

This increased conceptual nuance follows the broadening of the concept from its initial relatively narrow focus of what Holling (1973) referred to as 'engineering resilience') into a more elaborated concept which embraces the ability not simply to bounce back but also to adapt and to transform, captured by 'ecological resilience'. Recent advances to go beyond even this term with the notion of adaptive capacity. In the context of climate change, the IPCC defines adaptive capacity as 'the ability of a system to adjust to climate change (including climate variability and extremes) to moderate potential damages, to take advantage of opportunities, or to cope with the consequences' (IPCC, 2001). On the other hand, transform-ability is the 'capacity to create a fundamentally new system when ecological, economic or social structures make the existing system untenable' (Walker et al., 2004: 5). Along with the concept of absorptive capacity or persistence (that is, the various strategies by which individuals and/or households moderate or buffer the impacts of shocks on their livelihoods and basic needs), these three elements can be seen as the three core components of resilience in a develop-ment context. Drawing on this, Béné et al. (2014) employ absorptive adaptive, and transformative capacity as the three structuring ele-ments of an analytical framework aimed at understanding better what exactly 'strengthening resilience' means. The framework is presented in Figure 14.2.

Reasons for resilience...

Given the momentum behind resilience, in addition to its concep-tual sophistication and relevance, might we not see it as a bridge

Figure 14.2 *The 3D resilience framework*

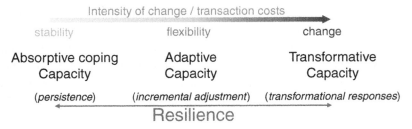

Source: Reproduced from Béné et al., 2012.

between conservation and development, a means through which to bring a sense of peace and solidity to an often uneasy relationship?

In one sense, the question can be answered with a straightforward yes, in that it is impossible, ultimately, to disentangle nature and society in such a way that the one is not influenced by and influencing the other, in a relationship of co-production which Chapter 3 has explored in further depth. Because resilience is so rooted in the concept of the social-ecological system, it provides deep theoretical insights into the interconnections and interdependencies characterising the relationship between conservation and development. It helps us to understand, in essence, why it is necessary to reconcile conservation imperatives with the demand for development. It allows us to see, in other words, how any development trajectory requires some level of conservation built into it, whilst the nature of what we want to conserve is fundamentally altered by the development trajectory that we choose. Think, for instance, of how different the impacts of the Asian Tsunami of 2004 might have been on humans and countries' entire development trajectories if the loss of coastal ecosystems, which would have provided a substantial buffer to the impacts, had not been quite so severe (Carpenter et al., 2006). Think also of the widely-acknowledged: relation between (rural) poverty and greater dependency on natural resources (Reddy and Chakravarty, 1999; Beck and Nesmith, 2001; Béné et al., 2009). Ultimately, it could be argued that without the holistic perspective that resilience thinking encourages, it will not be possible to fashion a healthy relationship between conservation and development.

One useful conceptual insight from resilience which is highly relevant both in conservation and development contexts is its emphasis on scale and levels within scales (think of local, regional and national levels within a geographical scale, for example). It has been well documented (for instance by Gunderson and Holling, 2002; Gunderson, 2008) that ecosystem processes happening at different levels within a geographical scale affect and are affected by each other (think of the way in which a fire that starts at a local level can engulf an entire forest, and the way in which the broader-level, structural characteristics of the forest provide a kind of 'memory' which permits regeneration). Within- and cross-scale effects are also relevant to development intervention. The debate around decentralisation and development effectiveness is grounded in considerations of whether more government functions should be given over to administrative units working at the local level, or whether they should remain concentrated within central government, because of its better understanding of national level processes and phenomena.

On a purely practical level, resilience may be seen not just as an organising principle but also as a 'mobilising metaphor' (Pain and Levine, 2012). That is, it is a term with such wide appeal that it can bring together a very disparate group of actors, organisations, communities of practice, etc. who probably should but don't always manage to coordinate and collaborate with each other. In fact, looking at the way in which resilience is used by policymakers and practitioners alike, it is clear that a range of meanings is attached to the term which go quite far beyond its disciplinary roots in ecology, engineering and psychology (Bahadur et al., 2010). Whilst this is unsatisfactory from the point of view of trying to establish a coherent definition of the term, the overlapping meanings imputed to it and the ambiguity which characterises its use is sometimes expedient in terms of bringing people together, especially when they have rather different objectives. This 'broker capacity' (Béné et al., 2014) has already been used in a development context to foster multi-sectoral collaboration. For instance, the World Food Program launched a 'Resilience Project' which uses the theme of resilience to encourage knowledge sharing, dialogue, and field level collaboration between food security, disaster risk reduction, climate change adaptation and social protection (SDC and WFP, 2011).

...But resilience is not a panacea

Sometimes, it can be very difficult to distinguish an opportunity from a risk, and this may be an appropriate way to think about current forms of engagement with resilience. One of the key risks of the current enthusiasm for resilience is that it stands in danger of being co-opted in the same way that 'sustainable development' has in the past. This is the other side of the coin to resilience as 'mobilising metaphor'. The key problem with 'sustainable development' is that it has been used (and abused) by so many different interest groups to refer to activities that are as apt to be mutually exclusive as they are to be complementary. In consequence, the term has sometimes been used to refer to forms of natural resource use which are anything but sustainable (Redclift, 1987). Duffy (2000) has gone so far as to propose that i has been used so diversely that it has in fact become meaningless. This is perhaps unfair, as it is possible to chart the different strands, meanings and discourses around the term (see Redclift, 2005 for an example). Ultimately, though, as with sustainability, the very neutrality of the term 'resilience' cannot resolve for us by itself what form of social-ecological system is worth trying to make resilient, or tell us much about the interests of different actors who favour one social-ecological system over another. Some commentators have argued that current development models are inadequate for dealing with climate change (Boyd et al., 2009). If this is the case, the risk is that resilience might itself become a justification for investing in approaches that a) may undermine long-term resilience and b) leave less space thereby for alternatives that might offer more hope for resilience. Boyd et al. (2009) has argued that this latter possibility is already happening with mainstream thinking on adaptation, which tends to favour market environmentalism and ecological modernisation, with an emphasis on maintaining rather than challenging the status quo (see chapters 2 and 5). Could the 'mainstreaming' of resilience follow a similar path? Recent work by Katrina Brown suggests that it might: she maintains that the rationale behind the World Bank's 'climate resilient development' policies is 'to protect economic growth from the ravages of climate change' (Brown, 2011: 28). However, we should not forget that Herman Daly, who famously labelled 'sustainable development' an 'oxymoron', worked at the World Bank. Perhaps, therefore, there is more space than we might imagine for unorthodoxy in some of the development world's biggest and most orthodox institutions. This would leave open more possibilities for using resilience to challenge, rather than to entrench, the modes of (economic) development that are causally linked to anthropogenic climate change.

Perhaps because of the fine line between using resilience for the purposes of mobilisation and/or co-optation, it is doubly important to understand that the conceptual roots of resilience are in fact descriptive rather than normative. In other words, resilience describes the likelihood of a particular state of affairs persisting, adapting to or otherwise responding to disturbance in ways which do not lead a social-ecological system to tip into an altogether different state. It has nothing to say about whether that is a good or a bad state to be in. It is up to the users of the term to decide which states are desirable or undesirable, and crucially, some enduringly resilient states are eminently undesirable. For instance, poor people can frequently be resilient almost by definition; that is, they have learned to survive in adverse conditions and become accustomed to poverty as a result. This is sometimes known as 'adaptive preference', which Amartya Sen (1999: 62) defines as follows:

> *Deprived people tend to come to terms with their deprivation because of the sheer necessity of survival, and they may, as a result … adjust their desires and expectations to what they unambitiously see as feasible.*

Becoming accustomed to conditions of poverty is in itself a resilience-building characteristic, which perpetuates, rather than challenges or allows people to escape those conditions of poverty (Béné et al., 2014).

Another example is the privileging of economic growth as a global policy objective, regardless of the fundamental and potentially irreversible character of the environmental externalities entailed. Given the threat to biodiversity of the accelerated rates of species extinction that we are witnessing, and given the dangers to economic growth of compromising the very natural capital upon which it depends, the continued influence of the idea of economic growth seems quite surprising, given its potential to become self-defeating. And yet the idea is enduringly resilient, to the chagrin of many working in the conservation or development sectors.

Considerations such as these have prompted some authors to argue that an uncritical use of the resilience concept risks obscuring the social and political factors which account for the persistence of enduringly undesirable resilient states (Cannon and Müller-Mahn, 2010; Davidson, 2010). Some are concerned that other concepts such as vulnerability, which have a stronger 'social/actor' focus, have been so far ahead of resilience in terms of emphasising issues around social justice or power distribution, they are at risk of being crowded out by resilience framings (Cannon and Müller-Mahn, 2010). It has even been suggested that using metaphors such as 'building resilience' is easier than engaging with these underlying issues of power and inequality which explain both environmental degradation and persistent poverty (Cannon and Müller-Mahn, 2010). Adopting a naïvely apolitical resilience framing poses the risk of a return to technocratic discourses under which social justice and transformative dimensions of conservation or development intervention are submerged or sidelined. Although some resilience scholars have taken such concerns on board (e.g. McLaughlin and Dietz, 2008; Miller et al., 2010), the very popularity of the term does not always militate in favour of using it in ways based on more subtle and contentious readings such as these. Without having ways to challenge and transform current social-ecological systems, we can only analyse and understand how resilient or vulnerable to disturbance and change they are likely to be. For this reason, following arguments laid out in Chapter 3, we suggest that in efforts to understand *whom* resilient system states best serve (and *dis-serve*) and through what processes they are constructed, political ecology is an essential conceptual resource. Efforts towards 'resilient' conservation and development could usefully take this on board.

Missing the wood for the trees – alternative relationships between conservation and development

Are we looking in the right places for conservation and development?

One of the most crucial issues for the future of conservation and development, and our understanding of the relationship between the two, is the extent to which efforts to reconcile conservation and development objectives map onto, or largely ignore, the processes which most fundamentally determine where people are on the planet and how they use it. Adams (2013) has argued forcefully that conservation has a rural and what we might call a 'biodiversity-centric' focus. Arguably, one of the consequences of this focus is that, whilst conservation has not really influenced broader development processes, efforts to undertake conservation and development simultaneously have likewise tended to have a rural focus, largely consisting in efforts in and adjacent to protected areas, or in efforts at community conservation which, whilst outside of protected areas, are mostly rural in their geographical location.

Against a background of disengagement with development processes amongst many conservationists in the twenty-first century (see Chapter 9), efforts to explore the relationship between poverty and conservation buck this trend (see Chapter 10). Yet even this literature is focused almost exclusively on rural and often remote settings. As such, it does not engage with perhaps the most important phenomenon of our time, when it comes to global environmental problems: the urban transition. This is the process through which the distribution of people (and poverty) increasingly tends towards a concentration in urban and peri-urban areas. Using FAOSTAT data, Borras (2009) charts the broad contours of this change. In 1970, the global population stood at 3.7 billion. Just over two thirds of this population was rural. At this time, the agricultural population amounted to 2 billion people and the non-agricultural population at 1.7 billion. By 2010, there were projected to be 6.8 billion people in the world, 3.3 billion of them in rural areas and 3.5 billion in urban areas. Of these, the 'non-agricultural' population had risen to 4.2 billion, whilst the agricultural population came in at 2.6 billion (Borras, 2009). The consequence of this dramatic change is that the majority of poor people in the world do not live in areas where much biodiversity is deemed to be worth conserving, because it is either transformed by peri-urban agriculture, or otherwise urbanised – formally and informally (Redford et al., 2008). There are strong environmental reasons for engaging in these kinds of contexts also. Cities are said to use two thirds of global energy and generate 70% of greenhouse gas emissions (IEA, 2011). Furthermore, whilst cities occupy perhaps 4% of the Earth's land surfaces, their inhabitants have been estimated to consume close to three-quarters of the world' natural resources, and to generate three-quarters of pollution and waste products (Redman and Jones, 2005).

Since the 1980s, one of the key bridges between conservation and development has been a focus on people (see chapters 2 and 9). But how can conservation and development be about people if there is comparatively little work done in the places where even now a majority of the world's population lives, and where ever greater numbers of people will come to live in the future? This is an argument made prominently by political ecologists such as Bill Adams (2013), and one which perhaps runs the risk of overlooking the exten of existing engagements with conservation and development – and indeed environment and development more broadly – in urban settings. Some of this we will touch upon below, in speculating what urban conservation and development might look like. However, we agree with his argument that overall, conservation is much more geared towards rural environments and, perhaps more problematically still, has become 'at ease with capitalism' (Adams, 2013: 312) as the basic system in which efforts to conserve biodiversity at the frontiers is most important. Whilst, moreover, there is an important engagement to be had with poverty in that context, it is not one which challenges the neoliberal, capitalist political and economic order which gives rise to global environmental problems – in fact, arguably, such forms of engagement – such as through ecotourism (see Chapter 11) – entrench it more than they reform (or transform) it. Adams concludes, in a sentiment we would share, that 'it should be a matter of great concern to biodiversity conservationists that so little of their effort is focused on the processes that drive biodiversity loss' (2013: 307).

Duffy (2010) makes a similar point in relation to efforts to protect gorillas in the Democratic Republic of Congo (DRC). She argues that the biggest threat to gorillas (and those people who happen to live close to them) is the global trade in coltan, a key material in the production of popular electronic devices such as mobile phones and games consoles. Up to 80% of coltan in the global market is sourced from mines within rebel-held territory, and contributes to sustaining the conflict within the DRC (Beswick, 2009). One consequence of the conflict has been that the many who have been displaced by it have had to abandon farming activities and to make charcoal to sell to urban centres in the Kivu region. This practice has been implicated in the destruction of gorilla habitat, and some conservation activities have sought to prohibit this practice, without providing comparably lucrative sources of alternative income. Duffy argues that this kind of intervention heaps conservation costs on already poor and distraught people, whilst failing to address the role of global consumption in fuelling the conflict dynamics which underlie conservation (and development) problems in the area.

As noted in Chapter 6, some have argued that this lack of engagement with the broader background of the ideological dominance of neo-liberal thinking is the 'elephant in the room' (Garland, 2008; Sullivan, 2006). Brockington et al. (2010) maintain that increasingly, conserva-tion is turning to capitalist modes of thinking and doing as the means through which to 'save' nature. This is paradoxical, they suggest, given that the biggest drivers of environmental degradation are rooted in the workings of a capitalist global economy (see Buscher and Arsel, 2012 for a synthesis of critiques of neo-liberal conservation). On this logic, Sullivan argues, both the 'preservationist' and commu-nity conservation approaches are actually two sides of the same coin, in that they both reproduce 'the value-frames that generate the biodiversity losses and other exclusions associated with modernity' (2006: 110). In other words, they are the product of, not the antidote to, a political and economic system which perpetuates environmental degradation as well as poverty and marginalisation.

In making this argument, Sullivan's critique brings into sharp focus perhaps the greatest legacy of integrated conservation and development: the tendency to privilege the economics of biodiversity conservation and development. This tendency continues to be seen in successor approaches such as payments for ecosystem services (Chapter 13), or leveraging private funding from transnational companies to fund the expansion of protected area networks (Chapter 6). Whilst not everyone working in such areas would subscribe to all of Sullivan's critique, ambivalence around using dominant modes of capitalist economic activity remains by no means uncommon. It also leaves those interventions which do attempt to reconcile conservation and development through mainstream economic means – such as carbon trading, payments for ecosystem services, etc. – open to the accusation that they make 'the wider problem just that little bit worse' (Adams, 2013: 309). This is perhaps especially the case to the extent that conservation organisations come to rely financially upon the support of transnational corporations whose resource use behaviour is fundamentally implicated in the production of global environmental crisis (Brockington et al., 2010).

Conservation, novel ecosystems and green cities?

It is difficult to see how conservation and development could move away from rural spaces and habitats entirely, and indeed that is not what is called for. However, perhaps we need to start to think in terms of *relationships* between conservation and development, acknowledging that these are likely to be quite different in diverse contexts. We would then be better placed to pose a question which currently, we feel, is receiving insufficient attention: what would conservation and development that engaged with an 'urban transition', unfolding against a neoliberal capitalist political and economic status quo, look like?

One part of this vision might entail changed attitudes and values towards urban ecosystems and the biodiversity contained within them

In recent years, and mostly within a northern country context, increasing attention has been paid to the idea of the 'novel ecosystem', perhaps most frequently associated with the work of Hobbs et al. (2006). These authors identify, broadly speaking, two defining characteristics of novel ecosystems:

a)   major changes in the abiotic environment as a result of human presence and activities;
b)   substantial changes in the species pool, owing to extinctions and/ or the introduction of novel biotic elements.

Both of these characteristics can lead to a fundamental shift in the functioning of ecosystems. Kowarik (2011) notes that on this definition, urban ecosystems very obviously qualify as examples of novel ecosystems, even though Hobbs et al. (2006) do not explicitly acknowledge them as such. A more careful investigation reveals the presence of four distinct categories of landscape relevant to urban transformation, resulting from different landscape histories, sometimes referred to as the 'four natures approach' (Kowarik, 2005): natural remnants, cultural landscapes, urban green spaces and urban-industrial spaces. A burgeoning literature is challenging the idea of urban ecosystems as being too degraded and insufficiently biodiverse to be worthy of much conservation attention. For some time, it has been recognised that plant species diversity can actually be higher in urban settings than in rural ones (Kühn et al., 2004). Some bird species whose populations were threatened by the expansion of agriculture have managed to re-establish populations in urban centres (BirdLife International, 2004). Insect species which specialise in open habitats and/or are capable of flight often do very well in cities (Niemelä and Kotze, 2009). Another concern sometimes raised is with the frequency of 'alien' or non-native species, which are often introduced in urban settings, or establish themselves there owing to the conditions.

There are a number of risks associated with non-native species, including their introduction to adjacent countryside (Sullivan et al., 2005), the displacement of native species (Didham et al., 2005), human health risks (Taramarcaz et al., 2005), impact on ecosystem services (Charles and Dukes, 2007) and the omission of volatile organic compounds (VOCs, which can increase the effect of some of smog and reduce air quality – see Niinemets and Peñuelas (2008). However, Kowarik

(2011) argues that there is a danger of over-generalising these risks, given that there often is uncertainty about the extent to which non-native species are responsible for perpetuating such risks, or indeed about their likelihood. For instance, examining the risk of a spread of an invasive species from urban to rural areas over time, through dispersal vectors such as birds, Botham et al. (2009) found that generally, little transfer from the one habitat to the other had occurred. Kowarik's broad point is that there are often benefits as well as risks from non-native species, but that these are not always recognised within conservation efforts because the fragmentation of urban ecosystems and other factors such as the prevalence of non-native species has given them a low conservation value. As a result, Kowarik contends that conservation strategies tend to focus on the conservation of 'natural remnants' in urban centres, without looking to engage with the 'other natures' which have emerged as a result of the urban transition. Some of these benefits may be of particular importance in relation to the development objectives in urban areas.

First, research has shown that species with higher tolerance of heat tend to do better in urban settings, to the point that Kowarik (2013) suggests that urban-industrial landscapes may have higher adaptive capacity in the face of climate impacts than natural remnants. This means that the ecosystem services provided by such landscapes may be less vulnerable to disruption, at least from warming effects. Second, a number of studies have found a correlation between increased levels of and access to biodiverse/'green' landscapes and human well-being (Miller, 2005), at least in northern urban contexts. It would be worth investigating the extent to which the same holds across a range of developing country contexts, not least because the concept of well-being is one which goes far beyond the narrow focus of development on economic growth, and one which is much more amenable to the kinds of intangible benefits to humans that urban ecosystems can provide. Third, the 'greening' of cities could be used to address some elements of the environmental burden of disease in urban contexts. Figure 14.3 is taken from a planning document relating to the city of Berlin's efforts to adapt to climate change ([German] Senate Department for Urban Development and the Environment, 2011). It demonstrates the ways in which climate change is likely to exacerbate the 'urban heat island' effect which in already exists in many cities, as well as the greening that measures could be used to lower the temperatures. Given the health impacts

Figure 14.3 *Potential of 'greening' measures to help with adaptation in urban centres*

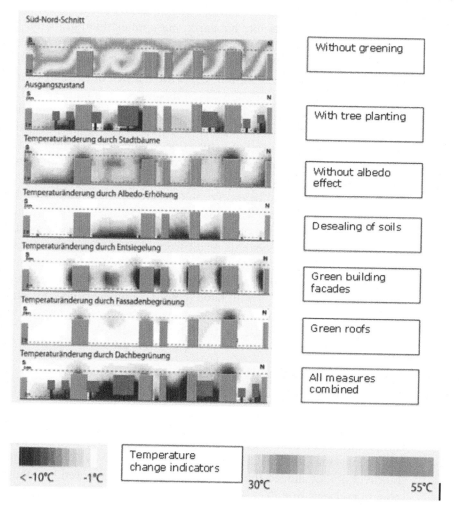

Source: Senate Department for Urban Development and the Environment (2011), 'Stadtentwicklungsplan Klima Urbane Lebensqualität im Klimawandel sichern'. Berlin, www.stadtentwicklung.berlin.de/planen/stadtentwicklungsplanung/download/klima/step_klima_broschuere.pdf

from higher temperatures and heat waves – increase in premature deaths, more amenable conditions for the spread of some diseases etc. – bringing down temperatures in urban settings could make an important contribution to development outcomes.

This re-evaluation of 'invasive' species and novel ecosystems has been gaining traction amongst ecologists and has recently been

brought to a broader audience by popular environmentalist, Fred
Pearce in his book *The New Wild* (2015). Pearce makes the same
arguments as ecologists like Kowarik, but applies them globally.
Alien species, he maintains, are likelier in the round to enrich ecosystems than to threaten them, reinvigorating nature as they spread. On
this view, trying to conserve the 'old wild' against the tide of change
represented by novel ecosystems is to misunderstand nature itself.
Perhaps unsurprisingly, making this argument, he recounts, gets him
'treated as a heretic in some conservation circles' (2015: 2). Yet this
reaction brings us back to the values underlying conservation and
raises the question of whose values most count when decisions are
made about what is worth conserving. As Chapter 9 argues, in
relation to conservation beyond protected areas, conservationists have
little value for the nature found in landscapes substantially modified
or transformed by human presence/activity – the ones where most
people live. Novel ecosystems result from a) the nature humans make
and b) the workings of nature autonomously from (and on) humans. A
relationship between conservation and development which genuinely
put people at its centre would have to confer higher value on such
ecosystems. It would have, in short, to value more of the relationship
between society and nature as it currently stands.

Conservation and development in urban, informal settlements?

The example from Berlin of the benefits of 'greening cities' would
come with trade-offs, inevitably. A focus on green spaces and necessary ecosystem services, for instance, does not always require higher
biodiversity, or indeed the kinds of biodiversity favoured by mainstream conservation. Yet it has clear relevance for a future of urban
development that will inevitably unfold in a changing climate. However, it is less applicable to many urban contexts across the 'global
South', simply because this kind of intervention has been conceived
of as part of a formal urban planning process. One of the biggest
characteristics of urban expansion in the global South, in contrast, is
informal settlements. Davis (2006) contends that since 1970, the
growth of informal settlements has dramatically outpaced formally
planned urbanisation. In one of the world's fastest-growing urban
frontiers, the Amazon, by the turn of the century, 80% of urban
growth took the form of shanty towns without access, in the round, to
standard formally planned utilities such as electricity and water

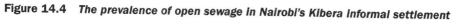

Figure 14.4 *The prevalence of open sewage in Nairobi's Kibera informal settlement*

infrastructure (Browder, 1997). In Kenya, 85% of population growth from 1989 to 1999 was found in the informal settlements of Nairobi and Mombasa (Herr and Karl, 2003). Many other examples – Beijing, Buenos Aires, Delhi, Guayaquil, Karachi, Kinshasa, Kolkata, Lima, Mexico City and São Paulo to name but a few – play host to similarly rapid informal urbanisation; to the point that, for Mike Davis (2006), the distinction between formal and informal urbanisation is moot. The latter is overwhelmingly more prevalent than the former. Crucially, furthermore, urbanisation of this kind across many parts of the 'global South' has become decoupled from the kinds of industrialisation which, in the nineteenth and twentieth centuries, provided migrants to urban areas with employment. It is driven, therefore, 'by the reproduction of poverty, not by the supply of jobs' (Davis, 2006: 16).

What would the relationship between conservation and development look like in an urban informal settlement in the 'global South'? It is unclear (to us, the authors of this book, at least) that there is as yet a comprehensive answer to this question, and the consideration given here is as incipient as it is speculative. However, two areas of priority suggest themselves. First is the propensity of informal settlements to arise in areas highly exposed to environmental hazards such as flooding (de Sherbinin, 2009), in urban areas in which safer, less exposed zones are already settled. The second priority area is the environmental burden of disease, that is, the adverse health effects of the exposure of informal settlement residents to pollutants and toxins. This is leading to a situation in which lower morbidity and mortality rates of urban centres, relative to rural ones, are being eroded in some cities (Montgomery, 2003). This tendency demonstrates just how much environment matters to development outcomes: a concrete example is provided by the East Kolkata Wetlands in Box 14.2.

The need to reconcile conservation and development objectives here remains imperative, as a means of reducing the environmental disease burden, of generating livelihoods and maintaining the existence of an ecosystem whose novelty is partly a function of human presence. A growing acknowledgement of humanity's role in transforming nature and the articulation of this role in the concept of the 'Anthropocene' opens up the opportunity to reconcile conservation and development.

Box 14.2

The East Kolkata Wetlands as the 'kidneys' of a metropolis

The East Kolkata Wetlands are a series of connected lakes and ponds which form part of the Gangetic Delta, designated in 2002 as of 'international importance' under the Ramsar Convention of Wetlands, of which India is a signatory (Bhattacharya et al., 2012). The wetlands were brought to prominence when Dhrubajyoti Ghosh, an engineer appointed by the West Bengal government, chose them when asked to identify the best model for a new sewage treatment system for the city (Parker, 2009). These wetlands currently serve to clean 250 million gallons of organic waste produced by Kolkata residents every day (Bhattacharya et al., 2012). This waste, courtesy of British sewage channels built in the colonial period, flows down to the wetlands (Parker, 2009). Once it has arrived, the amount of sewage entering each pond is regulated by a system of gates and bamboo sieves established by resident fishers who first settled in the wetlands over 100 years ago. The sewage entering the ponds is digested by bacteria, which produces carbon dioxide, in turn providing food for the fish in the ponds (Parker, 2009). Dhrubajyoti Ghosh (2005) claims that 74% of the people living in or adjacent to the East Kolkata Wetlands have livelihoods related to fish farming, agriculture and horticulture. Despite the substantial and indeed essential benefits provided by the wetlands both locally and to the city as a whole, urban expansion, industrial pollution, siltation and infestation are adversely affecting the health and functioning of the wetlands whose protected status is, to make matters worse, under threat precisely because parts of them have been and continue to be built over (Battacharya et al., 2012).

Figure 14.5 *The East Kolkata Wetlands, against a background of urban expansion*

There is much that environments like the Kolkata Wetlands can do for the ever increasing numbers of informal settlement residents.

If international conservation organisations put as much energy and resources into them as they currently put into combatting the international trade in illegal wildlife; and if urban planning processes could be oriented more towards using these resources to address the problems encountered by informal settlement residents, then we might have the makings of a relationship between conservation and development of a strikingly different complexion. Even this different geographical focus will not automatically address the broader concern that efforts to do conservation and development simultaneously contribute to, rather than provide an alternative to, the underlying are a global political and economic model which is responsible for global environmental problems. It would, however, put at the heart of our thinking about conservation and development processes which will fundamentally determine the prospects for both biodiversity and poverty in the twenty-first century.

Summary

- It is never easy to know what the future holds in any sphere of human or ecological activity. We have therefore focussed here on three themes/trends which could have profound ramifications for the relationship between conservation and development in coming decades: conflict connected to the illegal wildlife trade; resilience thinking as a potential 'bridge'; and alternative relationships between conservation and development.

- Conflict linked to illegal wildlife hunting has escalated, though the reasons for the escalation are still becoming clear. Some explanations look to greed and grievance motivations on the part of illegal hunters. Others note the vicious circle that results from beefed-up military capacity on the part of some hunters and conservation authorities. Others again are starting to wonder if conservation is being used as a proxy by actors pursuing military objectives which seek their justification in appeals to the global threat of terrorism. The concern from some commentators is that space and resources for community/participatory conservation approaches are squeezed
- Resilience, well-established in conservation circles, is now also going down a storm in the development arena. Could its popularity improve the prospects for reconciling conservation and development? This remains an open question. The conceptual utility of its focus on inter-connected social-ecological systems is appealing, as is its potential as a 'mobilising metaphor'. However the use (or co-optation?) of this term, by actors whose own way of strengthening their resilience weakens that of others, threatens to render it as oxymoronic as 'sustainable development' has, for some commentators, become.
- On balance, we agree with commentators who argue that conservation – but also efforts to achieve conservation and development simultaneously – remain unhelpfully fixated with rural settings high in the kinds of biodiversity traditionally prized by conservationists. There is reason to work in these areas, but an alternative vision for conservation and development would be to work more in rapidly-expanding, urban informal settlements; and to value novel ecosystems replete with 'alien' species that conservationists, in the round, would rather see eradicated.

Discussion questions

1. How should conservationists respond to the spike in the international trade in illegally procured wildlife products? How helpful is it to think in terms of 'resilient' conservation and development?
2. To what extent has 'big' conservation been co-opted by a global, political and economic order which is inherently environmentally destructive?
3. How necessary is it to rethink the relationship between conservation and development in terms of the 'urban transition'?

Note

1. 'Predator becomes prey: Google funded drones to hunt poachers in Africa', http://sys03-public.nbcnews.com/technology/predator-becomes-prey-google-funded-drones-hunt-poachers-africa-1C7456194 (accessed 24.07.14).

Further reading

Adams, W.M., 2013. Conservation in the Anthropocene: biodiversity, poverty and sustainability, in: Roe, D., Elliott, J., Sandbrook, C., and Walpole, Matt (eds), *Biodiversity Conservation and Poverty Alleviation: Exploring the evidence for a link*. John Wiley & Sons, pp. 304–315.

Béné, C., Newsham, A., Davies, M., Ulrichs, M., and Godfrey-Wood, R., 2014. Resilience, poverty and development. *Journal of International Development* n/a–n/a. doi:10.1002/jid.2992

Bhattacharya, S., Ganguli, A., Bose, S., and Mukhopadhyay, A., 2012. Biodiversity, traditional practices and sustainability issues of EastKolkata Wetlands: a significant Ramsar site of West Bengal (India). *Research and Reviews in Biosciences* 6, 340–7.

Borras, S.M., 2009. Agrarian change and peasant studies: changes, continuities and challenges – an introduction. *The Journal of Peasant Studies* 36, 5–31. doi:10.1080/03066150902820297

Duffy, R., 2014. Waging a war to save biodiversity: the rise of militarized conservation. *International Affairs* 90, 819–834. doi:10.1111/1468-2346.12142

Davis, M., 2006. *Planet of Slums*. Verso.

Kowarik, I., 2005. Wild urban woodlands: towards a conceptual framework, in: Kowarik, P.D.I., Körner, D.S. (Eds.), *Wild Urban Woodlands*. Springer Berlin Heidelberg, pp. 1–32.

Online resources

The Human Development Report – http://hdr.undp.org/en/content/human-development-report-2014 (accessed 30.01.2015)

The World Development Report – http://econ.worldbank.org/WBSITE/EXTERNAL/EXTDEC/EXTRESEARCH/EXTWDRS/EXTNWDR2013/0,,contentMDK:23459971~pagePK:8261309~piPK:8258028~theSitePK:8258025,00.html (accessed 30.01.2015)

The Nature Conservancy blogs on urban conservation: http://blog.nature.org/science/2013/10/07/new-report-what-will-the-urban-century-mean-for-nature/ (accessed 30.01.2015)

http://blog.nature.org/science/2014/11/18/urban-water-blueprint-security-freshwater-cities-supply-watershed/ (accessed 30.01.2015)

Bibliography

3iG (International Interfaith Investment Group) 2010. Available from: http:// www.3ignet.org/ [accessed December 2010].

Adams, W. M., 1995. *Future Nature: a Vision for Conservation*. London: Earthscan

Adams, W. M., Aveling, R., Brockington, D., Dickson, B., Elliott, J., Hutton, J. Roe, D., Vira, B., Wolmer, W., 2004. Biodiversity conservation and the eradication of poverty. *Science 2004*, 306: 1146–1149.

Adams, J. S. and McShane, T. O., 1992. *The Myth of Wild Africa: Conservation Without Illusion*. Berkeley: University of California Press.

Adams, W. M. and Hulme, D., 2001. If community conservation is the answer in Africa, what is the question? *Oryx*, 35 (3): 193–200.

Adams, W. M., 2004. *Against Extinction: the Story of Conservation*. London Earthscan.

Adams, W. M., 2009. *Green development: environment and sustainability in a developing world*. 3rd ed. London: Routledge.

Adams, W. M., 2013. Conservation in the Anthropocene: Biodiversity, Poverty and Sustainability. *In*: Roe, D., Elliott, J., Sandbrook, C., and Walpole, tt, eds. *Biodiversity Conservation and Poverty Alleviation: Exploring the Evidence for a Link* [online]. John Wiley & Sons, Ltd, 304–315. Available from: http:// onlinelibrary.wiley.com/doi/10.1002/9781118428351.ch19/summary [accessed 18 October 2013].

Adams, W. M., Aveling, R., Brockington, D., Dickson, B., Elliott, J., Hutton, J. Roe, D., Vira, B., and Wolmer, W., 2004. Biodiversity conservation and the eradication of poverty. *Science*, 306: 1146–1149.

Adams, W., Hulme, D., and Murphree, M., 2001. Changing Narratives, Policies and Practices in African Conservation'. *In*: *The Promise & Performance of Community Conservation*. Oxford: James Currey.

Agrawal, A. and Benson, C. S., 2011. Common property theory and resource governance institutions: strengthening explanations of multiple outcomes *Environmental Conservation*, 38 (2): 199–210.

Agrawal, A. and Redford, K. H., 2006. *Poverty in development and biodiversity conservation: shooting in the dark?* [online]. Wildlife Conservation Society. Working Paper No. 26. Available from: http://siteresources.worldbank.org/INTPOVERTYNET/Resources/Agrawal_Redford_WP26.pdf [accessed 14 October 2013].

Agrawal, A., 2005a. Environmentality – Community, intimate government, and the making of environmental subjects in Kumaon, India. *Current Anthropology*, 46 (2): 161–190.

Agrawal, A., 2005b. *Environmentality: Technologies of government and the Making Of Subjects.* Duke University Press.

Agrawal, A., Brown, D. G., Rao, G., Riolo, R., Robinson, D. T., and Bommarito II, M., 2013. Interactions between organizations and networks in common-pool resource governance. *Environmental Science & Policy*, 25: 138–146.

AKDN (Aga Khan Development Network) 2010. http://www.akdn.org [accessed December 2010].

Alcorn, J. B., 1993. Indigenous peoples and conservation. *Conservation Biology*, 7: 424–426.

Alcorn, J. B., Zarzycki, A., and de la Cruz, L. M., 2010. Poverty, governance and conservation in the Gran Chaco of South America. *Biodiversity (Ottawa)*, 11 (1–2, Sp. Iss. SI): 39–44.

Alix-Garcia, J. M., Shapiro, E. N., and Sims, K. R. E., 2012. Forest Conservation and Slippage: Evidence from Mexico's National Payments for Ecosystem Services Program. *Land Economics*, 88 (4): 613–638.

Andechs, 2010. Andechs Monastery www.andechs.de/index.asp?lng=en [accessed December 2010].

Anderson, A. B., May, P. H., and Balick, M. J., 1991. *The subsidy from nature: palm forests, peasantry, and development on an Amazon frontier.* New York: Columbia University Press.

Anderson, D., 1984. Depression, Dust Bowl, Demography, and Drought: The Colonial State and Soil Conservation in East Africa during the 1930s. *African Affairs*, 83 (332): 321–343.

Anderson, D. M., Salick, J., Moseley, R. K., and Ou, X. K., 2005. Conserving the sacred medicine mountains: a vegetation analysis of Tibetan sacred sites in Northwest Yunnan. *Biodiversity and Conservation*, 14: 3065–3091.

Anderson, K. and Bows, A., 2011. Beyond 'dangerous' climate change: emission scenarios for a new world. *Philosophical Transactions of the Royal Society A: Mathematical, Physical and Engineering Sciences*, 369: 20–44.

Anderson, S. and Cavanagh, J., 2000. Top 200: The Rise of Global Corporate Power. Global Policy Forum. https://www.globalpolicy.org/component/content/article/221/47211.html [accessed 03 February 2015].

Ansell, C. and Gash, A., 2008. Collaborative governance in theory and practice. *Journal of Public Administration Research and Theory*, 18 (4): 543–571.

Ansink, E., Hein, L., and Hasund, K. P., 2008. To value functions or services? An analysis of ecosystem valuation approaches. *Environmental Values*, 17: 489–503.

ARC (Alliance of Religions and Conservation) 2004. Case studies from Mongolia: 10 Selenge Aimag Monastery http://www.arcworld.org/projects.asp?projectID=424 [accessed December 2010].

ARC (Alliance of Religions and Conservation), 2005. Ethiopian Church Launche Organic Farming Programme www.arcworld.org/news.asp?pageID=9 [accessed December 2010].

ARC (Alliance of Religions and Conservation), 2006. Daoist monks and nuns t manage sacred mountains http://www.arcworld.org/projects.asp?projectID=25 [accessed December 2010].

ARC (Alliance of Religions and Conservation), 2010. http://www.arcworld.or [accessed December 2010].

Armitage, D., de Loe, R., and Plummer, R., 2012. Environmental governance an its implications for conservation practice. *Conservation Letters*, 5 (4): 245–255

Arriagada, R. A., Ferraro, P. J., Sills, E. O., Pattanayak, S. K., and Cordero-Sanchc S., 2012. Do Payments for Environmental Services Affect Forest Cover? Farm-Level Evaluation from Costa Rica. *Land Economics*, 88 (2): 382–399.

Arriagada, R. A., Sills, E. O., Pattanayak, S. K., and Ferraro, P. J., 2009 Combining Qualitative and Quantitative Methods to Evaluate Participation i Costa Rica's Program of Payments for Environmental Services. *Journal Sustainable Forestry*, 28 (3–5): 343–367.

Ashley, C. and Jones, B. T. B., 2001. Joint ventures between communities an tourism investors: experience in Southern Africa. *International Journal c Tourism Research*, 3.

Atsmon, Y., Dixit, V., Leibowitz, G., and Wu, C., 2011. *Understanding China growing love for luxury* [online]. McKinsey Insights China. Available from www.mkik.hu/download.php?id=9559 [accessed 17 Apr 2015].

Aylward, B. A., 2003. The actual and potential contribution of nature tourism i Zululand. *In*: Aylward, B. A. and Lutz, E., eds. *Nature Tourism, Conservation, an Development in Kwazulu-Natal, South Africa*. World Bank Publications, 1–40.

Bäckstrand, K. and Lövbrand, E., 2006. Planting trees to mitigate climate change Contested discourses of ecological modernization, green governmentality an civic environmentalism. *Global Environmental Politics*, 6 (1): 50.

Bäckstrand, K., 2006. Multi-stakeholder partnerships for sustainable developmen rethinking legitimacy, accountability and effectiveness. *European Environmen* 16 (5): 290–306.

Bäckstrand, K., 2008. Accountability of networked climate governance: The rise transnational climate partnerships. *Global Environmental Politics*, 8 (3): 74.

Baechler, G., 1999. *Violence Through Environmental Discrimination: Cause Rwanda Arena, and Conflict Model*. Springer Science & Business Media.

Bael, D. and Sedjo, R. A., 2006. "Toward Globalisation of the Forest Product Industry: Some Trends," Discussion Papers dp-06-35, Resources For the Futur

Bahadur, A. V., Ibrahim, M., and Tanner, T., 2010. *The Resilience Renaissance Unpacking of Resilience for Tackling Climate Change and Disasters*. [online Brighton: IDS. Strengthening Climate Resilience Programme. Available from http://community.eldis.org/.59e0d267/resilience-renaissance.pdf [accessed 2 February 2015].

Barnes, B., 1983. Social-Life as Bootstrapped Induction UR-://A1983RU1160000 *Sociology*, 17: 524–545.

Barnes, B., 1988. *The nature of power*. Cambridge: Polity Press.

Barnes, B., Bloor, D., and Henry, J., 1996. *Scientific knowledge: a sociologic analysis*. London: Athlone Press.

Barrow, E. and Murphree, M., 2001. Community Conservation: from concept to practice. *In*: *African Wildlife and Livelihoods: the promise and performance of community conservation*. Oxford: James Currey, 24–37.

Basil (Bhaktivedanta Academy for Sustainable Integrated Living) 2010. www.basilacademy.in/ [accessed December 2010].

Batisse, M., 1975. 'Man and the Biosphere', *Nature*, 256: 156–8.

Batisse, M., 1982. 'The biosphere reserve: a tool for environmental conservation and management', *Environmental Conservation*, 9: 101–11.

Batterbury, S., Forsyth, T., and Thomson, K., 1997. Environmental transformations in developing countries: Hybrid research and democratic policy. *The Geographical Journal*, 163 (2): 126–132.

BBC, 2009. Averting a perfect storm of shortages http://news.bbc.co.uk/1/hi/8213884.stm [accessed 20 January 2015].

Beck, T. and Nesmith, C., 2001. Building on Poor People's Capacities: The Case of Common Property Resources in India and West Africa. *World Development*, 29 (1): 119–133.

Becker, C. D., 2003. 'Grassroots to Grassroots: Why Forest Preservation Was Rapid at Loma Alta, Ecuador'. *World Development*, 31: 163–176.

Beckett, M., 2002. Looking Positively At The Jo'Burg Summit. *Guardian*, 2002.

Beinart, W. and McGregor, J., 2003. *Social history & African environments*. Oxford: Athens, OH: Cape Town: James Currey; Ohio Univesity Press; D. Philip.

Beinart, W., 1984. *Soil erosion, conservationism and ideas about development: a Southern African exploration, 1900–1960*. Abingdon: Carfax.

Beinart, W., Hughes, Lotte and Hughes, L., 2008. *Environment and empire*. Oxford: Oxford University Press.

Belshaw, D., Calderisi R., Sugden C., 2001. 'Faith in Development: Partnership Between the World Bank and the Churches of Africa' (Washington DC: World Bank).

Benavides Diaz, D., 2001. The Viability and Sustainability of International Tourism in Developing Countries. *In*: [online]. Presented at the World Trade Organisation Symposium on Tourism Services, 22–23 Feb 2001, Geneva, 21. Available from: https://iucn.oscar.ncsu.edu/mediawiki/images/2/29/Diaz_2001.pdf [accessed 16 March 2015].

Bendell, J., 2000. *Terms for Endearment: Business, NGOs and Sustainable Development*. Sheffield: Greenleaf Publishing.

Béné, C., Belal, E., Baba, M. O., Ovie, S., Raji, A., Malasha, I., Njaya, F., Na Andi, M., Russell, A., and Neiland, A., 2009. Power Struggle, Dispute and Alliance Over Local Resources: Analyzing 'Democratic' Decentralization of Natural Resources through the Lenses of Africa Inland Fisheries. *World Development*, 37 (12): 1935–1950.

Béné, C., Godfrey Wood, R., Newsham, A., and Davies, M., 2012. *Resilience: New Utopia or New Tyranny? Reflection about the Potentials and Limits of the Concept of Resilience in Relation to Vulnerability Reduction Programmes* [online]. Falmer, Sussex: Institute of Development Studies. No. 405. Available from: http://www.ids.ac.uk/publication/resilience-new-utopia-or-new-tyranny [accessed 3 April 2015].

Béné, C., Newsham, A., Davies, M., Ulrichs, M., and Godfrey-Wood, R., 2014. Resilience, Poverty and Development. *Journal of International Development*.

Bengtsson, J., Angelstam, P., Elmqvist, T., Emanuelsson, U., Folke, C., Ihse, M., Moberg, F., and Nystrom, M., 2003. Reserves, resilience & dynamic landscapes. *Ambio*, 32 (6): 389–396.

Benjaminsen, T. A., Aune, J., and Sidibe, D., 2010. A critical political ecology of cotton and soil fertility in Mali. *Geoforum*, 41 (4): 647–656.

Bennett et al., 2009. Understanding relationships among multiple ecosystem services. *Ecology Letters*, 12: 1394–1404.

Berkes, F., Hughes, T. P., Steneck, R. S., Wilson, J. A., Bellwood, D. R., Cron, B., Folke, C., Gunderson, L. H., Leslie, H. M., Norberg, J., Nyström, M., Olsson, P., Österblom, H., Scheffer, M., and Worm, B., 2006. Globalisation, roving bandits, and marine resources. *Science*, 311: 1557–1558.

Berkes, F., 2009. Evolution of co-management: role of knowledge generation, bridging organisations and social learning. *Journal of Environmental Management*, 90: 1692–1702.

Berkes, F. and Folke, C., 1998. Linking social and ecological systems for resilience and sustainability. *In*: Berkes, F. and Folke, C., eds. *Linking Social and Ecological Systems: Management Practices and Social Mechanisms for Building Resilience*. Cambridge: Cambridge University Press, 1–26.

Berkes, F., 1981. Fisheries of the James Bay Area and Northern Quebec: a case study in resource management. *In*: Freeman, M. M., ed. *Renewable Resources and the Economy of the North*. Ottawa: Association of Canadian Universities of Northern Studies/Man and the Biosphere Program, 143–60.

Berkes, F., 1989. *Common property resources: ecology and community-based sustainable development*. London: Belhaven Press.

Berkes, F., 2008. *Sacred Ecology*. New York: Routledge.

Berkes, F., 2009. Indigenous ways of knowing and the study of environmental change. *Journal of the Royal Society of New Zealand*, 39: 151–156.

Berkes, F., Colding, J., and Folke, C., 2000. Rediscovery of Traditional Ecological Knowledge as Adaptive Management. *Ecological Applications*, 10: 1251–1262.

Berkes, F., Colding, J., and Folke, C., 2008. *Navigating social-ecological systems: building resilience for complexity and change* [online]. Cambridge; New York: Cambridge University Press. Available from: http://site.ebrary.com/lib/aberdeenuniv/Doc?id=10069848.

Bernauer, T. and Betzold, C., 2012. Civil Society in Global Environmental Governance. *The Journal of Environment & Development*, 21 (1): 62–66.

Bernstein, H., 1971. 'Modernisation Theory and The Sociological Study of Development'. *Journal of Development Studies*, 7 (2): 141–160.

Berry, M. A. and Rondinelli, D. A., 1998. Proactive Corporate Environmental Management: A New Industrial Revolution. *The Academy of Management Executive (1993–2005)*, 12 (2): 38–50.

Beswick, D., 2009. Killing Neighbors: webs of violence in Rwanda. *Journal of Modern African Studies*, 47 (4): 631–632.

Betsill, M. and Corell, E., 2001. NGO Influence in International Environmental Negotiations: A Framework for Analysis. *Global Environmental Politics*, 1 (4): 65–85.

Betsill, M. M. and Corell, E., 2008. *NGO Diplomacy: The Influence of Nongovernmental Organizations in International Environmental Negotiations*. Cambridge, Mass.; London: MIT Press.

Beyer, P., 1992. 'The Global Environment As a Religious Issue: A Sociological Analysis'. *Religion*, 22: 1–19.

Beyers, R. L., Hart, J. A., Sinclair, A. R. E., Grossmann, F., Klinkenberg, B., and Dino, S., 2011. Resource wars and conflict ivory: the impact of civil conflict on elephants in the Democratic Republic of Congo – the case of the Okapi Reserve. *Plos One*, 6 (11): e27129.

Bhagwat S. A., Nogué, S., and Willis K. J., 2012. Resilience of an ancient tropical forest landscape to 7500 years of environmental change. *Biological Conservation*, 153: 108–117.

Bhagwat, S. A. and Palmer, M., 2009. 'Conservation: the World's Religions Can Help'. *Nature,* 461: 37.

Bhagwat, S. A. and Rutte, C., 2006. 'Sacred Groves: Potential for Biodiversity Management' *Frontiers in Ecology and the Environment*, 4 (10): 519–524.

Bhagwat, S. A., Kushalappa, C. G. Williams, P. H. and Brown, N. D., 2005. 'Landscape approach to biodiversity conservation of sacred groves in the Western Ghats of India'. *Conservation Biology*, 19: 1853–1862.

Bhagwat, S. A., Dudley, N. and Harrop S. R., 2011. 'Religious Following in Biodiversity Hotspots: Challenges and Opportunities for Conservation and Development' *Conservation Letters* in press.

Bhagwat, S., Willis, K. J., Birks, H., John, B. and Whittaker, R. J., 2008. Agroforestry: a refuge for tropical biodiversity? *Trends in Ecology & Evolution*, 23 (5): 261–267.

Bhattacharya, S., Ganguli, A., Bose, S., and Mukhopadhyay, A., 2012. Biodiversity, traditional practices and sustainability issues of EastKolkata Wetlands: A significance Ramsar site of West Bengal, (India). *Research and Reviews in Biosciences*, 6 (11): 340–347.

Bierbaum, R., Stocking, M., Bouwman, H., Cowie, A., Diaz, S., Granit, J., Patwardhan, A., Sims, R., Duron, G., Gorsevski, V. and others, 2014. Delivering Global Environ-mental Benefits for Sustainable Development. Report of the Scientific and Technical Advisory Panel (STAP) to the 5th GEF Assembly, México 2014. *Global Environment Facility*, Washington, DC [online]. Available from: http://www.stapgef.org/stap/wp-content/uploads/2014/05/STAP-GEF-Delivering-Global-Env.pdf [accessed 3 February 2015].

Biermann, F. and Gupta, A., 2011. Accountability and legitimacy in earth system governance: A research framework. *Ecological Economics*, 70 (11): 1856–1864.

Biermann, F. and Pattberg, P. H., 2012. *Global Environmental Governance Reconsidered*. Cambridge, Mass.; London: MIT Press.

BirdLife International, 2004. *Birds in the European Union: a status assessment.* [online]. Wageningen, The Netherlands: BirdLife International. Available from: http://www.birdwatchireland.ie/Portals/0/pdfs/BOCC_birds_in_the_eu.pdf [accessed 26 February 2015].

Blaikie, P. and Brookfield, H., 1987. *Land Degradation and Society*. London: Methuen.

Blaikie, P. M., 1985. *The Political Economy of Soil erosion in Developing Countries*. New York: Longman Scientific and Technical.

Bliss, K., 2009. *Trafficking in the Mesoamerican Corridor: A Threat to Regional and Human Security*. Washington DC: Center for Strategic and International Studies.

Bloor, D., 1991. *Knowledge and social imagery*. Chicago: University of Chicago Press.

Bloor, D., 1999. Anti-Latour. *Studies in the history and philosophy of science*, 30 (1) 81–112.

Boardman, R., 1981. *International organisations and the conservation of nature* Bloomington, IN: Indiana University Press.

Boivin, N. and Fuller, D. Q., 2009. Shell middens, ships and seeds: Exploring coastal subsistence, maritime trade and the dispersal of domesticates in and around the ancient Arabian peninsula, *Journal of World Prehistory*, 22: 113–180

Bond, I., Hulme, D., and Murphree, M., 2001. CAMPFIRE & the incentives for institutional change. *In: African Wildlife & Livelihoods*. Oxford: James Currey 227–243.

Bonham, C. A., Sacayon, E., and Tzi, E., 2008. Protecting imperiled 'paper parks' potential lessons from the Sierra Chinajai, Guatemala. *Biodiversity and Conservation*, 17 (7): 1581–1593.

Bookbinder, M. P., Dinerstein, E., Rijal, A., Cauley, H., and Rajouria, A., 1998 Ecotourism's support of biodiversity conservation. *Conservation Biology*, 1 (6): 1399–1404.

Borel-Saladin, J. M. and Turok, I. N., 2013. The impact of the green economy on jobs in South Africa. *South African Journal of Science*, 109 (9–10): a0033.

Borras, S. M., 2009. Agrarian change and peasant studies: changes, continuities and challenges – an introduction. *Journal of Peasant Studies*, 36: 5–31.

Borrini-Feyerabend, G., Pimbert, M., Taghi Farvar, M., Kothari, A., and Renard, Y., 2007. *Sharing Power: A Global Guide to Collaborative Management of Natural Resources*. London: Earthscan.

Boserup, E., 1965. *The Conditions of Agricultural Growth*. London: Earthscan.

Bossel, H., 1999. *Indicators for Sustainable Development: Theory, Method Applications*. Balaton Group, International institute for Sustainable Development Winnipeg, Canada.

Botham, M. S., Rothery, P., Hulme, P. E., Hill, M. O., Preston, C. D., and Roy D. B., 2009. Do urban areas act as foci for the spread of alien plant species? A assessment of temporal trends in the UK. *Diversity and Distributions*, 15 (2) 338–345.

Botkin, D. B., 1990. *Discordant Harmonies: A new ecology for the twenty-first century*. Oxford: Oxford University Press.

Bowles, I. A., Rice, R. E., Mittermeier, R. A., da Fonseca, G. A. B., 1998. Logging and Tropical Forest Conservation, *Science*, 280 (5371): 1899–1900.

Boyd, C. and Slaymaker, T., 2000. *Re-examining the 'More People, Less Erosion Hypothesis: Special Case or Wider Trend*. [online]. No. No 63. Available from http://www.odi.org.uk/resources/download/1703.pdf.

Boyd, E., Grist, N., Juhola, S., and Nelson, V., 2009. Exploring Development Futures in a Changing Climate: Frontiers for Development Policy and Practice *Development Policy Review*, 27: 659–674.

Boyd, J. and Banzhaf, S., 2007. 'What are ecosystem services? The need for standardized environmental accounting units'. *Ecological Economics*, 63 616–626.

Boyd, J. M., 1984. 'The Role of Religion in Conservation'. *The Environmentalist* 4: 40–44.

Brainard, L., Purvis, N., Jones, A., 2009. *Climate Change and Global Poverty: A Billion Lives in the Balance?* London: Brookings Institution Press.

Braithwaite, J. and Drahos, P., 2000. *Global Business Regulation*. Cambridge: Cambridge University Press.

Brandon, K. and Wells, M., 1992. Planning for People and Parks – Design Dilemmas. *World Development*, 20 (4): 557–570.

Brandon, K., Redford, K. H., and Sanderson, S. E., 1998. *Parks in Peril: People, Politics, and Protected Areas*. Washington, D.C.: Island Press.

Brenner, N. and Theodore, N., 2002. Cities and the Geographies of "Actually Existing Neoliberalism". *Antipode*, 34 (3): 349–379.

Briguglio, L., Cordina, G., Farrugia, N., and Vella, S., 2009. Economic Vulnerability and Resilience: Concepts and Measurements. *Oxford Development Studies*, 37 (3): 229–247.

Brock, K., Cornwall, A., and Gaventa, J., 2004. *Power, knowledge and political spaces in the framing of poverty policy*. Brighton: Institute of Development Studies.

Brockington, D. and Duffy, R., 2010. 'Conservation and Capitalism: an Introduction.' *Antipode*, 42 (3): 469–484.

Brockington, D. and Scholfield, K., 2010. The Conservationist Mode of Production and Conservation NGOs in sub-Saharan Africa. *Antipode*, 42: 551–575. doi: 10.1111/j.1467-8330.2010.00763.x

Brockington, D., 2002. *Fortress conservation: the preservation of the Mkomazi Game Reserve, Tanzania*. [London]: The International African Institute in association with James Currey, Oxford; Mkuki Na Nyota, Dar es Salaam; Indiana University Press, Bloomington & Indianapolis.

Brockington, D., Duffy, R., and Igoe, J., 2010. *Nature unbound: conservation, capitalism and the future of protected areas*. London: Earthscan.

Brockington, D., Homewood, K., Leach, M., and Mearns, R., 1996. Debates concerning Mkomazi Game Reserve Tanzania. *In*: *The lie of the land: Challenging received wisdoms on the African environment*. Oxford: James Currey.

Brockington, D., Igoe, J., and Schmidt-Soltau, K., 2006. Conservation, human rights, and poverty reduction. *Conservation Biology*, 20 (1): 250–252.

Brokensha, D., Warren, D. M., and Werner, O., 1980. *Indigenous knowledge systems and development*. Maryland: University of America.

Browder, J. O., 1997. *Rainforest Cities: Urbanization, Development, and Globalization of the Brazilian Amazon*. New York: Columbia University Press.

Brown, K., 2004. Trade-off analysis for integrated conservation and development. *In*: McShane, T. O. and Wells, M. P., eds. *Getting biodiversity projects to work*. Chichester; New York: University of Columbia Press, 232–255.

Brown, K., 2011. Sustainable adaptation: An oxymoron? *Climate and Development*, 3 (1): 21–31.

Brulle, R. J., 2000. *Agency, Democracy, and Nature: The U.S. Environmental Movement from a Critical Theory Perspective*. Cambridge, Mass.; London: MIT Press.

Brundtland, Gro Harlem (ed.), 1987. *Our common future: The World Commission on Environment and Development*. Oxford: Oxford University Press.

Bryant, D., Nielsen, D., and Tangley, L., 1997. *The Last Frontier Forests: Ecosystems and Economies on the Edge*. Washington DC: World Resources Institute.

Buckley, R., 2002. Tourism ecocertification in the International Year of Ecotourism. *Journal of Ecotourism*, 1 (2–3): 197–203.

Buckley, R., 2013. Three reasons for eco-label failure. *Nature*, 500 (7461): 151.

Budowski, G., 1976. Tourism and Environmental Conservation: Conflict Coexistence, or Symbiosis? *Environmental Conservation*, 3 (01): 27–31.

Buscher, B. and Arsel, M., 2012. Introduction: neoliberal conservation, uneven geographical development and the dynamics of contemporary capitalism *Tijdschrift Voor Economische En Sociale Geografie*, 103: 129–135.

Büscher, B. and Dressler, W., 2012. Commodity conservation: The restructuring of community conservation in South Africa and the Philippines. *Geoforum*, 4 (3): 367–376.

Buscher, B., Sullivan, S., Neves, K., Igoe, J., and Brockington, D., 2012 Toward a synthesized critique of neoliberal biodiversity conservation. *Capitalism Nature Socialism*, 23 (2): 4–30.

Butcher, J., 2007. *Ecotourism, NGOs and Development: A Critical Analysis* Abingdon: Taylor & Francis.

Butcher, J., 2011. Can ecotourism contribute to tackling poverty? The importance of 'symbiosis'. *Current Issues in Tourism*, 14 (3): 295–307.

Buxton, E. N., 1902. *Two African Trips: With Notes and Suggestions on Big Game Preservation in Africa*. London: Edward Stanford.

Callicott, J. B., 2008. 'What 'Wilderness' in Frontier Ecosystems?' *Environmental Ethics*, 30: 235–249.

Callon, M., 1986. Some Elements of a Sociology of Translation - Domestication of the Scallops and the Fishermen of St-Brieuc Bay. *Sociological Review Monograph*, 196–233.

Callon, M., Lascoumes, P., and Barthe, Y., 2009. *Acting in an uncertain world: a essay on technical democracy*. Cambridge, Mass.; London: MIT Press.

Callon, M., Latour, B., and Pickering, A., 1992. Don't Throw the Baby Out with the Bath School! A Reply to Collins and Yearley. *In: Science as Practice an Culture*. Chicago: University of Chicago Press: 343–368.

Campbell, L. M. and Vainio-Mattila, A., 2003. Participatory development an community-based conservation: Opportunities missed for lessons learned *Human Ecology*, 31: 417–437.

Campling, L. and Havice, E., 2014. The problem of property in industrial fisheries *Journal of Peasant Studies*, 41 (5): 707–727.

Cannon, T. and Müller-Mahn, D., 2010. Vulnerability, resilience and development discourses in context of climate change. *Natural Hazards*, 55 (3): 621–635.

Carlson, L. and Berkes, F., 2005. Co-management: concepts and methodological implications. *Journal of Environmental Management*, 75: 65–76. Available from: http://dx.doi.org/10.1016/j.jenvman.2004.11.008

Carpenter, S. R., DeFries, R., Dietz, T., Mooney, H. A., Polasky, S., Reid, W. V and Scholes, R. J., 2006. Millennium Ecosystem Assessment: Research Needs *Science*, 314 (5797): 257–258.

Carpenter, S. R., Brock, W., and Hanson, P., 1999. Ecological and social dynamic in simple models of ecosystem management. *Conservation Ecology*, 3 (2): 4.

Carr, D. L., 2002. The role of population change in land use and land cover change in rural Latin America: uncovering local processes concealed by macro-level data. *In*: Himiyama, Y., Hwang, M. and Ichinose, T. (eds.) *Land Use Change in Comparative Perspective*. Enfield, NH/Plymouth, UK: Science, 133–148.

Carrier, J. G., 2010. Protecting the Environment the Natural Way: Ethical Consumption and Commodity Fetishism. *Antipode*, 42: 672–689. doi 10.1111/j.1467-8330.2010.00768.x

Carrington, S. H. H., 2002. *The Sugar Industry and the Abolition of the Slave Trade, 1775–1810.* Gainesville: University of Florida Press.

Carson, R., 1963. *Silent spring.* London: Hamish Hamilton.

Cash, D. W., Adger, W. N., Berkes, F., Garden, P., Lebel, L., Olsson, P., Pritchard, L., and Young, O., 2006. Scale and cross-scale dynamics: Governance and information in a multilevel world. *Ecology and Society* [online], 11. Available from: ://WOS:000243280800003.

Cashore, B., 2002. Legitimacy and the privatization of environmental governance: How non-state market-driven (NSMD) governance systems gain rule-making authority. *Governance-an International Journal of Policy and Administration,* 15 (4): 503–529.

Castley, J. G., Patton, C., and Magome, H., 2012. Making 'conventional' parks relevant to all of society: the case of SAN Parks. *In:* Child, B., Jones, B. T. B., Murphree, M. W., Spenceley, A. and Suich, H., eds. Abingdon: Routledge.

Castree, N., 2008. Neoliberalising nature: processes, effects, and evaluations. *Environment and Planning A,* 40 (1): 153–171.

Cater, E. and Lowman, G., 1994. *Ecotourism: a sustainable option?* Oxford: Wiley.

CBD, 2010. The Convention on Biological Diversity Aichi Biodiversity Targets. Available from: http://www.cbd.int/sp/targets/ [accessed 20 January 2015].

Ceballos-Lascuráin, H., 1988. The Future of Ecotourism. *Mexico Journal,* 17: 13–14.

Ceballos-Lascurain, H., 1996. *Tourism, ecotourism, and protected areas: The state of nature-based tourism around the world and guidelines for its development.* IUCN, xiv + 301 pp.

Chambers, R. and Conway, G., 1992. *Sustainable rural livelihoods: practical concepts for the 21st century.* Brighton: Institute of Development Studies, University of Sussex. No. 296.

Chambers, R., 1983. *Rural Development: Putting the Last First.* Harlow: Longman.

Chambers, R., 1997. *Whose Reality Counts? Putting the First Last.* London: ITDG Publishing.

Chambers, R., Pacey, A., and Thrupp, L. A., 1989. *Farmer first: farmer innovation and agricultural research.* London; New York: Intermediate Technology Pub.; Bootstrap Press.

Chapin, M., 2004. A challenge to conservationists. *World Watch,* November/December.

Charles, H. and Dukes, J. S., 2007. Impacts of invasive species on ecosystem services. *In:* Nentwig, W., ed. *Biological Invasions.* Berlin; New York: Springer, 217–237.

Chee, Y. E., 2004. 'An ecological perspective on the valuation of ecosystem services'. *Biological Conservation* 120: 549–565.

Child, M. F., 2009. 'The Thoreau Ideal as a Unifying Thread in the Conservation Movement'. *Conservation Biology,* 23: 241–243.

Chimedsengee, U., Cripps, A., Finlay, V., Verboom, G., Batchuluun, V. M., and Byambajav Khunkhur, V. D. L., 2009. *Mongolian Buddhists Protecting Nature: A Handbook on Faiths, Environment and Development.* Ulaanbaatar: The Alliance of Religions and Conservation.

Christensen, N. L., Bartuska, A. M., Brown, J. H., Carpenter, S., DAntonio, C., Francis, R., Franklin, J. F., MacMahon, J. A., Noss, R. F., Parsons, D. J.,

Peterson, C. H., Turner, M. G., and Woodmansee, R. G., 1996. The report of the ecological society of America committee on the scientific basis for ecosystem management. *Ecological Applications*, 6 (3): 665–691.

CII (Conservation International Indonesia) 2005. 'Islamic Boarding Schools and Conservation: The World Bank Faith and Environment Initiative Agreement No 7133121' Final Report. Jakarta, Indonesia: Conservation International Indonesia

Clark, W. C., Jones, D. D., and Holling, C. S., 1979. Lessons for ecological policy design: a case study of ecosystem management. *Ecological Modelling*, 7: 2–53

Clarke, G., 2007. 'Agents of Transformation? Donors, Faith-based Organisation and International Development' *Third World Quarterly*, 28 (1): 77–96.

Clarke, R. and Timberlake, L., 1982. *Stockholm Plus Ten: Promises, Promises? The Decade since the 1972 UN Environment Conference*. London: Earthscan.

Coates, B., 2002. More Business As Usual. *Guardian*, 2002.

Cohen, M. P., 1984. *The pathless way: John Muir and American wilderness* Madison: University of Wisconsin Press.

Colding, J. and Folke, C., 1997. The relations among threatened species, their protection, and taboos. *Conservation Ecology (online)*, 1 (1 CITED JUNE 30, 1997)

Collier, P., 2008. *The Bottom Billion: Why the Poorest Countries are Failing and What Can Be Done About It*. Oxford: Oxford University Press.

Collier, P., 2010. *The Plundered Planet: Why We Must - and How We Can Manage Nature for Global Prosperity*. Oxford: Oxford University Press.

Collins, H. M., Yearley, S., and Pickering, A., 1992. Epistemological Chicken. *In Science as Practice and Culture*. Chicago: University of Chicago Press, 301–326

Colwell, M., Finlay, V., Hilliard, A., and Weldon, S., (Ed.) 2009. *Many Heavens One Earth: Faith Communities to Protect the Living Planet*. Bath: Alliance of Religions and Conservation and United Nations Development Program.

Conca, K., 1995. Greening the United-Nations: Environmental-Organizations and the UN System. *Third World Quarterly*, 16 (3): 441–457.

Constantine, J. and Pontual, M., 2015. *Shaping the Post 2015 development agenda: understanding the Rising Powers' contribution to the Sustainable Development Goals*. Falmer, Sussex: Institute of Development Studies, Rapid Response Briefing.

Corbridge, S., 1991. Third-World Development. *Progress in Human Geography* 15 (3): 311–321.

Coria, J. and Calfucura, E., 2012. Ecotourism and the development of indigenous communities: The good, the bad, and the ugly. *Ecological Economics*, 73: 47–55

Cornwall, A., 2002. *Making spaces, changing places: situating participation in development*. Brighton: Institute of Development Studies.

Cornwall, A., 2004. Introduction: New Democratic Spaces? The Politics and Dynamics of Institutionalised Participation. *IDS Bulletin*, 35 (2): 1–10.

Corson, C., 2010. Shifting Environmental Governance in a Neoliberal World: US AID for Conservation. *Antipode*, 42: 576–602. doi: 10.1111/j.1467-8330 2010.00764.x

Costanza, R., 2008. 'Ecosystem services: multiple classification systems are needed'. *Biological Conservation* 141: 350–352.

Costanza, R., de Groot, R., Sutton, P., van der Ploeg, S., Anderson, S. J Kubiszewski, I., Farber, S., and Turner, R. K., 2014. Changes in the global value of ecosystem services. *Global Environmental Change*, 26: 152–158.

Costanza, R., d'Arge, R., deGroot, R., Farber, S., Grasso, M., Hannon, B Limburg, K., Naeem, S., O'Neill, R. V., Paruelo, J., Raskin, R. G., Sutton, P and van den Belt, M., 1997. 'The value of the world's ecosystem services and natural capital', *Nature* 387: 253–260.

Côté, I. M., Darling, E. S., 2010. Rethinking Ecosystem Resilience in the Face of Climate Change. *PLoS Biol*, 8 (7): e1000438. doi:10.1371/journal.pbio.1000438

Crane, B. B. and Dusenberry, J., 2004. 'Power and Politics in International Funding for Reproductive Health: The US Global Gag Rule'. *Reproductive Health Matters*, 12 (24): 128–137.

Crosby, A. W., 2004. *Ecological Imperialism: The Biological Expansion of Europe, 900–1900*. Cambridge: Cambridge University Press.

Crush, J. S., 1995. *Power of development*. London; New York: Routledge.

Crutzen, P., 2000. The Anthropocene. *Global Change Newsletter*, 41 (1): 17–18, Springer.

Dahlberg, K. A., 1985. *Environment and the global arena: actors, values, policies, and futures*. Durham, N.C.: Duke University Press.

Daily, G. C., 1997. *Nature's Services: Societal Dependence on Natural Ecosystems*. Washington, D.C.: Island Press.

Daily, G. C., 2000. 'Management objectives for the protection of ecosystem services'. *Environmental Science and Policy*. 3: 333–339.

Daly, H., 2010. How a green economy grows: Remarks by Herman Daly, National Council for Science and Environment. *Social Issues, University of Maryland* [online], 2010. Available from: http://www.newsdesk.umd.edu/sociss/release. cfm?ArticleID=2065 [accessed 4 October 2014].

Darwin, C., 1859. *On the Origin of Species by Means of Natural Selection, or the Preservation of Favored Races in the Struggle for Life*. London: John Murray.

Davidson, D. J., 2010. The Applicability of the Concept of Resilience to Social Systems: Some Sources of Optimism and Nagging Doubts. *Society & Natural Resources*, 23 (12): 1135–1149.

Davis, M., 2006. *Planet of Slums*. London: Verso.

Davis, S. M., 1994. Phosphorus inputs and vegetation sensitivity. *In*: Davis, S. M. and Ogden, J., eds. *The Everglades: the ecosystem and its restoration*. Florida: St Lucie Press, 357–78.

DDVE (Development Dialogue on Values and Ethics), 2010. www.worldbank. org/developmentdialogue [accessed December 2010].

De Cordier, B., 2009. 'Faith-based Aid, Globalisation and the Humanitarian Frontline: an Analysis of Western-based Muslim Aid Organisations'. *Disasters*, 33: 608–628.

De Groot, J. I. M. and Steg, L., 2009. 'Mean or Green: Which Values Can Promote Stable Pro-environmental Behaviour?' *Conservation Letters*, 2: 61–66.

De Groot, R. S., 1992. *Functions of Nature: Evaluation of Nature in Environmental Planning, Management and Decision Making*. Groningen: Wolters-Noordhoff.

De Groot, R. S., Wilson, M. A., and Boumans, R. M. J., 2002. 'A typology for the classification, description and valuation of ecosystem functions, goods and services'. *Ecological Economics*, 41: 393–408.

De Sherbinin, A., 2009. Introduction: population-environment dynamics in the developing world. *In*: Rahman, A., Barbier, E. B., Fotso, J. C., and Zhu, Y., eds. *Urban Population-Development-Environment Dynamics in the Developing World: Case Studies and Lessons Learned*. [online]. Paris: Committee for International Cooperation in National Research in Demography (CICRED), 19–39. Available from: http://www.populationenvironmentresearch.org/ workshops.jsp#W2007.

De Sherbinin, A., Carr, D., Cassels, S., and Jiang, L., 2007. Population and environment. *Annual Review of Environmental Resources*, 32: 345–373.

De Sherbinin, A., Lopez-Carr, D., Cassels, S., Jiang, L., Lopez-Carr, A., and Miller K. M., 2014. *Population-environment theory and contemporary applications The Encyclopaedia of the Earth*. [online]. Available from: http://www.eoearth. org/view/article/537f8ed20cf2aafa2ccd987e/ [accessed February 2015].

De Soysa, I., 2013. Environmental security and the resource curse. *In*: Floyd, R and Matthew, R., eds. *Environmental Security: Approaches and Issues* Abingdon: Routledge, 64–81.

Defra. 2006. Ecosystem Services: Living with Environmental Limits. Available from http://www.ecosystemservices.org.uk/ecoserv.htm [accessed January 2009].

Deligiannis, T., 2013. The evolution of qualitative environment-conflict research moving towards consensus. *In*: Floyd, R. and Matthew, R. A., eds. *Environmenta Security: Approaches and Issues*. Abingdon: Routledge.

Denevan, W., 1992. The Pristine Myth - the Landscape of the America in 1492 *Annals of the Association of American Geographers*, 82 (3): 369–385.

Denman, R., 2001. *Guidelines for community-based ecotourism developmen* [online]. WWF. Available from: http://www.widecast.org/Resources/Docs WWF_2001_Community_Based_Ecotourism_Develop.pdf [accessed 1 March 2015].

Deveny, A., Nackoney, J., Purvis, N., Gusti, M., Kindermann, G., Macauley, M. Myers Madeira, E., Obersteiner, M., and Stevenson, A., 2009. *Forest Carbon Index: The Geography of Forests in Climate Solutions*. Resources for the Future

DFID, 2011. *Defining disaster resilience: a DFID approach paper*. [online] London: DFID. Available from: https://www.gov.uk/government/publications defining-disaster-resilience-a-dfid-approach-paper [accessed 25 February 2015]

Diamond, J., 2005. *Collapse: How Societies Choose to Fail or Succeed*. London Penguin.

Diamond, J. M., 1998. *Guns, Germs, and Steel: A Short History of Everybody fo the Last 13,000 Years*. London: Random House.

Didham, R. K., Tylianakis, J. M., Hutchison, M. A., Ewers, R. M., and Gemmell N. J., 2005. Are invasive species the drivers of ecological change? *Trends i Ecology & Evolution*, 20 (9): 470–474.

Dinham, A. and Lowndes, V., 2008. 'Religion, Resources, and Representation Three Narratives of Faith Engagement in British Urban Governance'. *Urba Affairs Review*, 43 (6): 817–845.

Dolsak, N. and Ostrom, E., 2003. *The commons in the new millennium: challenge and adaptation*. Cambridge, Mass.; London: MIT Press.

Dorward, A. R., 2014. Livelisystems: a conceptual framework integrating social ecosystem, development, and evolutionary theory. *Ecology and Society*, 19 (2): 44

Douglas, L. R. and Alie, K., 2014. High-value natural resources: Linking wildlif conservation to international conflict, insecurity, and development concerns *Biological Conservation*, 171: 270–277.

Dressler, W. and Roth, R., 2011. The Good, the Bad, and the Contradictory Neoliberal Conservation Governance in Rural Southeast Asia. *Worl Development*, 39 (5): 851–862.

Drexhage, J. and Murphy, D., 2010. *Sustainable Development: From Brundtlan to Rio 2012* [online]. IISD. Available from: http://www.un.org/wcm/webdav/site climatechange/shared/gsp/docs/GSP1-6_Background%20on%20Sustainable%2 Devt.pdf.

Drèze, J. and Sen, A., 1991. *The Political Economy of Hunger: Volume 1: Entitlement and Well-being.* Oxford: Clarendon Press.

Dublin, H. T., Sinclair, A. R. E., and McGlade, J., 1990. Elephants and fire as causes of multiple stable states in the Serengetti-Mara woodlands. *Journal of Animal Ecology,* 59: 1147–1164.

Dubois, J. C. L., 1990. Secondary Forests as a Land-Use Resource in Frontier Zones of Amazonia. *In:* Anderson, A. B., ed. *Alternative to Deforestation: Steps toward Sustainable Use of the Amazon Rainforest.* New York: Columbia University Press.

Dudley, J. P., Ginsberg, J. R., Plumptre, A. J., Hart, J. A., and Campos, L. C., 2002. Effects of war and civil strife on wildlife and wildlife habitats. *Conservation Biology,* 16: 319–329.

Dudley, N. (ed.), 2008. *Guidelines for Applying Protected Area Management Categories.* Gland, Switzerland: IUCN. x + 86pp.

Dudley, N., Higgins-Zogib, L., and Mansourian, S., 2009. 'The Links Between Protected Areas, Faiths, and Sacred Natural Sites'. *Conservation Biology,* 23: 568–577.

Duffy, R. and Moore, L., 2010. Neoliberalising Nature? Elephant-Back Tourism in Thailand and Botswana. *Antipode,* 42: 742–766. doi: 10.1111/j.1467-8330.2010.00771.x

Duffy, R., 2000. *Killing for Conservation: Wildlife Policy in Zimbabwe.* Oxford: James Currey.

Duffy, R., 2002. *A trip too far: ecotourism, politics, and exploitation.* London; Sterling, VA: Earthscan.

Duffy, R., 2008. Neoliberalising nature: Global networks and ecotourism development in Madagasgar. *Journal of Sustainable Tourism,* 16 (3): 327–344.

Duffy, R., 2010. *Nature Crime: How We're Getting Conservation Wrong.* New Haven, CT; London: Yale University Press.

Duffy, R., 2014. Waging a war to save biodiversity: the rise of militarized conservation. *International Affairs,* 90 (4): 819–834.

Duffy, R., Emslie, R. H., and Knight, M. H., 2013. *Rhino Poaching: How do we Respond?* UK Department for International Development (DFID). No. HD087.

Duffy, R., in review. War, by conservation, 30.

Duncan R. C., 2003. 'Three world oil forecasts predict peak oil production'. *Oil & Gas Journal,* (101/21): 18–21.

Dunlap, T. R., 1988. *Saving America's Wildlife.* Princeton, NJ: Princeton University Press.

Eakin, H., 2005. Institutional change, climate risk, and rural vulnerability: Cases from Central Mexico. *World Development,* 33: 1923–1938.

Easwaran, E., 2005. *The Upanishads.* Tomales, CA: Nilgiri Press.

Egoh, B., Rouget, M., Reyers, B., Knight, A. T., Cowling, R. M., van Jaarsveld, A. S., and Welz. A., 2007. 'Integrating ecosystem services into conservation assessments: a review'. *Ecological Economics,* 63: 714–721.

Ehrenfeld, D., 2003. Globalisation: Effects on Biodiversity, Environment and Society. *Conservation and Society,* 1: 99–111.

Ehrlich, P. R., 1975. *Population Bomb.* rev ed. [S.l.]: Rivercity, Mass.: Rivercity Press.

Eigenbrod et al., 2009. Ecosystem service benefits of contrasting conservation strategies in a human-dominated region. *Proceedings of the Royal Society B,* 276: 2903–2911.

Ekirch, A. A., 1963. *Man and nature in America.* New York,: Columbia University Press.

Ellen, R. F. and Harris, P., 2000. Introduction. *In*: Ellen, R. F., Bicker, A., and Parkes, P., eds. *Indigenous environmental knowledge and its transformations* Amsterdam: Harwood Academic Publishers, 213–51.

Ellen, R. F., 2007. *Modern crises and traditional strategies: local ecologica knowledge in island Southeast Asia* [online]. New York; Oxford: Berghah Books. Available from: http://prism.talis.com/sussex-ac/items/1068851 http:/ www.loc.gov/catdir/toc/ecip0715/2007012587.html.

Elliott, L. M., 2004. *The Global Politics of the Environment.* New York: Nev York University Press.

Ellis, S., 1994. Of elephants and men: politics and nature conservation in Sout Africa. *Journal of Southern African Studies*, 20 (1): 53–69.

Eltis, D., 2000. *The Rise of African Slavery in the Americas.* Cambridge Cambridge University Press.

Emerton, L., Hulme, D., and Murphree, M., 2001. The nature of benefits and th benefits of nature: why wildlife conservation has not economically benefitte communities in Africa. *In: African wildlife and livelihoods.* Oxford: James Currey 208–226.

ENDS, 2002. *Earth Summit ends in disillusion.* ENDS Report 332, 19–22 Available from: http://www.endsreport.com/9053/earth-summit-disillusio [accessed February 2015].

Engler, M., 2008. Value of the international trade in wildlife. *TRAFFIC Bulletir* 22 (1): 4–5.

Escobar, A., 1991. Anthropology and the development encounter: the making an marketing of the third world. *American Ethnologist*, 18 (4): 658–681.

Escobar, A., 1992. Imagining a Post-Development Era? Critical Though Development and Social Movements. *Social Text*, 20–56.

Escobar, A., 1995. *Encountering Development: The Making and Unmaking of th Third World.* Princeton: Princeton University Press.

Evans, J., 2012. *Environmental governance.* Abingdon; New York: Routledge.

Evans, S., 2010. *The Green Republic: A Conservation History of Costa Rica* University of Texas Press.

Fa, J. E., Seymour, S., Dupain, J., Amin, R., Albrechtsen, L., and Macdonald, D 2006. Getting to grips with the magnitude of exploitation: bushmeat in th Cross-Sanaga rivers region, Nigeria and Cameroon. *Biological Conservation* 129: 497–510.

Faccer, K., Nahman, A., and Audouin, M., 2014. Interpreting the green economy Emerging discourses and their considerations for the Global South. *Developme Southern Africa*, 31 (5): 642–657.

Fairhead, J. and Leach, M., 1998. *Reframing deforestation: global analysis an local realities: studies in West Africa.* London; New York: Routledge.

Fairhead, J. and Leach, M., 2009. Amazonian Dark Earths in Africa? *In*: Wood W. I., Teixeira, W. G., Lehmann, J., Steiner, C., Winkler Prins, A. M. G. A and Rebellato, L., eds. *Amazonian Dark Earths: Wim Sombroek's Visio* Amsterdam: Springer Netherlands, 265–278.

Falkner, R., 2008. *Business power and conflict in international environment politics.* Basingstoke: Palgrave Macmillan.

FAO, 2012. State of the World's Forests 2012 Food and Agriculture Organisatic of the United Nations, Rome.

FAO, 2015. Search Geographical Information: Fishery and Aquaculture Country Profiles. Available from: http://www.fao.org/fishery/countryprofiles/search/en [accessed 10 February 2015].

FAO, 2015. *FAO Soils Portal*. [online]. Food and Agriculture Organization of the United Nations. Available from: http://www.fao.org/soils-portal/soil-degradation-restoration/en/ [accessed 13 February 2015].

FAO-UNEP (Food and Agriculture Organisation of the United Nations and United Nations Environmental Program), 2008. Conference on Ecological Agriculture: Mitigating Climate Change, Providing Food Security and Self-reliance for Rural Livelihoods in Africa - Conclusions and Recommendations. African Union Headquarters Addis Ababa, Ethiopia. Available from: http://www.fao.org/docrep/012/al134e/al134e00.pdf [accessed December 2010].

Feeny, D., Berkes, F., McCay, B. J., M, A. J., and Baden, J. A. D. S. N., 1990. The tragedy of the commons twenty-two years later. *In: Managing the commons.* Indiana: Indiana University Press, 77–94.

Feng, K., Davis, S. J., Sun, L., Lie, X., Guan, D., Liuh, W., Fan, Z., and Hubacek, K., 2013. Outsourcing CO2 within China. *Proceedings of the National Academy of Sciences*, 110 (28): 11654–11659.

Ferguson, J., 1990. *The anti-politics machine: development, depoliticization, and bureaucratic power in Lesotho.* Cambridge: Cambridge University Press.

Ferguson, Mi. A. D., Viventsova, K., and Huntington, H. P., 2007. *Community-conserved areas in the circumpolar arctic: Preliminary description and assessment of their legal status, threats and needs* [online]. Report prepared for the IUCN Strategic Direction on Governance, Communities, Equity, and Livelihood Rights in Relation to Protected Areas. Available from: http://cmsdata.iucn.org/downloads/arctic_cca_study.pdf [accessed 13 February 2015].

Ferraro, P. J. and Kiss, A., 2002. Direct Payments to Conserve Biodiversity. *Science*, 298 (5599): 1718–1719.

Ferraro, P. J. and Simpson, R. D., 2002. The cost-effectiveness of conservation payments. *Land Economics*, 78 (3): 339–353.

Feurstein, G., 2001. *The Yoga Tradition: Its History, Literature, Philosophy and Practice.* Prescott, Arizona: Hohm Press.

Fiering, M., 1982. Alternative indices of resilience. *Water Resources Research*, 18: 33–39.

Fischer, A. and Kothari, U. (eds). 2011. Resilience in an unequal capitalist world – virtual issue. *Journal of International Development* [online], (Virtual issue). Available from: http://onlinelibrary.wiley.com/journal/10.1002/%28ISSN%291099-1328/homepage/jid_virtual_issue.htm [accessed 25 February 2015].

Fisher, B. and Christopher, T., 2007. 'Poverty and Biodiversity: Measuring the Overlap of Human Poverty and the Biodiversity Hotspots'. *Ecological Economics*, 62: 93–101.

Fisher, B., and Turner, R. K., 2008. 'Ecosystem services: Classification for valuation'. *Biological Conservation*, 141: 1167–1169.

Fitter, R. S. R. and Scott, J. C., 1978. *The penitent butchers*. London: Collins.

Fitzpatrick, Matthew C., Gove, A. D., Sanders, N. J., and Dunn R. R., 2008. Climate change, plant migration, and range collapse in a global biodiversity hotspot: the Banksia (Proteaceae) of Western Australia. *Global Change Biology*, 14 (6): 1337–1352.

Fletcher, R., 2009. Ecotourism discourse: Challenging the stakeholders theory. *Journal of Ecotourism*, 8 (3): 269–285.

Floresta 2008. *Plant With Purpose. 'Global Charity Floresta Announces Milestone Four Million Trees Planted Worldwide'*. Available from: http://floresta org/4MillionTrees.pdf [accessed December 2010].

Folke C., Colding, J., and Berkes, F., 2002. Building resilience for adaptive capacity in social-ecological systems. In: Berkes F., J. Colding, and C. Folke (eds). *Navigating Social-Ecological Systems: Building Resilience fo Complexity and Change*. Cambridge: Cambridge University Press.

Folke, C., Hahn, T., Olsson, P., and Norberg, J., 2005. Adaptive governance of social-ecological systems. *In: Annual Review of Environment and Resources* Palo Alto: Annual Reviews, 441–473.

Folke, C., Carpenter, S. R., Walker, B., Scheffer, M., Chapin, T., and Rockström J., 2010. Resilience thinking: integrating resilience, adaptability and transformability. *Ecology and Society*, 15 (4): 20. [online]. Available from http://www.ecologyandsociety.org/vol15/iss4/art2/

Foltz, R. and Saadi-nejad, M., 2007. 'Is Zoroastrianism an Ecological Religion?' *Journal for the Study of Religion, Nature & Culture*, 1: 413–430.

Ford, J., 1971. *The Role of Trypanosomiasis in African Ecology: a Study of th Tsetse Fly Problem*. Oxford: Clarendon Press.

Forsyth, T., 1996. Science, myth and knowledge: Testing Himalayan environmenta degradation in Thailand. *Geoforum*, 27 (3): 375–392.

Fox, S., 1985. *The American Conservation Movement: John Muir and his Legacy* Madison: The University of Wisconsin Press.

Francis, R. A. and Krishnamurthy, K., 2014. Human conflict and ecosystem services: finding the environmental price of warfare. *International Affairs*, 9 (4): 853–869.

Fraserburgh Herald, 2012. *Localisation: The way forward for fishing?* Availabl from: http://www.fraserburghherald.co.uk/news/local-news/localisation-the way-forward-for-fishing-1-2275380 [accessed 10 February 2015].

Fuller, D., Buultjens, J., and Cummings, E., 2005. Ecotourism and indigenou micro-enterprise formation in northern Australia opportunities and constraint *Tourism Management*, 26 (6): 891–904.

Gadgil, M. and Guha, R., 1995. *Ecology and equity: the use and abuse of natur in contemporary India*. London: Routledge.

Galaz, V., Crona, B., Österblom, H., Olsson, P., and Folke, C., 2012. Polycentri systems and interacting planetary boundaries – Emerging governance of climat change–ocean acidification–marine biodiversity. *Ecological Economics*, 81: 21–32

Galbraith, J. K., 1998. *The Affluent Society*. Boston, Mass.: Houghton Mifflin Harcour

Gallopin, G. C., Funtowicz, S., O'Connor, M., and Ravetz, J., 2001. Science fo the twenty-first century: from social contract to the scientific core. *Internationc Social Science Journal*, 53: 219–+.

Gaodi, X., Cao, S., Yang, Q., Xia, L., Fan, Z., Chen, B., Zhou, S., Chang, Y., G L., Cook, S., and Humphrey, S., 2012. *China Ecological Footprint Repor 2012: Consumption, Production and Sustainable Development*. [online] Available from: http://wwf.panda.org/?207011/China-needs-innovative solutions-to-reduce-footprint---2012-China-Ecological-Footprint-Report.

Garcia, C. A., Bhagwat, S. A., Ghazoul, G., Nath, C. D., Nanaya, K. M Kushalappa, C. G., Raghuramulu, Y., Nasi, R., and Vaast, P., 2010. Biodiversit

Conservation in Agricultural Landscapes: Challenges and Opportunities of Coffee Agroforests in the Western Ghats, India. *Conservation Biology*, 24 (2): 479–488.

Gardner, K. and Lewis, D., 1996. *Anthropology, development and the post-modern challenge*. London: Pluto.

Garland, E., 2008. The Elephant in the Room: Confronting the Colonial Character of Wildlife Conservation in Africa. *African Studies Review*, 51 (3): 51–74.

Gatta, J., 2004. *Making Nature Sacred: Literature, Religion, and Environment in America from the Puritans to the Present*. Oxford: Oxford University Press.

Geall, S., 2013. *China and the Environment: The Green Revolution*. London: Zed Books Ltd.

Geist, H. J. and Lambin, E. F., 2001. *What drives tropical deforestation?*, *BioScience* 52 (2): 143–150.

Gettleman, J., 2012. Elephants Dying in Epic Frenzy as Ivory Fuels Wars and Profits. *New York Times* [online], 2012. Available from: http://www.nytimes.com/2012/09/04/world/africa/africas-elephants-are-being-slaughtered-in-poaching-frenzy.html?pagewanted=all&_r=0 [accessed 10 July 2014].

Ghosh, D., 2005. *Ecology and traditional wetland practice: lessons from wastewater utilisation in the East Calcutta Wetlands*. Kolkata: Worldview.

Gilbert, G., 2008. Introduction to Malthus T.R. 1798. *An Essay on the Principle of Population*. Oxford World's Classics reprint. Oxford: Oxford Paperbacks.

Gleditsch, N. P. and Urdal, H., 2002. Ecoviolence? Links Between Population-Growth, Environmental Scarcity and Violent Conflict in Thomas Homer-Dixon's Work. *Journal of International Affairs*, 56 (1): 283–302.

Goarch, 2010. Greek Orthodox Archdiocese of America. Available from: http://www.goarch.org/ourfaith/ourfaith8046 [accessed December 2010].

Gomezpompa, A. and Kaus, A., 1992. Taming the Wilderness Myth. *Bioscience*, 42 (4): 271–279.

Gordillo, J., Hunt, C., and Stronza, A. L., 2008. An ecotourism partnership in the Peruvian Amazon: the case of Posada Amazonas. *In*: Stronza, A. L., ed. *Ecotourism and Conservation in the Americas*. Oxford; Cambridge, MA: CABI, 30–48.

Gorman, P., 2009. The National Religious Partnership for the Environment Congregational Programs. Available from: http://www.nrpe.org/whatisthepartnership/history02.htm#hist03 [accessed December 2010].

Gossling, S., Hansson, C. B., Horstmeier, O., and Saggel, S., 2002. Ecological footprint analysis as a tool to assess tourism sustainability. *Ecological Economics*, 43 (2–3): 199–211.

Gottlieb, R. S., 2007. 'Religious Environmentalism: What it is, Where it's Heading and Why We Should be Going in the Same Direction'. *Journal for the Study of Religion, Nature & Culture*, 1: 81–91.

Gowdy, J. M., 1997. 'The value of biodiversity: markets, society, and ecosystems'. *Land Economics*, 73: 25–41.

Gowdy, J. M., and Mayumi, K. 2001. 'Reformulating the foundations of consumer choice theory and environmental valuation'. *Ecological Economics*, 39: 223–237.

Graham, A., 1973. *The Gardeners of the Eden*. Hemel Hempstead: Allen and Unwin.

Green, S. J., White, A. T., Flores, J. O., Carreon, M. F. III., and Sia, A. E., 2003. *Philippine fisheries in crisis: A framework for management*. Coastal Resource Management Project of the Department of Environment and Natural Resources, Cebu City, Philippines. 77 p.

Greenquist, P., 1997. Tourism and biodiversity preservation in Honduras. *Planeta com* [online]. Available from: http://www.planeta.com/planeta/98/0298mosquitia html [accessed 16 March 2015].

GRN, 1996. *Nature Conservation Amendment Act*. Government Gazette No. 1333.

Groenfelt, D., 1991. Building on tradition: indigenous irrigation knowledge and sustainable development in Asia. *Agriculture and Human Values*, 8: 114–120.

Grossmann, D. and Koch, E., 1995. *Ecotourism Report: Nature Tourism in South Africa: Links with the Reconstruction and Development Program*. Pretoria, South Africa: SATOUR.

Grove, R., 1995. *Green imperialism: colonial expansion, tropical island Edens and the origins of environmentalism, 1600–1860*. Cambridge; New York: Cambridge University Press.

Grubb, M., Koch, M., Munson, A., Sullivan, F., and Thomson, K., 1993. *The Earth Summit Agreements: A Guide and Assessment*. London: Earthscan.

Guan, D., Hubacek, K., Weber, C. L., Peters, G. P., and Reiner, D. M., 2008. The drivers of Chinese CO_2 emissions from 1980 to 2030. *Global Environmental Change*, 18 (4): 626–634.

Guha, R. and Martínez-Alier, J., 1997. *Varieties of Environmentalism: Essays North and South*. London: Routledge.

Guha, R., 2000. *Environmentalism: a global history*. New York; Harlow: Longman.

Gullison, R. E., Frumhoff, P. C., Canadell, J. G., Field, C. B., Nepstad, D. C., Hayhoe, K., Avissar, R., Curran, L. M., Friedlingstein, P., Jones, C. D., and Nobre, C., 2007. Tropical Forests and Climate Policy. *Science*, 316 (5827): 985–986.

Gunderson L. H. and Holling, C. S. (eds), 2002. *Panarchy: understanding transformations in human and natural systems*. Washington, D.C.; London: Island Press.

Gunderson, L. H., 2008. Adaptive dancing: interactions between social resilience and ecological crises. *In*: Berkes, F., Colding, J. and Folke, C. *Navigating Social-Ecological Systems: Building Resilience for Complexity and Change*. Cambridge: Cambridge University Press.

Gundlach, E. R., 1977. Oil tanker disasters. *Environment*, 19 (9): 16–27.

Gunningham, N. and Grabosky, P. N., 1998. *Smart Regulation: Designing Environmental Policy*. Oxford: Clarendon Press.

Gustafsson, B., 1998. Scope and limits of the market mechanism in environmental management. *Ecological Economics*, 24 (2–3): 259–274.

Haar, G. T. and Ellis, S., 2006. The Role of Religion in Development: Towards a New Relationship Between the European Union and Africa. *The European Journal of Development Research*, 18: 351–367.

Hall, M., Grim, J., and Tucker, M. E., 2009. Need for Religions to Promote Values of Conservation. *Nature*, 462: 720.

Hamel, G., and Valikangas, L. The Quest for Resilience. *Harvard Business Review*, (September 2003): 52–63.

Hansen, A. J., and R. DeFries., 2007. Ecological mechanisms linking protected areas to surrounding lands. *Ecological Applications*, 17: 974–988.

Hanson, T., Brooks, T. M., da Fonseca, G. A. B., Hoffmann, M., Lamoreux, J. F., Machlis, G., Mittermeier, C. G., and John, D. P., 2009. Warfare in biodiversity hotspots. *Conservation Biology*, 23: 578–587.

Harbinson, R., 2007. *Development recast? A review of the impact of the Rio Tinto ilmenite mine in southern Madagascar* [online]. Friends of the Earth and Panos. Available from: http://www.foe.co.uk/sites/default/files/downloads/development_recast.pdf [accessed 2 March 2014].

Hardesty, D. L., 1977. *Ecological anthropology*. New York: Wiley.

Hardin, G. J., 1974. Lifeboat Ethics: the Case Against Helping the Poor. *Pschology Today* [online], (September 1974). Available from: http://www.garretthardinsociety.org/articles/art_lifeboat_ethics_case_against_helping_poor.html [accessed 3 May 2015].

Hardin, G., 1968. The Tragedy of the Commons. *Science*, 162: 1243–1248.

Hardoon, D., 2015. *Wealth: Having it all and wanting more.* Research report, Oxfam International. ISBN 978-1-78077-795-5.

Harvey, D., 2005. *A Brief History of Neoliberalism.* Oxford: Oxford University Press.

Hassan, R. M., Scholes, R., and Ash, N., 2005. *Ecosystems and Human Well-Being: Current State and Trends: Findings of the Condition and Trends Working Group.* Washington, D.C.: Island Press.

Hay, P. R., 2002. *Main Currents in Western Environmental Thought.* Bloomington, IN: Indiana University Press.

Heede, R., 2013. Tracing anthropogenic carbon dioxide and methane emissions to fossil fuel and cement producers, 1854–2010. *Climatic Change*, 122 (1–2): 229–241.

Hegde, R. and Bull, G. Q., 2011. Performance of an agro-forestry based Payments-for-Environmental-Services project in Mozambique: A household level analysis. *Ecological Economics*, 71: 122–130.

Hemmati, M., 2002. *Multi-stakeholder Processes for Governance and Sustainability: Beyond Deadlock and Conflict.* London: Earthscan.

Herr, H. and Karl, G., 2003. *Estimating Global Slum Dwellers: Monitoring the Millenium Development Goal 7, Target 11* [online]. Nairobi, Kenya: UN-HABITAT. Available from: http://unstats.un.org/unsd/accsub/2002docs/mdg-habitat.pdf [accessed 23 March 2015].

Heske, F., 1938. *German forestry: [Trans.]: Published for the Oberlaender trust of the Carl Schurz memorial foundation.* New Haven: Yale University Press.

Hildyard, N., 1993. Foxes in charge of the chickens. *In*: Sachs, W., ed. *Global ecology.* London: Zed Books, 22–35.

Hiro, D., 2012. *After Empire: The Birth of a Multipolar World.* New York: Nation Books.

Hobbs, R. J., Arico, S., Aronson, J., Baron, J. S., Bridgewater, P., Cramer, V. A., Epstein, P. R., Ewel, J. J., Klink, C. A., Lugo, A. E., Norton, D., Ojima, D., Richardson, D. M., Sanderson, E. W., Valladares, F., Vilà, M., Zamora, R., and Zobel, M., 2006. Novel ecosystems: theoretical and management aspects of the new ecological world order. *Global Ecology and Biogeography*, 15 (1): 1–7.

Hobsbawm, E. J., 1994. *The Age of Extremes: The Short Twentieth Century, 1914–1991.* London: Michael Joseph.

Hochstetler, K. and Keck, M. E., 2007. *Greening Brazil: Environmental Activism in State and Society.* Durham, N.C.: Duke University Press.

Holdgate, M., 1999. *The Green Web: A Union for World Conservation.* London: Routledge.

Holling, C. S., 1973. Resilience and Stability of Ecological Systems. *Annua Review of Ecology and Systematics*, 4 (1): 1–23.

Holling, C. S., 1986. The resilience of terrestrial ecosystems: local surprise an global change. *In*: Clark, W. C. and Munn, R. E., eds. *Sustainable developmen of the biosphere*. Cambridge: Cambridge University Press, 292–317.

Holling, C. S., Berkes, F., and Folke, C., 1998. Science, sutainability and resourc management. *In*: *Linking Social and Ecological Systems*. New York: Cambridg University Press, 343–362.

Holmberg, J., Thomson, K., Timberlake, L., and Edwards, M., 1993. *Facing th Future: Beyond the Earth Summit*. London: Earthscan.

Holmes, G., 2010. The Rich, the Powerful and the Endangered: Conservatio Elites, Networks and the Dominican Republic. *Antipode*, 42 (3): 624–646.

Homer-Dixon, T. F., 1999. *Environment, Scarcity and Violence*. Princeton, NJ Princeton University Press.

Homewood, K., Chevenix Trench, P., and Brockington, D., 2013. Pastoralism an conservation - who benefits? *In*: Roe, D., Elliott, J., Sandbrook, C., an Walpole, M., eds. *Biodiversity conservation and poverty alleviation: explorin the evidence for a link*. Chichester: Wiley-Blackwell, 239–52.

Honey, M., 2001. Certification programmes in the tourism industry. *Industry an environment*, 24 (3): 28–29.

Honey, M., 2008. *Ecotourism and Sustainable Development, Second Edition: Wh Owns Paradise?* Washington, D.C.: Island Press.

Honey-Rosés, J., López-García, J., Rendón-Salinas, E., Peralta-Higuera, A., an Galindo-Leal, C., 2009. To pay or not to pay? Monitoring performance an enforcing conditionality when paying for forest conservation in Mexic *Environmental Conservation*, 36 (2): 120–128.

Hopkin, R., 2008. *The Transition Handbook: From Oil Dependency to Loc Resilience*. Cambridge: Green Books.

Horton, L. R., 2009. Buying Up Nature Economic and Social Impacts of Cost Rica's Ecotourism Boom. *Latin American Perspectives*, 36 (3): 93–107.

Hughes, R., and Flintan, F., 2001. Integrating Conservation and Developmen Experience: A Review and Bibliography of the ICDP Literature. Londo International Institute for Environment and Development.

Hughes, T. P., Bellwood, D. R., Folke, C., et al., 2005. New paradigms f supporting the resilience of marine ecosystems. *Trends in Ecology & Evolutio* 20: 380–386.

Humphreys, J. and Smith, M. L. R., 2011. War and wildlife: the Clausewit connection. *International Affairs*, 87: 121–142.

Hunt, T. L. and Lipo, C. P., 2006. Late Colonisation of Easter Island. *Scienc* March 9, 2006.

Hurrell, A. and Sengupta, S., 2012. Emerging powers, North-South relations an global climate politics. *International Affairs*, 88: 463–484.

Hutton, J., Adams, W. M., and Murombedzi, J. C., 2005. Back to the Barriers Changing Narratives in Biodiversity Conservation. *Forum for Developmer Studies*, 32 (2): 341–370.

Hviding, E. and Baines, G. B. K., 1992. *Fisheries management in the Pacifi tradition and challenges of development in Marovo, Solomon Islands*. Genev United Nations Research Institute for Social Development.

IAASTD, 2009. *International Assessment of Agricultural Knowledge, Science and Technology for Development: Sub-Saharan Africa (SSA) Report.* McIntyre, BD; HR Herren; J Wakhungu; RT Watson (Ed.). Island Press, Washington DC.

IEA, 2011. *World Energy Outlook 2011* [online]. Paris: OECD, International Energy Agency. Available from: http://www.worldenergyoutlook.org/publications/weo-2011/ [accessed 23 March 2015].

IFOAM (International Federation of Organic Agriculture Movements), 2010. Available from: http://www.ifoam.org/ [accessed December 2010].

Igoe, J., Neves, K., and Brockington, D., 2010. A Spectacular Eco-Tour around the Historic Bloc: Theorising the Convergence of Biodiversity Conservation and Capitalist Expansion. *Antipode,* 42: 486–512. doi: 10.1111/j.1467-8330.2010.00761.x

ILO (International Labor Organisation), 2000. Tripartite Meeting on Moving to Sustainable Agricultural Development through the Modernisation of Agriculture and Employment in a Globalized Economy. Geneva, 18–22 September 2000, International Labor Office Geneva. Available from: http://www.ilo.org/public/english/dialogue/sector/techmeet/tmad00/ [accessed December 2010].

Inglehart, R., 1977. *The Silent Revolution: Changing Values and Political Styles Among Western Publics.* Princeton: Princeton University Press.

IPCC, 2007. *Climate Change 2007: The Physical Science Basis. Contribution of Working Group I to the Fourth Assessment Report of the Intergovernmental Panel on Climate Change* [Solomon, S., D. Qin, M. Manning, Z. Chen, M. Marquis, K. B. Averyt, M. Tignor and H. L. Miller (eds.)]. New York; Cambridge: Cambridge University Press.

IPCC, 2001. *2001 A Report of Working Group II, Climate Change 2001: Impacts, Adaptation, and Vulnerability.* [online]. Cambridge: Cambridge University Press. Available from: http://www.grida.no/publications/other/ipcc_tar/.

Isaac J. L., 2009. 'Effects of climate change on life history: implications for extinction risk in mammals'. *Endangered Species Research,* 7 (2): 115–123.

Iskandar, J., 2007. Responses to Environmental Stress in the Baduy Swidden System, South Banten, Java. *In*: Ellen, R. F., ed. *Modern crises and traditional strategies: local ecological knowledge in island Southeast Asia.* New York; Oxford: Berghahn Books, 112–132.

IUCN, 2003. International Union for Conservation of Nature. World Parks Congress recommendation V.26. Available from: http://cmsdata.iucn.org/downloads/recommendationen.pdf [accessed 11 February 2015].

IUCN, 2007. International Union for Conservation of Nature Protected Areas Programme, World Commission on Protected Areas, IUCN, Gland, Switzerland.

IUCN, 1980. *World Conservation Strategy: Living Resource Conservation for Sustainable Development.* Gland, Switzerland: IUCN/UNEP/WWF.

IUCN, 1997. *Resolutions and Recommendations: World Conservation Congress, Montreal, Canada, 13–23 October 1996* [online]. IUCN. Available from: http://books.google.co.uk/books?id=iU7y1NJviOkC&dq=1996%20world%20conservation%20Congress%20resolutions%20on%20indigenous%20people&source=gbs_similarbooks [accessed 3 March 2014].

IUCN, 2003. *Durban Accord: Our Global Commitment for People and the Earth's Protected Areas Including the Durban Action Plan: Draft of 7 September 2003.* IUCN--The World Conservation Union.

IUCN, 2008. Protecting the Sacred Natural Sites of the World. Available from http://cms.iucn.org/about/work/programmes/social_policy/sp_themes/sp_themes_sns/index.cfm [accessed January 2009].

J. D. Power & Associates, 2011. *Global automotive outlook for 2011 appears positive as mature or auto markets recover, emerging markets continue to expand. Press release.* [online]. Available from: http://businesscenter.jdpower com/news/pressrelease.aspx?ID=2011018.

Jackson, T., 2009. *Prosperity without growth: economics for a finite planet* London: Earthscan.

Janssen, M. A., 2007. An update on the scholarly networks on resilience vulnerability, and adaptation within the human dimensions of global environmental change. *Ecology and Society*, 12 (2): 9.

Janssen, M. A., Bodin, O., Anderies, J. M., Elmqvist, T., Ernstson, H., McAllister R. R. J., Olsson, P., and Ryan, P., 2006. Toward a network perspective of the study of resilience in social-ecological systems. *Ecology and Society*, 11 (1): 15

Janssen, M. A., Schoon, M. L., Ke, W., and Borner, K., 2006. Scholarly network on resilience, vulnerability and adaptation within the human dimensions of global environmental change. *Global Environmental Change-Human and Policy Dimensions*, 16 (3): 240–252.

Jasanoff, S., 2004. *States of knowledge: the co-production of science and social order* [online]. London: Routledge. Available from: http://www.loc.gov/catdir enhancements/fy0651/2003017603-d.html http://www.loc.gov/catdir/toc ecip047/2003017603.html.

Jessop, B., 2002. *The Future of the Capitalist State*. Cambridge: Polity.

Jhamtani, H., 2010. *The Green Revolution in Asia: Lessons for Africa. Food and Agriculture Organisation of the United Nations.* Available from: http://www fao.org/docrep/012/al134e/al134e02.pdf [accessed December 2010].

Johannesen, A. B. and Skonhoft, A. 2005. 'Tourism, Poaching and Wildlife Conservation: What Can Integrated Conservation and Development Project Accomplish?' *Resource and Energy Economics*, 27: 208–226.

Jones, B. T. B., Davis, A., Diggle, R., and Diez, L., 2013. Community-based natural resource management and reducing poverty in Namibia. *In*: Roe, D Elliott, J., Sandbrook, C., and Walpole, M., eds. *Biodiversity conservation and poverty alleviation: exploring the evidence for a link*. Chennai: India & Malaysia: Wiley-Blackwell, 191–205.

Jones, P. G. and Thornton, P. K., 2003. 'The potential impacts of climate change in tropical agriculture: the case of maize in Africa and Latin America in 2055' *Global Environmental Change*, 13: 51–59.

Jordan, A., 2008. The governance of sustainable development: taking stock and looking forwards. *Environment and Planning C-Government and Policy, 2* (1): 17–33.

Kaplan, B. J., 2007. *Divided by Faith: Religious Conflict and the Practice of Toleration in Early Modern Europe*. Cambridge, MA: Harvard University Press.

Karamanolis, G. E., 2006. *Plato and Aristotle in Agreement? Platonists on Aristotle From Antiochus to Porphyry*. Oxford: Oxford University Press.

Kartha, S. and Erickson, P., 2011. *Comparison of Annex 1 and non - Annex pledges under the Cancun Agreements* [online]. Stockholm: Stockholm Environment Institute. No. WP - US - 1107. Available from: http://www

sei-international.org/mediamanager/documents/Publications/Climate/sei-workingpaperus-1107.pdf [accessed 5 March 2015].

Katz, E., 1997. Artifacts and functions: a note on the value of nature. *In*: *Nature as Subject: Human Obligation and Natural Community* E. Katz, ed. Lanham, MD: Rowman & Littlefield Publishers. Lanham, MD: Rowman & Littlefield, 120–131.

Keeley, J. and Scoones, I., 2003. *Understanding environmental policy processes: cases from Africa*. London; Sterling, VA: Earthscan.

Kepe, T. and Scoones, I., 1999. Creating grasslands: social institutions and environmental change in Mkambati area, South Africa. *Human Ecology*, 27 (1): 29–54.

Khanna, P., 2009. *The Second World: How Emerging Powers Are Redefining Global Competition In The Twenty-First Century*. London: Random House.

Kinsley, D. R., 1994. *Ecology and Religion: Ecological Spirituality in Cross-cultural perspective*. London: Prentice Hall.

Kiple, K. F., 1984. *The Caribbean Slave: Biological History*. Cambridge: Cambridge University Press.

Kirkby, C. A., Giudice-Granados, R., Day, B., Turner, K., Velarde-Andrade, L. M., Duenas-Duenas, A., Lara-Rivas, J. C., and Yu, D. W., 2010. The Market Triumph of Ecotourism: An Economic Investigation of the Private and Social Benefits of Competing Land Uses in the Peruvian Amazon. *Plos One*, 5 (9): e13015.

Kirkby, C. and Yu, D. W., 2010. Ecotourism case study. *In*: Ghazoul, J. and Sheil, D., eds. *Tropical Rain Forest Ecology, Diversity, and Conservation*. Oxford; New York: Oxford University Press.

Kiss, A., 2004. Is community-based ecotourism a good use of biodiversity conservation funds? *Trends in Ecology & Evolution*, 19 (5): 232–237.

Klein, R. J. T., Nicholls, R. J., and Thomalla, F., 2003. Resilience to natural hazards: How useful is this concept? *Environmental Hazards*, 5 (1): 35–45.

Klijn, E.-H. and Skelcher, C., 2007. Democracy and governance networks: Compatible or not? *Public Administration*, 85 (3): 587–608.

Knorr-Cetina, K., 1985. Germ Warfare. *Social Studies of Science*, 15: 577–586.

Koch, E., 1994. *Reality or Rhetoric? Ecotourism and rural reconstruction in South Africa*. Geneva: United Nations Research Institute for Social Development.

Koens, J. F., Dieperink, C., and Miranda, M., 2009. Ecotourism as a development strategy: experiences from Costa Rica. *Environment Development and Sustainability*, 11 (6): 1225–1237.

Kowarik, I., 2005. Wild Urban Woodlands: Towards a Conceptual Framework. *In*: Kowarik, P. D. I. and Körner, D. S., eds. *Wild Urban Woodlands* [online]. Springer Berlin Heidelberg, 1–32. Available from: http://link.springer.com/chapter/10.1007/3-540-26859-6_1 [accessed 4 March 2015].

Kowarik, I., 2011. Novel urban ecosystems, biodiversity, and conservation. *Environmental Pollution*, 159 (8–9): 1974–1983.

Kowarik, I., 2013. *Urban ecosystems helping cities to adapt to a changing climate*. [online]. [online]. Available from: https://www.bfn.de/fileadmin/MDB/documents/themen/biologischevielfalt/Klimaseite/25.06-16.50_Kowarik.pdf [accessed 26 February 2015].

Krasner, S. D., 1983. *International Regimes*. Ithaca, N.Y.: Cornell University Press.

Kreg, L., 2003. Tourism's contribution to conservation in Zululand. *In*: Aylward B. A. and Lutz, E., eds. *Nature Tourism, Conservation, and Development in Kwazulu-Natal, South Africa*. World Bank Publications, 203–240.

Kühn, I., Brandl, R., and Klotz, S., 2004. The flora of German cities is naturally species rich. *Evolutionary Ecology Research*, 6 (5): 749–764.

Kula, E., 2001. 'Islam and Environmental Conservation'. *Environmental Conservation*, 28 (1): 1–9.

Kumar, S., 2002. Does participation in common pool resource management help the poor? A social cost-benefit analysis of joint forest management in Jharkhand, India. *World Development*, 30 (5): 763–782.

Kumar, M., and P. Kumar., 2008. 'Valuation of the ecosystem services: a psycho cultural perspective'. *Ecological Economics*, 64: 808–819.

Kumleben, M. E., 1996. *Report of the Commission of Inquiry into the Alleged Smuggling of and Illegal Trade in Ivory and Rhinoceros Horn in South Africa*. Durban: Government of South Africa. Commission Report.

Lander, E., 2011. *The Green Economy: the Wolf in Sheep's clothing* [online]. Transnational Institute. Available from: http://www.tni.org/report/green economy-wolf-sheeps-clothing [accessed 5 March 2015].

Lapeyre, R., 2010. Community-based tourism as a sustainable solution to maximise impacts locally? The Tsiseb Conservancy case, Namibia. *Development Southern Africa*, 27 (5): 757–772.

Larkin, P. A., 1977. An epitaph for the concept of maximum sustained yield. *Transactions of the American Fish Society*, 106: 1–11.

Latour, B., 1987. *Science in action: how to follow scientists and engineers through society*. Milton Keynes: Open University Press.

Latour, B., 1993. *We have never been modern*. Harlow: Longman.

Latour, B., 1996. On actor-network theory - A few clarifications. *Soziale Welt Zeitschrift Fur Sozialwissenschaftliche Forschung Und Praxis*, 47 (4): 369–8.

Latour, B., 2007. *Reassembling the Social: An Introduction to Actor-Network Theory*. Oxford: Oxford University Press.

Lawrence, D., 2000. *Kakadu: The making of a national park*. Melbourne: The Miegunyah Press, Melbourne University Press.

Le Billon, P., 2005. *Geopolitics of Resource Wars: Resource Dependence Governance and Violence*. London: Frank Cass Publishers.

Leach, M. and Fairhead, J., 2000. Challenging neo-Malthusian deforestation analyses in West Africa's dynamic forest landscapes. *Population and Development Review*, 26 (1): 17–43.

Leach, M. and Mearns, R., 1996. Challenging Received Wisdom in Africa. *In*: The lie of the land. *Challenging received wisdoms on the African environment*. Oxford: Heinmann/James Currey, 240–240.

Leach, M., Fairhead, J., and Fraser, J., 2012. Green grabs and biochar: Revaluing African soils and farming in the new carbon economy. *Journal of Peasant Studies*, 39 (2): 285–307.

Lehmann, J., Kern, D., German, L., McCann, J., Martins, G. V. C., and Moreira, A., 2003. Soil fertility and production potential. *In*: Lehmann, J., Kern, D., Glaser, B., and Woods, W. I., eds. *Amazonian Dark Earths: Origin, Properties and Management*. Dordrecht: Kluwer.

Leichenko, R. and O'Brien, K., 2008. *Environmental Change and Globalization Double Exposures*. Oxford: Oxford University Press.

Lemos, M. C. and Agrawal, A., 2006. Environmental governance. *In: Annual Review of Environment and Resources*, 297–325.

Lepper, C. M. and Goebel, J. S., 2010. Community-based natural resource management, poverty alleviation and livelihood diversification: A case study from northern Botswana. *Development Southern Africa*, 27 (5): 725–739.

Levy, D. L. and Newell, P. J., 2005. *The Business of Global Environmental Governance*. Cambridge, Mass.; London: MIT Press.

Lindblom, C., 1979. Still Muddling, Not yet Through. *Public Administration Review*, 39 (6): 517–526.

Liu, C., Lu, J., and Yin, R., 2010. An Estimation of the Effects of China's Priority Forestry Programs on Farmers' Income. *Environmental Management*, 45 (3): 526–540.

Lobell, D., and Burke, M. (eds), 2010. *Climate change and food security: Adapting agriculture to a warmer world*. New York: Springer.

Lockwood, J. A. and Lockwood, D. R., 1993. Catastrophe theory: a unified paradigm for Rangelands ecosystem dynamics. *Journal of Range Management*, 46 (4): 282–288.

Long, S. A., 2004. Introduction. *In: Livelihoods and CBNRM in Namibia: The Findings of the WILD Project, Final Technical Report of the Wildlife Integration for Livelihood Diversification Project (WILD), prepared for the Directorates of Environmental Affairs and Parks and Wildlife Management, the Ministry of Environment and Tourism, the Government of the Republic of Namibia. Windhoek, March 2004*. Windhoek: MET/DfID, 1–12.

Lovejoy, P. E., 2000. *Transformations in Slavery: A History of Slavery in Africa*. Cambridge: Cambridge University Press.

Lovejoy, T. E., and Lee, H., (eds). 2005. *Climate Change and Biodiversity*. New Haven, CT: Yale University Press.

Lowe, P. and Ward, S., 1998. *British Environmental Policy and Europe: Politics and Policy in Transition*. London: Routledge.

Lowood, H., 1990. The calculating forester: quantification, cameral science and the emergence of scientific forestry management in Germany. *In*: Frängsmyr, T., Heilbron, J. L., and Rider, R. E., eds. *The Quantifying spirit in the 18th century*. Berkeley; Oxford: University of California Press, 315–42.

Ludwig, D., Jones, D., and Holling, C., 1978. Qualitative-Analysis of Insect Outbreak Systems – Spruce Budworm and Forest. *Journal of Animal Ecology*, 47 (1): 315–332.

Lujala, P. and Rustad, S. A., 2011. High-value natural resources: a blessing or a curse for peace? *Sustainable Development Law Policy*, 12 (19–22): 56–57.

Lynas, M., 2006. *Six Degrees: Our Future on a Hotter Planet*. London: Harper Perennial.

Maass, A. and Anderson, R. L., 1986. *--and the desert shall rejoice: conflict, growth, and justice in arid environments*. Malabar, Fla.: R. E. Krieger.

MacDonald, K. I., 2010. The Devil is in the (Bio)diversity: Private Sector "Engagement" and the Restructuring of Biodiversity Conservation. *Antipode*, 42: 513–550. doi: 10.1111/j.1467-8330.2010.00762.x

MacKenzie, J. M., 1988. *The empire of nature: hunting conservation and British imperialism*. Manchester: Manchester University Press.

Magome, H., Grossman, D., Fakir, S., and Stowell, Y., 2000. *Partnerships in conservation: the State, private sector and community at Madikwe Game*

Reserve, North-west Province, South Africa [online]. London: IIED. No. 7 Available from http://pubs.iied.org/pdfs/7802IIED.pdf [accessed 16 March 2015]

Mahony, K. and Van Zyl, J., 2001. *Practical strategies for pro-poor tourism, Cas studies of Makuleke and Manyeleti tourism initiatives.* CRT, IIED and OD Pro-poor tourism working paper No. 2.

Malthus, T. R., 1999. *An Essay on the Principle of Population.* Oxford: Oxford University Press.

Mamdani, M., 2002. African States, citizenship and war. *International Affairs*, 7 (3): 493–506.

Marcoux, A., 1999. *Population and environmental change: from linkages to polic issues.* [online]. Available from: http://www.fao.org/sd/wpdirect/WPre0089.htm

Marsh, J., 1982. *Back to the land: the pastoral impulse in England, from 1880 t 1914.* London: Quartet.

Martin, R. B., 1994. *Alternative Approaches to Sustainable Use.* Presented at th Conservation Through Sustainable Use of Wildlife Conference.

Marvier, M., 2013. New Conservation Is True Conservation. *Conservatio Biology*, 28 (1): 1–3.

Marzouki, M., Froger, G., and Ballet, J., 2012. Ecotourism versus Mass Tourism A Comparison of Environmental Impacts Based on Ecological Footprin Analysis. *Sustainability*, 4 (1): 123–140.

McAfee, K., 1999. Selling nature to save it? Biodiversity and gree developmentalism. *Environment and Planning D: Society and Space*, 17 (2) 133–154.

McCarthy, J., 2012. The Financial Crisis and Environmental Governance After Neoliberalism. *Tijdschrift Voor Economische En Sociale Geografie*, 103 (2 180–195.

McCauley, D. J., 2006. 'Selling out on nature'. *Nature*, 443: 27–28.

McCormick, J., 1989. *The global environmental movement: reclaiming paradise* London: Belhaven Press.

McKibben, B., 1989. *The End of Nature.* New York: Random House.

McLaughlin, P. and Dietz, T., 2008. Structure, agency and environment: Towar an integrated perspective on vulnerability. *Global Environmental Change*, 1 (1): 99–111.

McShane, T. O. and Wells, M. P. (eds), 2004. *Integrated conservation an development?* New York; Chichester: Columbia University Press.

McShane, T. O., Hirsch, P. D., Trung, T. C., Songorwa, A. N., Kinzig, A Monteferri, B., Mutekanga, D., Thang, H. V., Dammert, J. L., Pulga Vidal, M., Welch-Devine, M., Peter Brosius, J., Coppolillo, P., an O'Connor, S., 2011. Hard choices: Making trade-offs between biodiversit conservation and human well-being. *Biological Conservation*, 144 (3 966–972.

MEA, 2005. *Ecosystems and Human Well-Being: Synthesis.* Washington, D.C Island Press.

Meadows, D. H., Randers, J., and Meadows, D. L., 1972. *The Limits to growth: report for the Club of Rome's project on the predicament of mankind.* Londo Earth Island Ltd.

Mellor, J. 1992. *Agriculture on the Road to Industrialisation*, Baltimore: John Hopkins University Press.

Meza, L. E., 2009. Mapuche Struggles for Land and the Role of Private Protected Areas in Chile. *Journal of Latin American Geography*, 8 (1): 149–163.

Milburn, R., 2012. Mainstreaming the environment into postwar recovery: the case for 'ecological development'. *International Affairs*, 88: 1083–1100.

Miles, L. and Kapos, V., 2008. Reducing greenhouse gas emissions from deforestation and forest degradation: global land-use implications. *Science*, 320 (5882): 1454–1455.

Miller, T. R., Minteer, B. A., Malan, L-C., 2011. The new conservation debate: The view from practical ethics. *Biological Conservation*, 144: 948–957.

Miller, C., 2013. *Gifford Pinchot and the Making of Modern Environmentalism.* Washington: Island Press.

Miller, F., Osbahr, H., Boyd, E., Thomalla, F., Bharwani, S., Ziervogel, G., Walker, B., Birkmann, J., van de Leeuw, S., Rockström, J., Hinkel, J., Downing, C., Folke, C., and Nelson, D., 2010. Resilience and vulnerability: complementary or conflicting concepts? *Ecology and Society*, 15 (3): 11.

Miller, J. R., 2005. Biodiversity conservation and the extinction of experience. *Trends in Ecology & Evolution*, 20 (8): 430–434.

Miller, K., 1982. *Planning National Parks for Ecodevelopment: Methods and Cases from Latin America.* Fundación para la Ecologia y para la Protección del Medio Ambiente.

Minteer, B. A. and Miller, T. R., 2011. The New Conservation Debate: Ethical foundations, strategic trade-offs, and policy opportunities. *Biological Conservation*, 144 (3): 945–947.

Misc., 2005. Readers' responses to '"A challenge to conservationists"' — additions. *World Watch Magazine* [online], 18 (2). Available from: Http:// www.worldwatch.org/node/1832 [accessed 10 November 2013].

Mitchell, R. C., Dunlap, R. E., and Mertig, A. G., 2014. Twenty Years of Environmental Mobilization: Trends among National Environmental Organizations. *In*: Dunlap, R. E. and Mertig, A. G., eds. *American Environmentalism: The US Environmental Movement, 1970–1990.* London: Taylor & Francis, 11–26.

Mohai, P., Pellow, D., and Roberts, J. T., 2009. Environmental Justice. *In*: *Annual Review of Environment and Resources.* Palo Alto: Annual Reviews, 405–430.

Montgomery, M. R., 2003. *Cities transformed; demographic change and its implications in the developing world* [online]. Washington DC/Europe: The National Academy Press/Earthscan. Available from: http://www.nap.edu/ catalog.php?record_id=10693.

Moran, E. F., 2009. *Human Adaptability: An Introduction to Ecological Anthropology.* Boulder, CO: Westview Press.

Morgan, P. and Lawton, C., 2007. *Ethical Issues in Six Religious Traditions.* Edinburgh: Edinburgh University Press.

Mortimore, M., 2005. Dryland development: success stories from West Africa. *Environment*, 47 (1): 8–21.

Moseley, W., 2005. Global cotton and local environmental management: the political small-hold farmers ecology of rich and poor in southern Mali. *Geographical Journal*, 171: 36–55.

Moser, C., Norton, A., Stein, A., and Agbola, S. B., 2010. *Pro-Poor Adaptation to Climate Change in Urban Centres: Case Studies of Vulnerability and Resilience in Kenya and Nicaragua.* Washington, DC: Social Development Department The World Bank.

Mount Athos. 2008. Mount Athos. Available from: http://www.inathos.gr [accessed January 2009].

Mowforth, M. and Munt, I., 2009. *Tourism and Sustainability: Development Globalisation and New Tourism in the Third World*, 3rd Edition. London Taylor & Francis.

Muir, J., 1901. Our National Parks. Excerpted from Callicott, J. B. and M. P. Nelson eds. 1998. *The Great New Wilderness Debate*, p. 48–62. Athens: University o Georgia Press.

Muir, K., Bojö, J., and Cunliffe, R., 1996. Economic policy, wildlife and land use in Zimbabwe. *In: The Economics of Wildlife: Case Studies from Ghana, Kenya Namibia and Zimbabwe.* World Bank, 154.

Munasinghe, M., 1999. Is environmental degradation an inevitable consequence o economic growth: tunneling through the environmental Kuznets curve *Ecological Economics*, 29 (1): 89–109. doi:10.1016/S0921-8009(98)00062-7.

Murombedzi, J., 1992. *Decentralisation or Recentralisation? Implementing Campfire in the Omay Communal Lands of the Nyaminyami District.* Harare University of Zimbabwe.

Murombedzi, J., Hulme, D., and Murphree, M., 2001. Natural resource stewardship & community benefits in Zimbabwe's Campfire programme. *In: African Wildlife and Livelihoods: the promise and performance of community conservation.* Oxford: James Currey.

Murphree, M., 1997. *Congruent Objectives, Competing Interests and Strategic Compromise: Concept and Process in the Evolution of Zimbabwe's Campfire Programme.* Manchester: Institute for Development Policy and Management University of Manchester.

Nabhan, G. P., House, D., Humberto, S. A., Hodgson, W., Luis, H. S., and Guadalupe M., 1991. Conservation and Use of Rare Plants by Traditional Cultures of the US Mexico Borderlands. *In*: Oldfield, F. and Alcorn, J. B., eds. *Biodiversity: Culture Conservation and Ecodevelopment.* Boulder, Colorado: Westview Press.

NACSO, 2011. *Namibia's communal area conservancies: a review of progress 2010* [online]. Windhoek, Namibia: Namibian Association of CBNRM Suppor Organisations. Available from: http://www.nacso.org.na/SOC_2010/index.php [accessed 10 November 2013].

NACSO, 2013. *Namibia's communal conservancies: a review of progress an challenges - 2011* [online]. Windhoek, Namibia: Namibian Association o CBNRM Support Organisations. Available from: http://www.nacso.org.na SOC_2011/index.php [accessed 10 November 2013].

Nash, R., 1973. *Wilderness and the American Mind.* New Haven: Yale University Press.

Nash, R. F., 2001. Wilderness and the American Mind (fourth edition). New Haven: Yale University Press.

Nature Conservancy, 2014. *Livelihoods Approaches as a Conservation Too* Available from: http://www.reefresilience.org/pdf/LivelihoodsApproac ShortVersion.pdf [accessed 11 February 2015].

Nelson, D., Adger, W. N., and Brown, K., 2007. Adaptation to Environmenta Change: Contributions of a Resilience Framework. *Global Environmenta Change*, 32: 395–419.

Neumann, R. P., 1996. Dukes, earls and ersatz Edens: aristocratic nature preservationists in colonial Africa. *Environment and Planning D: Society and Space*, 14: 79–98.

Neumann, R. P., 1997. Primitive ideas: protected area buffer zones and the politics of land in Africa. *Development and Change*, 28: 559–82.

Neves, K., 2010. Cashing in on Cetourism: A Critical Ecological Engagement with Dominant E-NGO Discourses on Whaling, Cetacean Conservation, and Whale Watching. *Antipode*, 42: 719–741. doi: 10.1111/j.1467-8330.2010.00770.x

Neves, K. and Igoe, J., 2012. Uneven development and accumulation by dispossession in nature conservation: comparing recent trends in the Azores and Tanzania. *Tijdschrift Voor Economische En Sociale Geografie*, 103: 164–179.

New Carbon Finance, 2009. Price Forecasting. Available from: http://www.newcarbonfinance.com/?p=services&i=fundamentals [accessed June 2009].

Newell, P. and Mulvaney, D., 2013. The political economy of the 'just transition'. *The Geographical Journal*, 179 (2): 132–140.

Newell, P., 2008. The political economy of global environmental governance. *Review of International Studies*, 34 (3): 507–529.

Newell, P., 2008b. Trade and biotechnology in Latin America: Democratization, contestation and the politics of mobilization. *Journal of Agrarian Change*, 8 (2–3): 345–376.

Newell, P., Pattberg, P., and Schroeder, H., 2012. Multiactor Governance and the Environment. *In*: Gadgil, A. and Liverman, D. M., eds. *Annual Review of Environment and Resources, Vol 37*. Palo Alto: *Annual Reviews*, 365–387.

Newing, S. and Wahl, 2004. Benefiting Local Communities? Communal Reserves in Peru. *Cultural Survival Quarterly* [online], 28 (1). Available from: http://www.culturalsurvival.org/publications/cultural-survival-quarterly/peru/benefiting-local-populations-communal-reserves-peru [accessed 1 February 2014].

Newmark, W. D. and Hough, J. L., 2000. Conserving wildlife in Africa: integrated conservation and development projects and beyond. *BioScience*, 50.

Newsham, A. J., 2002. *Participation: processes, problematics and Campfire*. CAS Occasional Paper No 70. University of Edinburgh.

Newsham, A. J., 2004. *Who's involved and in what? Participation in natural resource management institutions and 'community based tourism enterprises' in two conservancies in Northwest Namibia*. Windhoek: University of Namibia.

Newsham, A. J., 2007. *Knowing and deciding: Participation in conservation and development initiatives in Namibia and Argentina*. PhD thesis. University of Edinburgh.

Niamir-Fuller, M., 1990. *Herders' decision making in natural resource management in arid and semi-arid Africa*. Rome: FAO.

Niemelä, J. and Kotze, D. J., 2009. Carabid beetle assemblages along urban to rural gradients: A review. *Landscape and Urban Planning*, 92 (2): 65–71.

Niinemets, Ü. and Peñuelas, J., 2008. Gardening and urban landscaping: significant players in global change. *Trends in Plant Science*, 13 (2): 60–65.

Nijman, V., 2010. An overview of international wildlife trade from Southeast Asia. *Biodiversity and Conservation*, 19:1101–1114.

Norton-Griffiths, M. and Said, M. Y., 2009. The Future for Wildlife on Kenya's Rangelands: An Economic Perspective. *In*: Head, J. T. du T., Kocknager, R., and Deutsch, J. C., eds. *Wild Rangelands* [online]. John Wiley & Sons, Ltd, 367–392. Available from: http://onlinelibrary.wiley.com/doi/10.1002/9781444317091.ch14/summary [accessed 11 October 2013].

Norton-Griffiths, M., and Southey, C., 1995. The Opportunity Costs of Biodiversity Conservation in Kenya. *Ecological Economics*, 12 (2): 125–139.

Norton-Griffiths, M., 2007. How Many Wildebeest do You Need? *World Economics*, 8 (2): 41–64.

Nunez, D., Nahuelhual, L., Oyarzun, C., 2006. Forests and water: the value of native temperate forests in supplying water for human consumption. *Ecological Economics*, 58: 606–616.

O'Brien, J. and Palmer M., 2007. *The Atlas of Religion*. London: Earthscan.

Oates, J. F., 1999. *Myth and Reality in the Rain Forest: How Conservation Strategies are Failing in West Africa*. Oakland, CA: University of California Press.

Oelschlaeger, M., 1991. *The Idea of Wilderness: From Prehistory to the Age of Ecology*. New Haven, CT: Yale University Press.

Orlove, B. S. and Brush, S. B., 1996. Anthropology and the conservation of biodiversity. *Annual Review of Anthropology*, 25: 329–352.

Ormaza, P. and Bajana, F., 2008. *Territorios A'i Cofan, Siekóya pâi, Siona, Shuar y Kichwa–zona baja de la Reserva de Producción Faunística Cuyabeno* [online]. IUCN/CEESP and GTZ. Available from: http://iccaforum.org/images/stories/Database/ea%20icca%20english.pdf [accessed 13 February 2015].

Ormsby, A. and Mannle, K., 2006. Ecotourism Benefits and the Role of Local Guides at Masoala National Park, Madagascar. *Journal of Sustainable Tourism* 14 (3): 271–287.

Ormsby, A. A. and Bhagwat, S. A., 2010. 'Sacred Forests of India: A Strong Tradition of Community-based Natural Resource Management' *Environmental Conservation*, 37 (1): 1–7.

Orrnert, A., 2006. The Heart (and Soul?) of International Development: The Role of Faith in Global Poverty Reduction *Public Administration and Development* 26: 185–189.

Ostrom, E., 1990. *Governing the Commons: the evolution of institutions for collective action*. Cambridge: Cambridge University Press.

Ostrom, E., Gardner, R., and Walker, J., 1994. *Rules, games and common-pool resources*. Ann Arbor: University of Michigan Press.

Oxfam, 2010. Available from: http://www.oxfam.org.uk/oxfam_in_action/history/index.html [accessed December 2010].

Pacheco, P., 2012. *Soybean and oil palm expansion in South America: A review of main trends and implications*. Working Paper 90. CIFOR, Bogor, Indonesia. Available from: http://www.cifor.org/publications/pdf_files/WPapers/WP90Pacheco.pdf [accessed 21 January 2015].

Pagiola, S., 2008. Payments for environmental services in Costa Rica. *Ecological Economics*, 65 (4): 712–724.

Pagiola, S., Arcenas, A., and Platais, G., 2005. Can Payments for Environmental Services Help Reduce Poverty? An Exploration of the Issues and the Evidence to Date from Latin America. *World Development*, 33 (2): 237–253.

Pain, A. and Levine, S., 2012. *A conceptual analysis of livelihoods and resilience addressing the insecurity of agency*. London: Overseas Development Institute (ODI).

Palmer, M. and Finlay, V., 2003. *Faith in Conservation: New Approaches to Religions and the Environment*. Washington, D.C.: The World Bank.

Paquette, A. and Messier, C., 2010. The role of plantations in managing the world's forests in the Anthropocene. *Frontiers in Ecology and the Environment*, 8: 27–34. dx.doi.org/10.1890/080116.

Parker, T., 2009. The kidneys of Kolkata. *Geographical*, 81 (12): 38–43.

Pearce, D. W., 1991. *Blueprint 2: greening the world economy*. London: Earthscan Publications, in association with the London Environmental Economics Centre.

Pearce, F., 2015. *The New Wild: Why Invasive Species Will Be Nature's Salvation*. Boston, Mass.: Beacon Press.

Peet, R., Robbins, P., and Watts, M., 2010. *Global Political Ecology*. London; New York: Routledge.

Peiser, B., 2005. From Genocide to Ecocide: The Rape of Rapa Nui. *Energy & Environment*, 16 No. 3&4.

Pelling, M., 2011. *Adaptation to Climate Change: From Resilience to Transformation*. London; New York: Routledge.

Peluso, N. L. and Watts, M., eds., 2001. *Violent Environments*. Ithaca, NY.: Cornell University Press.

Peters, J., 1998. Transforming the Integrated Conservation and Development Project (ICDP) Approach: Observations from the Ranomafana National Park Project, Madagascar. *Journal of Agricultural and Environmental Ethics*, 11 (1): 17–47.

Peterson, G., Allen, C. R., and Holling, C. S., 1998. Ecological resilience, biodiversity, and scale. *Ecosystems*, 1 (1): 6–18.

Peterson, M. N. and Liu, J. G., 2008. Impacts of Religion on Environmental Worldviews: The Teton Valley Case. *Society and Natural Resources*, 21: 704–718.

Phillips, D. C., 1990. Postpositivistic Science – Myths and Realities. *In*: Guba, E. G., ed. *The Paradigm Dialogue*. Newbuery Park, London: SAGE.

Pimbert, M. P. and Pretty, J. N., 1995. *Parks, People and Professionals: Putting 'Participation' into Protected Area Management* [online]. Available from: http://www.ibcperu.org/doc/isis/6931.pdf.

Pinchot, G., 1998. *Breaking new ground* [online]. Commemorative. Washington, D.C.: Island Press. Available from: http://www.loc.gov/catdir/toc/fy052/98012418.html http://www.loc.gov/catdir/enhancements/fy0666/98012418-d.html [accessed 24 November 2013].

Plant, S. J., 2009. International Development and Belief in Progress. *Journal of International Development*, 21: 844–855.

Plummer, R. and Arai, S., 2005. Co-management of natural resources: opportunities for and barriers to working with citizen volunteers. *Environmental Practice*, 7: 221–234. Available from: http://dx.doi.org/10.1017/S1466046605050362

Plummer, R., Armitage, D. R., and De Loë, R. C., 2013. Adaptive Comanagement and Its Relationship to Environmental Governance. *Ecology & Society*, 18 (1): 1–15.

Pottier, J., Bicker, A., Sillitoe, P., and Association of Social, A., 2003. *Negotiating local knowledge: power and identity in development*. London: Pluto Press.

Poulsen, J., Clark, C. J., Mavah, G. and Elkan, P. W., 2009. Bushmeat supply and consumption in a tropical logging concession in Northern Congo. *Conservation Biology*, 23: 1597–1608.

Poultney, C. and Spenceley, A., 2001. *Practical strategies for pro-poor tourism, Wilderness Safaris South Africa: Rocktail Bay and Ndumu Lodge* [online]. Available from: www.propoortourism.co.uk.

Power, M. E. and Chapin, F. S., 2009. Planetary Stewardship. *Frontiers in Ecolog* *and the Environment*, 7 (8): 399.

Prendergast, D. and Adams, W. M., 2003. Colonial wildlife conservation and th origins of the society for the preservation of the wild fauna of the Empir (1903–1914). *Oryx*, 37: 251–260.

Pretty, J. (ed.), 2007. *Sustainable Agriculture and Food – History of Agricultur* *and Food*. London: Earthscan.

Pretty, J. 2002. *Agri-Culture - Reconnecting People, Land and Nature*. Londor Earthscan.

Protected Planet, 2015. *Discover and learn about protected areas*. Available from http://www.protectedplanet.net [accessed 11 February 2015].

Pyle, R. M., 2003. 'Nature matrix: reconnecting people and nature'. *Oryx*, 37 206–214.

Qu, G. and Li, J., 1994. *Population and the Environment in China*. Boulder, CO. Lynne Rienner.

Ramadori, D. E., Brown, A. D., and Grau, H. R., 1995. Agricultura migratoria e el valle del Río Barití, Santa Victoria, Salta. *In*: *Investigación, conservación* *desarrollo en selvas subtropicales de montaña*. San Miguel de Tucumár Universidad de Tucumán/Laboratorio de Investigaciones Ecológicas en la Yungas, 205–214.

Ramankutty, N., Gibbs, H. K., Achard, F., Defries, R., Foley, J. A., and Houghtor R. A., 2007. Challenges to estimating carbon emissions from tropical deforestatior *Global Change Biology*, 13: 51–66. doi: 10.1111/j.1365-2486.2006.01272.x

Ramutsindela, M. F., 2002. The perfect way to ending a painful past? Makulek land deal in South Africa. *Geoforum*, 33 (1): 15–24.

Randall, A., 1988. 'What mainstream economists have to say about the value c biodiversity'. In W.E.O. (ed.), *Biodiversity*. Washington, D.C.: Nationa Academy Press, 217–223.

Rangarajan, M. and Shahabuddin, G., 2006. Displacement and relocation fror protected areas: Towards a biological and historical synthesis. *Conservatio* *and Society*, 4 (3): 359–78.

Raworth, K., 2012. *A Safe and Just Space for Humanity: Can we live within th* *doughnut?* Discussion paper, Oxfam International.

Raymond, L., 2006. Environmentality: Technologies of government and th making of subjects. *Comparative Political Studies*, 39 (2): 261–265.

Redclift, M. and Sage, C., 1995. *Strategies for sustainable development: loca* *agendas for the Southern Hemisphere*. Chichester; New York: J. Wiley & Son

Redclift, M., 1987. *Sustainable development: exploring the contradictions*. Londor Methuen.

Redclift, M., 2005. Sustainable development (1987–2005): An oxymoron come of age. *Sustainable Development*, 13: 212–227.

Reddy, S. R. C. and Chakravarty, S. P., 1999. Forest Dependence and Incom Distribution in a Subsistence Economy: Evidence from India. *Worl* *Development*, 27 (7): 1141–1149.

Redford, K. H., and Adams W. M., 2009. Payment for ecosystem services and th challenge of saving nature. *Conservation Biology*, 23: 785–787.

Redford, K. H., Brandon, K., and Sanderson, S. E., 1998. Holding Ground. *I* *Parks in Peril: People, Politics and Protected Areas*. Washington, D.C.: Islan Press, 455–464.

Redford, K. H., Levy, M. A., Sanderson, E. W., and de Sherbinin, A., 2008. What is the role for conservation organizations in poverty alleviation in the world's wild places? *Oryx*, 42 (4): 516–528.

Redman, C. and Jones, N. S., 2005. The environmental, social and health dimensions of urban expansion. *Population and Environment*, 26 (6): 505–520.

Rees, W. E., 2002. 'An Ecological Economics Perspective on Sustainability and Prospects for Ending Poverty'. *Population and Environment*, 24: 15–46.

Reeve, R. and Ellis, S., 1995. An insider's account of the South African security force's role in the ivory trade. *Journal of Contemporary African Studies*, 13 (2): 222–243.

Reij, C., 1991. *Indigenous soil and water conservation in Africa*. London: IIED.

Reis, A. Heitor and Miguel, Antonio F., 2009. Sustainable Energy Sources and Developing Clean Energy Technologies. *International Journal of Green Energy*, 6 (3): 229.

Resilience Alliance, 2002. *Resilience*. Available from: http://www.resalliance.org/index.php/resilience [accessed 21 January 2015].

Resnick, D., Tarp, F., and Thurlow, J., 2012. The Political Economy of Green Growth: Cases from Southern Africa. *Public Administration and Development*, 32 (3): 215–228.

Richards, P., 1985. *Indigenous agricultural revolution: ecology and food production in West Africa*. London: Hutchinson.

Richards, P. D., Myers, R. J., Swinton, S. M., and Walker, R. T., 2012. Exchange rates, soybean supply response, and deforestation in South America. *Global Environmental Change*, 22 (2): 454–462.

Richardson, B. C., 1997. *Economy and Environment in the Caribbean: Barbados and the Windwards in the late 1800s*. Barbados: University of West Indies Press.

Ridder, B., 2007. The naturalness versus wildness debate: ambiguity, inconsistency, and unattainable objectivity. *Restoration Ecology*, 15 (1): 8–12.

Rigg, J., 2006. Land, farming, livelihoods, and poverty: Rethinking the links in the Rural South. *World Development*, 34: 180–202.

Rio Tinto, 2012. *Sustainable development report 2012* [online]. Madagascar: Rio Tinto Madagascar and QIT Madagascar Minerals SA. Available from: http://www.riotintomadagascar.com/pdf/rdd12en.pdf [accessed 1 March 2014].

Rist, L., Shaanker, R. U., Milner-Gulland, E. J., and Ghazoul, J., 2010. The Use of Traditional Ecological Knowledge in Forest Management: an Example from India. *Ecology and Society* [online], 15. Available from: ://WOS:000277933900002.

Robalino, J. and Pfaff, A., 2013. Ecopayments and Deforestation in Costa Rica: A Nationwide Analysis of PSA's Initial Years. *Land Economics*, 89 (3): 432–448.

Robalino, J., Pfaff, A., Sánchez-Azofeifa, A., Alpízar, F., León, C., and Rodríguez, C., 2008. *Deforestation Impacts of Environmental Services Payments: Costa Rica's PSA Program 2000–2005* [online]. Resources for the Future. Environment for Development Discussion Paper Series. Available from: http://www.rff.org/RFF/Documents/EfD-DP-08-24.pdf [accessed 24 February 2015].

Robbins, P., 2006. Environmentality: Technologies of government and the making of subjects. *Geographical Review*, 96 (4): 715–718.

Robbins, P., 2012. *Political Ecology: A Critical Introduction*. London: John Wiley & Sons.

Robert, M., Hulten, P. and Frostell, B., 2007. Biofuels in the energy transition beyond peak oil. A macroscopic study of energy demand in the Stockholm transport system 2030. *Energy*, 32 (11): 2089–2098.

Roberts, C., 2009. *The Unnatural History of the Sea*. Washington: Island Press.

Roberts, J. T., 2011. Multipolarity and the new world (dis)order: US hegemonic decline and the fragmentation of the global climate regime. *Global Environmental Change-Human and Policy Dimensions*, 21 (3): 776–784.

Robins, S. and van der Waal, K., 2008. Model tribes' and iconic conservationists? The Makuleke restitution case in Kruger National Park. *Development and Change*, 39 (1): 53–72.

Rocheleau, D., 1995. More on Machakos. *Environment*, 37 (7): 3–5.

Rockström, J., Steffen, W., Noone, K., Persson, Å., Chapin, III, F. S., Lambin, E. F., Lenton, T. M., Scheffer, M., Folke, C., Schellnhuber, H. J., Nykvist, B., de Wit, C. A., Hughes, T., van der Leeuw, S., Rodhe, H., Sörlin, S., Snyder, P. K. Costanza, V., Svedin, U., Falkenmark, M., Karlberg, L., Corell, R. W., Fabry V. J., Hansen, J., Walker, B., Liverman, D., Richardson, K., Crutzen, P., and Foley, J. A., 2009. A safe operating space for humanity. *Nature*, 461: 472–475. doi:10.1038/461472a.

Roe D., Nelson, F., Sandbrook, C. (eds.), 2009. Community management of natural resources in Africa: Impacts, experiences and future directions. *Natural Resource Issues No. 18*, London: International Institute for Environment and Development.

Roe, D. and Elliott, J., 2010. *The Earthscan Reader in Poverty and Biodiversity Conservation*. London: Earthscan.

Roe, D. and Elliott, J., 2004. Poverty reduction and biodiversity conservation: rebuilding the bridges. *Oryx*, 38 (2): 137–139.

Roe, D. and Elliott, J., 2010. Biodiversity conservation and poverty reduction: an introduction to the debate. *In*: Roe, D. and Elliott, J., eds. *The Earthscan reader in poverty and biodiversity conservation*. London: Earthscan, 1–11.

Roe, D., 2008. The origins and evolution of the conservation-poverty debate: a review of key literature, events and policy processes. *Oryx*, 42 (4): 491–503.

Rohde, R. F., 1997. Looking into the past: interpretations of vegetation change in Western Namibia based on matched photography. *Dinteria*, 24: 121–149.

Rome, A., 2003. 'Give Earth a Chance': The Environmental Movement and the Sixties. *The Journal of American History*, 90 (2): 525–554.

Rootes, C., 2004. Environmental Movements. *In*: Snow, D. A., Soule, S. A., and Kriesi, H. (eds), *The Blackwell Companion to Social Movements*. Oxford Wiley.

Rosenzweig, M., 2003. Reconciliation ecology and the future of species diversity. *Oryx*, 37: 194–205.

Rosset, P. M. and Martínez-Torres, M. E., 2012. Rural social movements and agroecology: context, theory, and process. *Ecology and Society*, 17 (3): 17. Available from: http://dx.doi.org/10.5751/ES-05000-170317.

Ros-Tonen, M. A. F., 2003. Globalisation, localisation and tropical forest management: introducing the challenge of new markets and partnerships. *ETFRN News*, 39–40 (Autumn/Winter 2003):7–9.

Roy, S. D., Jackson, P., and Kemf, E., 1993. Mayhem in Manas: The threats to India's wildlife reserves. *In*: *Indigenous peoples and protected areas - the law of Mother Earth*. London: Earthscan.

Rozzi, R., Massardo, F., Cruz, F., Crenier, C., Muñoz, A., Mueller, E., and Elbers, J., 2010. Galápagos and Cape Horn: Ecotourism or Greenwashing in Two Iconic Latin American Archipelagoes? *Environmental Philosophy*, 7 (2): 1–32.

Ruddle, K., 1994. Local knowledge in the folk management of fisheries and coastal marine environments. *In*: Dyer, C. L. and McGoodwin, J. R., eds. *Folk management in the world's fisheries: lessons for modern fisheries management*. Niwot: University Press of Colorado, 161–206.

Rudel, T. K. and Horowitz, B., 2013. *Tropical Deforestation: Small Farmers and Land Clearing in Ecuadorian Amazon*. New York: Columbia University Press.

Runte, A., 2010. *National parks: the American experience*. 4th ed. Lanham, Md.: Taylor Trade Pub.

Ruse, M., 2005. *The Evolution-Creation Struggle*. Cambridge, MA: Harvard University Press.

Rutherford, P., 1999. The entry of life into history. *In*: *Discourses of the Environment*. Oxford: Blackwell, 37–62.

Rutten, M. M. E. M., 2002. *Parks beyond parks: genuine community-based wildlife eco-tourism or just another loss of land for Maasai pastoralists in Kenya?* [online]. 27. Available from: https://openaccess.leidenuniv.nl/handle/1887/9462 [accessed 16 March 2015].

Rydin, Y., 2010. *Governing for Sustainable Urban Development*. Earthscan Canada.

Saad Filho, A., 2010. Neoliberalism in Crisis: A Marxist Analysis. *Marxism*, 21: 235–257.

Sachs, J. D. and Warner, A. M., 2000. Natural resources and economic development: the curse of natural resources. *European Economic Review*, 45: 827–838.

Sachs, J., Woo, W. T., Fischer, S., and Hughes, G., 1994. Structural factors in the economic reforms of China, Eastern Europe, and the former Soviet Union. *Economic Policy*, 9: 101–145.

Sachs, W., 1992. *The development dictionary: a guide to knowledge as power*. London; Atlantic Highlands, N.J: Zed Books.

Sagoff, M., 2008. On the economic value of ecosystem services. *Environmental Values*, 17: 239–257.

Sahlins, M., Bargatzky, T., Bird-David, N., Clammer, J., Hamel, J., Maegawa, K., and Siikala, J., 1996. The sadness of sweetness: the native anthropology of Western cosmology. *Current Anthropology*, 37: 395–428.

Sakata, H. and Prideaux, B., 2013. An alternative approach to community-based ecotourism: a bottom-up locally initiated non-monetised project in Papua New Guinea. *Journal of Sustainable Tourism*, 21 (6): 880–899.

Salafsky, N., 2011. Integrating development with conservation: A means to a conservation end, or a mean end to conservation? *Biological Conservation*, 144: 973–978.

Salafsky, N., Cauley, H., Balachander, G., Cordes, B., Parks, J., Margoluis, C., Bhatt, S., Encarnacion, C., Russell, D., and Margoluis, R., 2001. A systematic test of an enterprise strategy for community-based biodiversity conservation. *Conservation Biology*, 15 (6): 1585–1595.

Salisbury, D. S. and Schmink, M., 2007. Cows versus rubber: Changing livelihoods among Amazonian extractivists. *Geoforum*, 38 (6): 1233–1249.

Salles-Reese, V., 1997. *From Viracocha to the Virgin of Copacabana: Representation of the Sacred at Lake Titicaca*. Austin, Texas: University of Texas Press.

Samii, C., Lisiecki, M., Kulkarni, P., Paler, L., and Chavis, L., 2014. Effects of Payment for Environmental Services (PES) on Deforestation and Poverty in Low and Middle Income Countries: A Systematic Review. *Campbell Systematic Reviews*, 11, DOI: 10.4073/csr.2014.11.

Samset, I., 2002. Conflict of interests or interests of conflict? Diamonds & war in the DRC. *Review of African Political Economy*, 93/94: 463–480.

Sandbrook, C. G., 2010a. Putting leakage in its place: The significance of retained tourism revenue in the local context in Rural Uganda. *Journal of International Development*, 22 (1): 124–136.

Sandbrook, C. G., 2010b. Local economic impact of different forms of nature-based tourism. *Conservation Letters*, 3 (1): 21–28.

Sant'Egidio, 2010. *Community of Sant'Egidio*, Available from: http://www.santegidio.org/ [accessed December 2010].

Santilli, M., Moutinho, P., Schwartzman, S., Nepstad, D., Curran, L., and Nobre, C., 2005. Tropical Deforestation and the Kyoto Protocol. *Climatic Change*, 7 (3): 267–276.

Schellhorn, M., 2010. Development for whom? Social justice and the business of ecotourism. *Journal of Sustainable Tourism*, 18 (1): 115–135.

Scherr, S. J., White, A., Kaimowitz, D., 2004. *A new agenda for forest conservation and poverty alleviation: making markets work for low-income producers.* Washington, DC: Forest Trends and CIFOR. 160p. ISBN: 0-9713606-6-9.

Schmitt, P. J., 1969. *Back to Nature: The Arcadian Myth in Urban America.* New York: Oxford University Press.

Scholte, P., 2003. Immigration: A Potential Time Bomb under the Integration of Conservation and Development *Ambio*, 32 (1): 58–64.

Schuurman, F. J., 2000. Paradigms Lost, Paradigms Regained? Development Studies in The Twenty-first Century, *Third World Quarterly*, 21: 7–20.

Schwartzman, S. and Zimmerman, B., 2005. Conservation alliances with indigenous peoples of the Amazon. *Conservation Biology*, 19 (3): 721–727.

Scialabba, N. E. H., 2007. Organic Agriculture and Food Security. *Food and Agriculture Organisation of the United Nations*, Available from: ftp://ftp.fao.org/docrep/fao/meeting/012/ah952e.pdf [accessed December 2010].

Scwartz, D., Deligiannis, T., and Dixon, H. T., 2001. The environment and violent conflict. *In*: Diehl, P. F. and Gleditsch, N. P., eds. *Environmental Conflic.* Boulder, CO.: Westview Press.

SDC and WFP, 2011. *Building resilience: bridging food security, climate change adaptation and disaster risk reduction.* [online]. Swiss Agency for Development and Cooperation (SDC) and World Food Programme (WFP). Available from http://www.preventionweb.net/files/24163_workshopbuildingresiliencecasestud. pdf [accessed 26 February 2015].

Seales, L. and Stein, T., 2012. Linking commercial success of tour operators and agencies to conservation and community benefits in Costa Rica. *Environmental Conservation*, 39 (1): 20–29.

Seeland, K., 2000. National park policy and wildlife problems in Nepal and Bhutan. *Population Environment*, 22: 43–62.

Seixas, C. S. and Berkes, F., 2008. Dynamics of social–ecological changes in a lagoon fishery in southern Brazil. *In*: Berkes, F., Colding, J., and Folke, C., eds. *Navigating social–ecological systems: building resilience for complexity and change.* Cambridge: Cambridge University Press, 271–298.

Selinger, L. 2004. The Forgotten Factor: Relationship Between Religion and Development. *Social Compass*, 51: 523–543.

Sen, A., 1999. *Development as freedom*. Oxford: Oxford University Press.

Senate Department for Urban Development and the Environment, 2011. *Stadtentwicklungsplan Klima Urbane Lebensqualität im Klimawandel sichern.* Berlin: der Senat von Berlin.

Seyfang, G., 2003. Environmental mega-conferences—from Stockholm to Johannesburg and beyond. *Institutions / Global Environmental Change*, 13: 223–228.

Shambaugh, J., Oglethorpe, J., and Ham, R., 2001. *The Trampled Grass: Mitigating the impacts of armed conflict on the environment.* Washington DC: Biodiversity Support Program.

Shapiro, J., 2012. *China's Environmental Challenges*. Cambridge; Malden, MA: Polity Press.

Sheikh, K. M., 2006. Involving Religious Leaders in Conservation Education in the Western Karakorum, Pakistan. *Mountain Research and Development*, 26: 319–322.

Shiva, V., 1991. *The Violence of the Green Revolution: Third World Agriculture, Ecology, and Politics*. London: Zed Books.

Siipi, H., 2008. Dimensions of naturalness. *Ethics & the Environment*, 13 (1): 71–103.

Simmie, J. and Martin, R., 2010. The economic resilience of regions: towards an evolutionary approach. *Cambridge Journal of Regions, Economy and Society* 2010, 3 (1): 27–43. doi:10.1093/cjres/rsp029

Simon, H. A., 1957. *Models of Man, Social and Rational*. New York: Wiley.

Simon, J. L. and Kahn, H., 1984. *The Resourceful earth: a response to Global 2000*. Oxford: Blackwell.

Singh, S., 2002. Contracting Out Solutions: Political Economy of Contract Farming in the Indian Punjab. *World Development*, 30: 1621–1638.

Singh, V. S., Pandey, D. N., and Prakash, N. P., 2011. What determines the success of joint forest management? Science-based lessons on sustainable governance of forests in India. *Resources, Conservation and Recycling*, 56 (1): 126–133.

Sklair, L., 2001. *The Transnational Capitalist Class*. Oxford: Wiley.

Slikkerveer, L. J., Brokensha, D., Dechering, W., and Warren, D. M., 1995. *The cultural dimension of development: indigenous knowledge systems*. London: Intermediate Technology Publications.

Sørensen, E. and Torfing, J., 2008. *Theories of democratic network governance.* Basingstoke: Palgrave Macmillan.

Soule, M., 2013. The "New Conservation". *Conservation Biology*, 27 (5): 895–897.

Spash, C. L., 2000. 'Ecosystems, contingent valuation and ethics: the case of wetlands re-creation'. *Ecological Economics*, 34 (2): 195–215.

Spash, C. L., 2006. Non-economic motivation for contingent values: rights and attitudinal beliefs in the willingness to pay for environmental improvements. *Land Economics*, 82 (4): 602–622.

Spash, C. L., 2008. 'Contingent valuation design and data treatment: if you can't shoot the messenger, change the message'. *Environment & Planning C: Government & Policy*, 26 (1): 34–53.

Spash, C. L., 2008a. 'How much is that ecosystem in the window? The one with the bio-diverse trail'. *Environmental Values*, 17: 259–284.

Spenceley, A., and Seif, J., 2003. *Strategies, Impacts and Costs of Pro-Poo Tourism Approaches in South Africa* [online]. Available from: www propoortourism.org.uk.

Spenceley, A., 2010. Local impacts of community-based tourism in Southern Africa. *In*: Spenceley, A., ed. *Responsible Tourism: Critical Issues fo Conservation and Development*. London; Sterling, VA: Routledge, 285–304.

Spierenburg, M. and Wels, H., 2010. Conservative Philanthropists, Royalty and Business Elites in Nature Conservation in Southern Africa. *Antipode*, 42 647–670. doi: 10.1111/j.1467-8330.2010.00767.x

Stacey, N., Izurieta, A., and Garnett, S. T., 2013. Collaborative measurement o performance of jointly managed protected areas in northern Australia. *Ecolog and Society*, 18 (1): 19.

Standish, A., 2009. A Review of Ecotourism, NGOs and Development. *Professiona Geographer*, 61 (2): 270–272.

Steinberg, P. F., 2001. *Environmental Leadership in Developing Countries Transnational Relations and Biodiversity Policy in Costa Rica and Bolivia* Cambridge, Mass.; London: MIT Press.

Stevens, S. (ed.), 1997. *Conservation through Cultural Survival: indigenou peoples and protected areas*. Washington, D.C.: Island Press.

Stocks, A., 2005. Too much for too few: Problems of indigenous land rights i Latin America. *In: Annual Review of Anthropology*, 85–104.

Strassburg et al., 2010. Global Congruence of Carbon Storage and Biodiversity i Terrestrial Ecosystems *Conservation Letters*, doi: 10.1111/j.1755 263X.2009.00092.x

Stringer, L. C., Dyer, J. C., Reed, M. S., Dougill, A. J., Twyman, C., an Mkwambisi, D., 2009. Adaptations to climate change, drought an desertification: local insights to enhance policy in southern Africa *Environmental Science & Policy*, 12 (7): 748–765.

Stronza, A. L., 2009. Commons management and ecotourism: Ethnographi evidence from the Amazon. *International Journal of the Commons*, 4 (1): 56–7.

Suich, H., 2013. Evaluating the household level outcomes of community base natural resource management: the Tchuma Tchato Project and Kwand Conservancy. *Ecology and Society*, 18 (4): 25.

Sullivan, J. J., Timmins, S. M., and Williams, P. A., 2005. Movement of exoti plants into coastal native forests from gardens in northern New Zealand. *Ne Zealand Journal of Ecology*, 29 (1): 1–10.

Sullivan, S., 1999. The Impacts of People and Livestock on Topographicall Diverse Open Wood- and Shrub-Lands in Arid North-West Namibia. *Globa Ecology and Biogeography*, 8: 257–277.

Sullivan, S., 2000. Getting the science right, or, introducing the science in the fir place? *In*: Stott, P. and Sullivan, S., eds. *Political Ecology: Science, Myth an Power*. New York: Arnold.

Sullivan, S., 2006. Elephant in the Room? Problematising 'New' (Neolibera Biodiversity Conservation. *Forum for Development Studies*, 33 (1): 105–135

Sullivan, S., Burgland, E., and Anderson, D., 2003. Protest, conflict and litigatio dissent or libel in resistance to a conservancy in North-West Namibia. *In Ethnographies of environmental under-privilege: Anthropological encounter with conservation*. Oxford: Berghahn Press.

Sunderlin, W. D., Angelsen, A., Belcher, B., Burgers, P., Nasi, R., Santoso, L., and Wunder, S., 2005. Livelihoods, forests, and conservation in developing countries: An Overview. *World Development*, 33 (9): 1383–1402.

Taramarcaz, P., Lambelet, C., Clot, B., Keimer, C., and Hauser, C., 2005. Ragweed (Ambrosia) progression and its health risks: will Switzerland resist this invasion? *Swiss Medical Weekly*, 135: 538–548.

Tarrow, S., 2005. *The New Transnational Activism*. Cambridge: Cambridge University Press.

Taylor, B., 1995. *Ecological Resistance Movements*. New York: State University of New York Press.

Taylor, B., 2005. *Encyclopedia of Religion and Nature*. New York: Continuum.

Taylor, B., 2010. *Dark green religion: nature spirituality and the planetary future*. Berkeley, CA: University of California Press.

TEEB, 2010. *The Economics of Ecosystems and Biodiversity: Mainstreaming the Economics of Nature: A Synthesis of the Approach, Conclusions and Recommendations of TEEB*. Available from: http://www.teebweb.org/publication/mainstreaming-the-economics-of-nature-a-synthesis-of-the-approach-conclusions-and-recommendations-of-teeb/ [accessed 08 October 2015].

Terborgh, J., 2004. *Requiem for Nature*. New York: Island Press.

Terborgh, J. C., Van Schaik, C., Davenport, L., and Rao, M. (eds.) 2002. *Making Parks Work: strategies for preserving tropical nature*. Washington DC: Island Press.

Terraviva, 2012. *Green Economy, the New Enemy | TERRAVIVA Rio + 20* [online]. Available from: http://www.ips.org/TV/rio20/green-economy-the-new-enemy/ [accessed 26 February 2015].

Thomas, S., 1995. *The legacy of dualism in decision-making within CAMPFIRE*. London: IIED. No. 4.

Thompson, B. and Coyle, J., 2005. *Trade Issues in Sustainable Tourism Certification: An examination of the constraints imposed by international trade rules and organizations' (NAFTA, WTO, etc.), barriers to trade* [online]. Stanford: The International Ecotourism Society. Available from: http://www.google.co.uk/url?sa=t&rct=j&q=&esrc=s&source=web&cd=2&cad=rja&uact=8&ved=0CCgQFjAB&url=http%3A%2F%2Fwww.gstcouncil.org%2Fdocs%2Fcategory%2F22-iso-technical-committee-228-on-tourism.html%3Fdownload%3D178%3Ainternational-trade-restraints-on-sustainable-tourism&ei=0JkGVdXEBIzUave0gcAE&usg=AFQjCNHbphHq0hnbd2E6KrcoFoGi-skp9Q&bvm=bv.88198703,d.d2s [accessed 16 March 2014].

Thynne, I., 2008. Symposium introduction - climate change, governance and environmental services: institutional perspectives, issues and challenges. *Public Administration and Development*, 28 (5): 327–339.

Tienhaara, K., Orsini, A., and Falkner, R., 2012. Global corporations. *In:* Biermann, F. and Pattberg, P., eds. *Global environmental governance reconsidered*. Cambridge (Mass.); London: M I T Press, 45–68.

Tietenberg, T., 1991. *Environmental and Natural Resource Economics*. New York: Harper Collins.

Tiffen, M., Mortimore, M., and Gichuki, F., 1994. *More people, less erosion: environmental recovery in Kenya*. Chichester: J. Wiley.

Tilman, D., Cassman, K. G., Matson, P. A., Naylor, R., and Polasky, S., 2002 Agricultural Sustainability and Intensive Production Practices. *Nature*, 418 671–677.

Timko, J. A. and Satterfield T., 2008. Seeking Social Equity in National Parks Experiments with Evaluation in Canada and South Africa. *Conservation & Society*, 6 (3): 238–254.

Toledo, V. M., 1992. Biodiversity and indigenous peoples. *Etnoecologica*, 1 (1) 5–21.

Tomalin, E., 2002. The Limitations of Religious Environmentalism for India *Worldviews*, 6: 12–30.

Tomalin, E., 2004. Bio-divinity and Biodiversity: Perspectives on Religion and Environmental Conservation in India. *Numen*, 51: 265–295.

Tomalin, E., 2009. *Bio-divinity and Biodiversity: the Limits of Religiou. Environmentalism*. Aldershot: Ashgate.

Trejos, B. and Chiang, L.-H. N., 2009. Local economic linkages to community based tourism in rural Costa Rica. *Singapore Journal of Tropical Geography* 30 (3): 373–387.

Trương, T.-Đ., 1990. *Sex, money, and morality: prostitution and tourism in Southeast Asia /*. London: Zed Books.

Tscharntke, T., Clough, Y., Bhagwat, S. A., Buchori, D., Faust, H., Hertel, D. Hölscher, D., Juhrbandt, J., Kessler, M., Perfecto, I., Scherber, C., Schroth, G. Veldkamp, E., and Wanger, T. C., 2011. Multifunctional shade-tree managemen in tropical agroforestry landscapes - a review. *Journal of Applied Ecology*, 48 619–629.

Tucker, M. E. and Grim, J., 2004. *Overview of World Religions and Ecology Forum on Religion and Ecology*, Available from: http://fore.research.yale.edu religion/ [accessed December 2010].

Turner, B. L., Matson, P. A., McCarthy, J. J., Corell, R. W., Christensen, L. Eckley, N., Hovelsrud-Broda, G. K., Kasperson, J. X., Kasperson, R. E., Luers, A Martello, M. L., Mathiesen, S., Naylor, R., Polsky, C., Pulsipher, A., Schiller A., Selin, H., and Tyler, N., 2003. Illustrating the coupled human-environmen system for vulnerability analysis: Three case studies. *Proceedings of th National Academy of Sciences of the United States of America*, 100 (14) 8080–8085.

Turner, N. J., 2007. *The earth's blanket: traditional teachings for sustainabl living*. Seattle: University of Washington Press.

Turner, R. K. and Daily, G. C., 2008. The ecosystem services framework and natura capital conservation. *Environmental and Resource Economics*, 39: 25–35.

Turner, W. R., Brandon, K., Brooks, T. M., Gascon, C., Gibbs, H. K., Lawrence K., Mittermeier, R. A., and Selig, E. R., 2013. The potential, realised an essential ecosystem service benefits of biodiversity conservation. *In*: Roe, D and Elliott, J., eds. *Biodiversity conservation and poverty alleviation: explorin the evidence for a link*. Chennai, India & Malaysia: Wiley-Blackwell, 21–35.

Tversky, A., and Kahneman, D., 1991. 'Loss aversion in riskless choice a reference dependent model'. *The Quarterly Journal of Economics*, 10(1039–1061.

UN, 2000. *The Millennium Development Goals*. Available from: http://www unmillenniumproject.org/goals/ [accessed 20 January 2015].

UN, 2002a. *Frequently Asked Questions about the Johannesburg Summit* [online]. Available from: http://www.johannesburgsummit.org/html/basicinfo/faqs.html.

UN, 2002b. *Johannesburg Declaration on Sustainable Development* [online]. Available from: http://www.johannesburgsummit.org/html/documents/summit docs/1009wssdpoldeclaration.doc [accessed 4 October 2006].

UNDESA, 2012. *Sustainable Development in the 21st century (SD21) Review of Implementation of Agenda 21 and the Rio Principles* [online]. United Nations Department of Economic and Social Affairs Division for Sustainable Development. Available from: https://sustainabledevelopment.un.org/content/documents/641Synthesis_report_Web.pdf [accessed 3 May 2015].

UNDESA, 2013. *World Population Prospects: The 2012 Revision, Highlights and Advance Tables.* [online]. Washington DC: United Nations Department of Economic and Social Affairs, Population Division. Available from: http://esa.un.org/wpp/documentation/publications.htm.

UNEP (United Nations Environmental Program, 2010. Available from: http://www.un-foodsecurity.org/node/447 [accessed December 2010].

UNEP, 1978. *Review of the areas of environment and development and environmental management.* UNEP. No. 3.

UNEP, 2011. *Towards a Green Economy: Pathways to Sustainable Development and Poverty Eradication - A Synthesis for Policy Makers* [online]. Available from: www.unep.org/greeneconomy.

UNEP, 2012. *Geo 5: Global Environment Outlook, Environment for the Future We Want* [online]. Nairobi, Kenya: United Nations Environment Programme. Available from: http://www.unep.org/geo/geo5.asp [accessed 5 March 2015].

UNEP, 2013. *Green Economy and Trade – Trends, Challenges and Opportunities* [online]. UNEP. Available from: http://www.unep.org/greeneconomy/Portals/88/GETReport/pdf/FullReport.pdf [accessed 16 March 2015].

UNEP, CITES, IUCN, and TRAFFIC, 2013. *Elephants in the Dust – the African Elephant Crisis.* [online]. United Nations Environment Programme (UNEP). A Rapid Response Assessment. Available from: http://www.unep.org/pdf/RRAivory_draft7.pdf.

UNFPA, 2007. *State of world population 2007: unleashing the potential of urban growth.* United Nations Population Fund. Available from: http://www.unfpa.org/sites/default/files/pub-pdf/695_filename_sowp2007_eng.pdf [accessed 20 January 2015].

UNU, 2008. World Income Inequality Database V2.0C May 2008. Available from: http://www.wider.unu.edu/research/Database/en_GB/database/ [accessed 26 January 2015].

UNWTO, 2006. *UNWTO Annual Report 2005* [online]. Madrid: UNWTO. Available from: https://www.wto.org/english/res_e/booksp_e/anrep_e/anrep06_e.pdf [accessed 15 March 2015].

UNWTO, 2013. *UNWTO Annual Report 2012* [online]. Madrid: UNWTO. Available from: http://www2.unwto.org/publication/unwto-annual-report-2012 [accessed 15 March 2015].

USAID, 2012. *Building resilience to recurrent crisis USAID policy and program guidance.* [online]. US Agency for International Aid. Available from: http://www.usaid.gov/sites/default/files/documents/1866/Policy%20%26%20

Program%20Guidance%20-%20Building%20Resilience%20to%2(
Recurrent%20Crisis_Dec%202012.pdf [accessed 12 April 2013].

Van Houtan, K. S., 2006. Conservation as Virtue: a Scientific and Social Proces.
for Conservation Ethics. *Conservation Biology*, 20: 1367–1372.

Van Kersbergen, K. and van Waarden, F., 2004. 'Governance' as a bridge betweei
disciplines: Cross-disciplinary inspiration regarding shifts in governance an(
problems of governability, accountability and legitimacy. *European Journal o,
Political Research*, 43 (2): 143–171.

Van Noordwijk, M. and Leimona, B., 2010. Principles for fairness and efficiency
in enhancing environmental services in Asia: payments, compensation, o(
co-investment? *Ecology and Society*, 15 (4): 17. [online] Available from: http:/
www.ecologyandsociety.org/vol15/iss4/art17/

Van Noordwijk, M., Leimona, B., Jindal, R., Villamor, G. B., Vardhan, M.,
Namirembe, S., Delia, C., Kerr, J., Minang, P. A., and Tomich, T. P., 2012
Payments for Environmental Services: Evolution Toward Efficient and Fai
Incentives for Multifunctional Landscapes. *In*: Gadgil, A. and Liverman, D. M.
eds. *Annual Review of Environment and Resources, Vol 37* [online]. 389–420
Available from: ://WOS:000310224900017.

Ver Beek, K. A., 2002. 'Spirituality: a Development Taboo' In Eade D. (ed.
Development and Culture: Selected Essays from Development in Practice. Oxford
Oxfam GB in association with World Faiths Development Dialogue: 58–75.

Verschuuren, B., 2007. *Believing is Seeing: Integrating Cultural and Spiritua
Values in Conservation Management*. Gland, Switzerland: Foundation fo(
Sustainable Development; EarthCollective; IUCN.

Verstraete, M. M., Brink, A. B., Scholes, R. J., Beniston M. and Smith M. S.
2008. Climate change and desertification: Where do we stand, where should w
go? *Global and Planetary Change*, 64: 105–110.

Victor, P. A. and Jackson, Ti., 2012. A commentary on UNEP'S green econom.
scenarios. *Ecological Economics*, 77: 11–15.

Vigne, L. and Martin E., 2000. Price for Rhino Horn Increases in Yeme
Pachyderm, 28: 91–100.

Vilaca, A. and Wright, R. M., 2009. *Native Christians: Modes and Effects c
Christianity Among Indigenous Peoples of the Americas*. VT: Burlington.

Vira, B. and Kontoleon, A., 2013. Dependence of the poor on biodiversity: whic
poor, what biodiversity? *In*: Roe, D. and Elliott, J., eds. *Biodiversity conservatio,
and poverty alleviation: exploring the evidence for a link*. Chennai, India &
Malaysia: Wiley-Blackwell, 52–84.

Vitali, S., Glattfelder, J. B., and Battiston, S., 2011. The Network of Globa
Corporate Control. *PLoS ONE* 6 (10): e25995. doi:10.1371/journal.pone.002599

Vitousek, P. M., Mooney, H. A., Lubchenco, J., and Melillo, J. M., 1997. Huma
domination of earth's ecosystems. *Science*, 277: 494–504.

Vogel, S., 2003. The nature of artifacts. *Environmental Ethics*, 25: 149–168.

Von Frantzius, I., 2004. World Summit on Sustainable Development Johannesbur
2002: A Critical Analysis and Assessment of the Outcomes. *Environmentc
Politics*, 13: 467–473.

Votrin, V., 2005. The Orthodoxy and Sustainable Development: A Potential fc
Broader Involvement of the Orthodox Churches in Ethiopia and Russi:
Environment Development and Sustainability, 7: 9–21.

Wackernagel, M., and Rees, W., 1998. *Our Ecological Footprint: Reducing Human Impact on the Earth*. Vancouver: New Society Publishers.

Wackernagel, M., Kitzes, J., Moran, D., Goldfinger, S., and Thomas, M., 2006. The ecological footprint of cities and regions: comparing resource availability with resource demand. *Environment and Urbanisation*, 18 (1): 103–112.

Waithaka, J., 2002. *The Role of Community Wildlife-based Enterprises in Reducing Human Vulnerability and Environmental Degradation: The Case Of Il Ngwesi Ecotourism Project, Kenya*. African Conservation Center.

Walker, B., Carpenter, S., Anderies, J., Abel, N., Cumming, G. S., Jansson, M., Lebel, L., Norberg, J., Peterson, G. D., and Pritchard, R., 2002. Resilience management in social-ecological systems: a working hypothesis for a participatory approach. *Conservation Ecology*, 6 (1): 14. [online]. Available from: http://www.consecol.org/vol6/iss1/art14/. *Conservation Ecology*, 6 (1): 14.

Walker, B., Holling, C. S., Carpenter, S. R., and Kinzig, A., 2004. Resilience, adaptability and transformability in social-ecological systems. *Ecology and Society*, 9 (2): 5.

Wallace, D. R., 1992. *The Quetzal and the Macaw: The Story of Costa Rica's National Parks*. San Francisco, CA: Sierra Club Books.

Wallace, K. J., 2007. Classification of ecosystem services: problems and solutions. *Biological Conservation*, 139: 235–246.

Wallace, K. J., 2008. Ecosystem services: multiple classifications or confusion? *Biological Conservation*, 141: 353–354.

Walls, L. D., 2009. *The passage to Cosmos: Alexander von Humboldt and the shaping of America*. Chicago, Ill.: University of Chicago Press; Bristol: University Presses Marketing [distributor].

Waltner-Toews, D., Kay, J. J., Neudoerffer, C., and Gitau, T., 2003. Perspective changes everything: managing ecosystems from the inside out. *Frontiers in Ecology and the Environment*, 1 (1): 23–30.

Wapner, P. K., 1996. *Environmental Activism and World Civic Politics: The Cultural and Political Landscape of Anishinaabe Anti-Clearcutting Activism*. SUNY Press.

Watts, D., 1987. *The West Indies: Patterns of Development, Culture and Environmental Change since 1492*. Cambridge: Cambridge University Press.

WCC (World Council of Churches), 2010. Available from: http://www.oikoumene.org/ [accessed December 2010].

WCED, 1987. *Our common future*. Oxford: Oxford University Press.

WCRP (World Conference on Religions for Peace), 2010. *Religions for Peace*. Available from: http://www.wcrp.org/ [accessed December 2010].

Weaver, D. B., 1999. Magnitude of ecotourism in Costa Rica and Kenya. *Annals of Tourism Research*, 26 (4): 792–816.

Weber, N. and Christophersen, T., 2002. The influence of non-governmental organisations on the creation of Natura 2000 during the European Policy process. *Forest Policy and Economics*, 4 (1): 1–12.

Wells, M., 1996. The role of economics in critical African wildlife policy debates. Manchester: Institute for Development Policy and Management, University of Manchester.

Wells, M., Brandon, K., and Hanna, L., 1992. *People and parks: linking protected area management with local communities*. Washington, D.C.: The World Bank.

Wells, M., Guggenheim, S., Khan, A., Wardojo, W., and Jepson, P., 1999 *Investing in biodiversity: a review of Indonesia's conservation and developmen projects*. Washington, D.C.: The World Bank.

West, P., 2010. Making the Market: Specialty Coffee, Generational Pitches, an Papua New Guinea. *Antipode*, 42: 690–718. doi: 10.1111/j.1467-8330 2010.00769.x

West, P., 2008. Tourism as Science and Science as Tourism Environmen Society, Self, and Other in Papua New Guinea. *Current Anthropology*, 49 (4) 597–626.

West, P., Igoe, J., and Brockington, D., 2006. Parks and peoples: the social impac of protected areas. *Annual Review of Anthropology*, 35: 251–277.

WFDA (Women, Faith, and Development Alliance), 2010. Available from: http:/ www.wfd-alliance.org/ [accessed December 2010].

WFDD (World Faiths Development Dialogue), 2009. *History and Objectives c the World Faiths Development Dialogue*. Available rom: http://berkleycente georgetown.edu/wfdd/about [accessed December 2010].

White, S. and Tiongco R., 1997. *Doing Theology and Development: Meeting th Challenge of Poverty*. Edinburgh: St. Andrew's University Press.

WHO, 2007. *Country Profiles of Environmental Burden of Disease*. [online] 2007. Available from: http://www.who.int/quantifying_ehimpacts countryprofiles/en/ [accessed 23 March 2015].

Wild, R., and McLeod C. (eds), 2009. *Sacred Natural Sites: Guidelines fo Protected Area Managers*. Gland, Switzerland: IUCN and Paris, France UNESCO.

Wilkie, D. S., and Carpenter, J. F., 1999. Bushmeat hunting in the Congo Basi an assessment of impacts and options for mitigation. *Biodiversity an Conservation*, 8: 927–955.

Wilkinson, R., and Pickett, K., 2010. *The Spirit Level: Why Equality is Better fc Everyone*. London: Penguin.

Williams, R., 1973. *The country and the city*. London: Hogarth.

Willis, K. J. and Bhagwat, S. A., 2009. Biodiversity and Climate Change. *Scienc* 326.5954: 806–807.

Wilshusen, P. R., 2010. The Receiving End of Reform: Everyday Responses Neoliberalisation in Southeastern Mexico. *Antipode*, 42: 767–799. do 10.1111/j.1467-8330.2010.00772.x

Wilshusen, P. R., Brechin, S. R., Fortwangler, C. L., and West, P. C., 200. Reinventing a square wheel: Critique of a resurgent "protection paradigm" i international biodiversity conservation. *Society & Natural Resources*, 15 (1 17–40.

Wilson, E. O., 1984. *Biophilia*. Cambridge, MA: Harvard University Press.

Wilson, E. O., 2006. *The Creation: An Appeal to Save Life on Earth*. New Yor W.W. Norton and Company.

Winkler, T., 2008. When God and Poverty Collide: Exploring the Myths of Fait sponsored Community Development. *Urban Studies*, 45: 2099–2116.

Wittemyer, G., Elsen, P., Bean, W. T., Burton, A. C. O., and Brashares, J. S., 200 Accelerated human population growth at protected area edges. *Science*, 32 123–126.

Wittmer, H., Berghofer, A., and Sukhdev, P., 2013. Poverty reduction and biodiversity conservation: using the concept of ecosystem services to understand the linkages. *In*: *Biodiversity conservation and poverty alleviation: exploring the evidence for a link*. Chennai, India & Malaysia: Wiley-Blackwell, 36–51.

Wolmer, W., 2007. *From wilderness vision to farm invasions: conservation & development in Zimbabwe's South-east Lowveld*. Oxford: James Currey.

Wood, D., 1993. Forests to Fields - Restoring Tropical Lands to Agriculture. *Land Use Policy*, 10 (2): 91–107.

Woods, W. I., and Denevan, W. M., 2009. Amazonian dark earths: the first century of reports. *In*: Woods, W. I., Teixeira, W. G., Lehmann, J., Steiner, C., WinklerPrins, A., and Rebellato, L., eds. *Amazonian Dark Earths: Wim Sombroek's Vision* [online]. Dordrecht: Springer Netherlands. Available from: http://link.springer.com/10.1007/978-1-4020-9031-8 [accessed 10 April 2015].

World Bank, 2001. *World Development Report 2000/2001: Attacking Poverty*. [online]. Washington D.C: World Bank. Available from: https://publications.worldbank.org/index.php?main_page=product_info&cPath=0&products_id=21654 [accessed 10 November 2013].

World Bank, 2011. *China data*. [online]. Available from: http://data.worldbank.org/country/china. [accessed 16 March 2015]

Worm B., Barbier, E. B., Beaumont, N., Duffy, J. E., Folke, C., Halpern, B. S., Jackson, J. B. C., Lotze, H. K., Micheli, F., Palumbi, S. R., Sala, E., Selkoe, K. A., Stachowicz, J. J. and Watson, R., 2006. Impacts of biodiversity loss on ocean ecosystem services. *Science*, 314.5800: 787–790.

WTO, 2011. World Trade Organization International Trade Statistics 2011 Available from: http://www.wto.org/english/res_e/statis_e/its2011_e/its11_charts_e.htm. [accessed 26 January 2015].

Wunder, S., 2000. Ecotourism and economic incentives – an empirical approach. *Ecological Economics*, 32 (3): 465–479.

Wunder, S., 2005. *Payments for environmental services: Some nuts and bolts*. CIFOR Occasional Paper, 42. Jakarta: The Centre for International Forestry Research (CIFOR). Available from: http://www.cifor.org/pes/publications/pdf_files/OP-42.pdf [accessed 16 January 2015].

Wunder, S., Engel, S., and Pagiola, S., 2008. Taking stock: A comparative analysis of payments for environmental services programs in developed and developing countries. *Ecological Economics*, 65 (4): 834–852.

WWF (World Wildlife Fund), 2009. Eco-Islam: Malaysia's Imams to Preach Against Poaching. Available from; http://www.panda.org/wwf_news/news/?162082/Eco-Islam-Malaysias-Imams-to-preach-against-poaching [accessed December 2010].

WWF, 2012. *Fighting Illicit Wildlife Trafficking: A Consultation with Governments*. Gland, Switzerland: WWF International.

Wyler, L. S., and Sheikh, P. A., 2008. *International illegal trade in wildlife: threats and U.S. Policy. United States Congressional Research Service*. Washington, DC: United States Congressional Research Service.

Wynne, B., Benton, T., and Redclift, M., 1994. *Scientific knowledge and the global environment*. London: Routledge.

Ye, L. and van Ranst, E., 2009. Production scenarios and the effect of soil degradation on long-term food security in China. *Global Environmental Change*, 19: 464–481.

You, L., Spoor, M., Ulimwengu, J., and Zhang, S., 2011. Land use change and environmental stress of wheat, rice and corn production in China. *China Economic Review*, 22: 461–473.

Young, O. R., 1989. *International Cooperation: Building Regimes for Natural Resources and the Environment*. Ithica, N.Y.: Cornell University Press.

Young, O. R., 1999. *Governance in World Affairs*. Ithica, N.Y.: Cornell University Press.

Ysseldyk, R., Matheson, K., and Anisman, H., 2010. Religiosity as Identity: Toward an Understanding of Religion from a Social Identity Perspective. *Personality and Social Psychology Review*, 14: 60–71.

Zbinden, S. and Lee, D. R., 2005. Paying for Environmental Services: An Analysis of Participation in Costa Rica's PSA Program. *World Development*, 33 (2): 255–272.

Zeppel, H., 2006. *Indigenous Ecotourism: Sustainable Development and Management*. CABI.

Zimmerer, K. S., 2011. 'Conservation Booms' with Agricultural Growth? Sustainability and Shifting Environmental Governance in Latin America, 1985–2008 (Mexico, Costa Rica, Brazil, Peru, Bolivia). *Latin American Research Review*, 46: 82–114.

Index